2250

STATISTICAL ENERGY ANALYSIS OF DYNAMICAL SYSTEMS:
THEORY AND APPLICATIONS

STATISTICAL ENERGY ANALYSIS OF DYNAMICAL SYSTEMS:

THEORY AND APPLICATIONS

Richard H. Lyon

The MIT Press
Cambridge, Massachusetts, and London, England

ACKNOWLEDGMENT
This book was written as a result of research performed for the Aerospace
Dynamics Branch, Vehicle Dynamics Division, Wright-Patterson Air Force
Base, Ohio.

PUBLISHER'S NOTE
This format is intended to reduce the cost of publishing certain works in
book form and to shorten the gap between editorial preparation and final
publication. The time and expense of detailed editing and composition in
print have been avoided by photographing the text of the book directly
from the author's typescript.

Printed in the United States of America

Library of Congress Cataloging in Publication Data

Lyon, Richard H.
    Statistical Energy Analysis of Dynamical Systems.

    Bibliography: p.
    Includes index.
    1. Structural dynamics. 2. Vibration. I. Title.
TA654.L9   624'.176   75-19074
ISBN 0-262-12071-2

CONTENTS

PART I BASIC THEORY

CHAPTER 1. THE DEVELOPMENT OF STATISTICAL ENERGY ANALYSIS

1.0   Introduction

This book is a presentation of the basic theory and procedure for application of a branch of study of dynamical
systems called Statistical Energy Analysis, which we shall
refer to as "SEA". The name SEA was coined in the early
1960's to emphasize certain aspects of this new field of
study. _Statistical_ emphasizes that the systems being studied
are presumed to be drawn from statistical populations having
known distributions of their dynamical parameters. _Energy_
denotes the primary variable of interest. Other dynamical
variables such as displacement, pressure, etc., are found from
the energy of vibration. The term _Analysis_ is used to emphasize that SEA is a framework of study, rather than a particular technique.

Statistical Vibration Analysis.   Statistical approaches
in dynamical analysis have a long history. In mechanics, we
are most familiar with their application to the vibration that
is random in time of a deterministic system. We shall use
this analysis in parts of Chapters 2 and 3. It is useful to
emphasize here that the important feature of SEA is the
description of the vibrating system as a member of a statistical
population or ensemble, not whether or not the temporal behavior is random.

Traditional analyses of mechanical system vibration of
machines and structures have been directed at the lower few
resonant modes because these modes tend to have the greatest
displacement response in many instances, and the frequencies
of excitation were fairly low. Of course, the vibration of
walls and floors and their high frequency sound radiation have
been of interest for a long time, but mechanical and structural
engineers have been generally unaware of or unconcerned with
this work. The advent of fairly large and lightweight aerospace structures, and their attendant high frequency broad-
band loads, has meant much more attention to higher order
modal analysis for purposes of predicting structural fatigue,
equipment failure and noise production.

A characteristic of higher order mode analysis, however,
is basic uncertainty in modal parameters. The resonance
frequencies and mode shapes of these modes show great sensi-

tivity to small details of geometry and construction. In
addition, the computer programs used to evaluate the mode shapes
and frequencies are known to be rather inaccurate for the
higher order modes, even for rather idealized systems. In
light of these uncertainties, a statistical model of the
modal parameters seems quite natural and appropriate.

If there is cause for statistical approaches from the
nature of the dynamical problem, there is equal motivation
from the viewpoint of application. Mechanical and structural
designers are often faced with making environmental and
response estimates at a stage in a project where structural
detail is not known. These estimates are made for the
qualification of equipment and the design of isolation,
damping, or structural configurations to protect equipment
and protect the integrity of the structure. Highly detailed
analyses requiring specific knowledge of shape, construction,
loading functions, etc., are not appropriate. Simpler
statistical analytical estimates of response to environment
that preserve parameter dependence (such as damping, average
panel thickness, etc.) are appropriate to the designer's
need at this stage.

Inspirations for the SEA Approach. There is experience
in dealing with dynamical systems described by random parameters. Two notable examples that have served as "touch-
stones" in early developments of SEA are the theory of room
acoustics, and statistical mechanics. Room acoustics deals
with the excitation of systems of very many degrees of freedom (there may be over a million modes of oscillation of a
good sized room in the audible frequency range) and the interactions between such systems (sound transmission through a
wall is an example). The analyses are carried out using
both modal and wave models. The very large number of degrees
of freedom is an advantage from a statistical viewpoint --
it tends to diminish the fluctuations in prediction of
response. We shall show why this is so in Chapter 4.

Statistical mechanics deals with the random motion of
systems with either a few or very many degrees of freedom.
However, it is random motion of a very special type, which
we may call "maximally disordered." In this state of
vibration, all modes, whether they resonate at frequencies
near each other or are far apart, tend to have equal energy
of vibration and to have incoherent motion. We must add the
proviso, "ignoring quantum effects." The energy of the
modes is, aside from a universal constant, the system temperature. The state of equal modal energy is spoken of as

"equipartition of energy". In SEA, we sometimes make the
equipartition assumption for modes that resonate in the
same frequency band, but not for all modes.

Statistical mechanics, and its related science, heat
transfer, also teach us that thermal (random vibration)
energy flows from hotter to cooler systems, and that the rate
of flow is proportional to temperature (modal energy) difference. In Chapter 3, we show that this result also applies
to dynamical systems excited by broad band noise sources.
Not only that, but since we also show in Chapters 3 and 4,
that narrow band sources are equivalent to broad band sources
when system averages are taken, the result can be generalized,
with proper care, to pure tones.

The study of the statistical mechanics of electrical
circuits shows that a resistor at known temperature is
equivalent to a thermal reservoir (or temperature bath). The
interaction of the dynamical system with this reservoir is
represented as a white noise generator in circuit theory. We
can turn the argument around and say that a damper (mechanical
resistance element), in conjunction with a noise generator,
represents a thermal reservoir, and we should not be surprised
when certain "thermal" results develop from its analysis.

The Advantages and Limitations of Statistical Analysis.
One advantage of statistical analysis of systems may be seen
from the practical aspects of room acoustics. If one truly
has 10^6 modes to deal with, even the m.s. pressure associated
with each, changing with time as the flute gives way to the
tympani, would be a hopeless mass of information to assimilate.
What one does instead is to describe the field by a few
coherent features of the modal pattern (direct field and a
few early reflections) and the incoherent energy (reverberant
field) totalled into a few frequency bands. Thus, instead of
10^6 measures of the sound field (which would be incomplete in
any case without the coherence data), we are able to describe
the sound field effectively by 10-20 measures.

The statistical analysis also allows for much simpler
description of the system, whether one describes the field
by modes or waves. In the former case, modal density,
average modal damping, and certain averages of modal impedance
to sound sources are required. In a wave description, such
parameters as mean free path for waves, surface and volume
absorption, and general geometric configuration are required.
The number of input parameters is generally in balance with
the number of measures (10-20) to be taken.

The most obvious disadvantage of statistical approaches
is that they give statistical answers, which are always subject
to some uncertainty. In very high order systems, this is not
a great problem. Many of the systems we may wish to apply
SEA to, however, may not have enough modes in certain frequency
bands to allow predictions with an acceptable degree of cer-
tainty. To keep track of this, we may calculate the variance
as well as the mean, and also calculate the confidence for
prediction intervals. This is discussed in Chapter 4.

In addition to hard and fast computational problems,
there are certain difficulties in the psychology of statistical
methods. A designer is not dealing with a gas of complicated
molecules in random collision -- he is concerned about predicting the structural response of a wing, for which he has
engineering drawings, to a loading environment, for which he
has flight data. Instead of following a deterministic calculation (probably computer based) it is suggested that he
will get a "better" estimate if he represents the wing as a
flat plate of a certain average thickness and total area!
His incredulity may be imagined. But he must remember that
his knowledge of the wing at the 50th mode of vibration may
be just as well represented by the flat plate as it is by
his drawings. Also, the answers he gets by SEA will be in a
form that is usable to him, generally retaining parameter
dependence that will allow him to interpret the effect of
certain simple design changes on response level.

To close this introductory section, it may be revealing
to quote from M.L. Mehta, a theoretical physicist concerned
with applying statistical methods to large nuclei.
"Statistical theory does not predict the detailed level sequence [resonance frequencies] of any one nucleus, but it
does describe the...degree of irregularity...that is expected
to occur in any nucleus....Here we shall renounce knowledge
of the system itself...There is a reasonable expectation,
though no rigorous mathematical proof, that a system under
observation will be described correctly by an ensemble
average.... If this particular [system] turns out to be far
removed from the ensemble average, it will show that...[the
system]...possesses specific properties of which we are not
aware. This, then, will prompt us to try to discover the
nature and origin of these properties." The problems we
face seem to be universal.

1.1 A Brief Historical Survey

Beginnings. The earliest work in the development of
SEA as an identifiable entity were two independent calculations
in 1959 by R. H. Lyon and P. W. Smith, Jr. In England, on
an NSF postdoctoral fellowship, Lyon calculated the power
flow between two lightly coupled, linear resonators excited
by independent white noise sources. He found that the power
flow was proportional to the difference in uncoupled energies
of the systems and that it always flowed from the system of
higher to lower modal energy.

The other calculation was by Smith at BBN working under
U.S. Air Force support. Smith calculated the response of a
resonator excited by a diffuse, broad band sound field, and
found that the response of the system reached a limit when
the radiation damping of the resonator exceeded its internal
damping, but that this limit did not depend on the precise
value of the radiation damping.

This result of Smith's was somewhat surprising since
many workers regarded an acoustic noise field simply as a
source of broad band random excitation. When a resonator,
excited by broad band noise, has its internal damping reduced
to zero, the response diverges, i.e., goes to infinity. The
limit involved in Smith's result was of course due to the
reaction of the sound field itself on the resonator.

After Lyon joined BBN in the fall of 1960, it developed
that Smith's limiting vibration amounted to an equality of
energy between the resonator and the average modal energy of
the sound field. The two calculations were consistent, and
power would flow between resonators until equilibrium would
develop. If the coupling were strong enough compared to
internal damping, equipartition would result. But how
specifically did the wave-field result of Smith relate to
the two-mode interaction that Lyon had studied?

To answer this question, Lyon and Maidanik wrote the
first paper that may be said to be an SEA publication, although
the name SEA had not then been coined. This paper combined
Lyon's work in England with extensions to make it able to deal
with the kind of problem Smith had analyzed. Formulas for
the interaction of a single mode of one system with many
modes of another were developed, and experimental studies of
a beam (few mode system) with a sound field (multi modal

system) were reported. This work also showed the importance
of the basic SEA parameters for response prediction: modal
density, damping, and coupling loss factor.

   Early Extensions and Improvements of the Theory.
Almost immediately, the activity in SEA split along two
lines. One line was the clarification of basic assumptions
and improvement in the range of approximation to real
system performance. The second line was the application of
SEA to other systems. The earliest application was to sound-
structure interaction, largely a result of Smith's work, but
also because it seemed "obvious" that SEA would work best
when a sound field, with all of its many degrees of freedom
was involved. Very soon, however, applications were also made
to structure-structure interactions.

   One basic assumption in SEA had to deal with light
coupling. How much of a restriction did this represent?
Also, what were the uncoupled systems? Two independent
studies, by Ungar and Scharton, showed that the light
coupling assumption was unnecessary if the uncoupled systems
were defined as the blocked system, meaning that the other
system was held fixed while the system being considered was
allowed to vibrate. Also, the energy flow relations were
valid whether one was using the "blocked" energies of the
system as the driving force or the actual energy of each
subsystem with the coupling intact. Of course, the constant of proportionality would be different for these two
calculations.

   The question of quantifying the uncertainty in the
prediction of energy flow was examined by Lyon, who developed
a theory of response variance and prediction intervals for
SEA calculations. The calculation of variance for structure-
structure interactions in which relatively few modes
participate in the energy sharing process was first included
in a paper by Lyon and Eichler.

   An important extension of the two system theory was
made by Eichler, who developed predictions from energy
distribution for three systems connected in tandem. The
practical application of SEA often involves the flow of
vibratory energy through several intervening "substructures"
before it gets to a vibration sensitive area. It is
important, therefore, to be able to predict the energy distribution for such cases.

<u>Improvements in Range of Application.</u>   The earliest
work on structure-structure vibration transmission was undertaken with Air Force sponsorship and was concerned with
electronic package vibration.   An early paper on this topic
by Lyon and Eichler dealt with plate and beam interaction,
very similar to an example discussed in Chapter 4, and two
plates connected together.   A subsequent paper on a three
element, plate-beam-plate system by Lyon and Scharton made
use of the earlier formulation of three element systems by
Eichler and some plate edge-admittance calculations, also
by Eichler.

The basic SEA theory was pretty much directly applicable
to these new systems.   The major problem was evaluation of
modal densities and coupling loss factors for various interacting junctions between systems.   For example, the
radiation of sound by reinforced plates was evaluated by
Maidanik, and a similar study of the radiation of sound by
cylinders was undertaken by Manning and Maidanik.   We have
already mentioned the plate-edge admittance calculations of
Eichler.   These, along with earlier (pre-SEA) calculations
of force and moment impedances of beams and plates have
allowed a wide variety of structural coupling loss factors
to be evaluated.   Quite recently, a series of soil-foundation
impedance evaluations by Kurzweil allows one to apply SEA to
certain structure-ground vibration problems.

Modal densities of acoustical spaces have been studied
for quite a long time.   Also, the modal densities of some
flat and curved panel structures pre-date SEA.   However, the
activity in SEA has motivated work in modal density evaluation.
For example, the modal density of cylinders has been studied
by Heckl, Manning, Chandiramani, Miller, and Szechenyi.   The
modal density of cones has been calculated by Chandiramani,
and by Wilkinson for curved sandwich panels.   Generally,
modal density prediction has not been as difficult as the
calculation of coupling loss factors.

Understanding and prediction of system damping has
improved very little over the years since SEA began.   In most
part, the improvements that have taken place were not
particularly related to SEA work, although the work by Heckl
on plate boundary absorption and by Ungar and Maidanik on
air pumping along riveted beams were generally related to
SEA.   Despite this work, our ability to predict damping in
built-up structures is still based largely on empiricism.

Other Developments.   We should also note certain
developments that might be termed "sword into plowshares"
activity.   Most early applications of SEA have been aerospace
related because that is where the problems have been (and
still are) and DOD and NASA were paying the bills.   There
have been recent applications of SEA to certain architectural
and building problems that are worthy of note.   In England,
Crocker and his associates have carried out studies of sound
transmission through double septum walls, following up on
some earlier work by White and Powell.   More recently, Ver has
applied SEA to the prediction of sound transmission through
floors consisting of a main slab with another lighter slab
floated on many small point springs.   Rinsky has also studied
sound transmission through double and single stud-reinforced
walls using SEA.

Finally, the attempts to better understand the theoretical
basis for SEA and the limiting effect of its assumptions on
the range of applications has continued.   The most notable
effort along these lines has been by Bogdanoff and Zeman.
Other work includes a recent thesis by Lotz and calculations
comparing ideal deterministic with averaged systems by
Scharton, Manning and Remington.   So far, the attempts at
further reducing the number of assumptions that one must make
in SEA have not been very effective, but we must not be dis-
couraged by this.   Every step, even the small ones taken to
this point, have been very productive in extending the reach
and usefulness of SEA.

1.2   The General Procedures of SEA

In the following chapters, we will derive the basic
equations of SEA and give motivation for the modeling and
computational procedures.   To provide an overview of this
process, however, in this section we describe the way that
SEA calculations are made, and the steps that are necessary
to arrive at a prediction of response.   Hopefully, this
will provide a framework for a better understanding of the
purpose of the later chapters and how the various elements
of SEA fit together.

Model Development.   In its simplest elements, SEA
results in a procedure for calculating the flow and storage
of vibrational energy in a complex system.   The energy
storage elements are   groups of "similar modes". Energy

input to each of the storage elements comes from a set of
external (usually random) sources. Energy is dissipated by
mechanical damping in the system, and transferred between
the storage elements. The analysis is essentially that of
linear R-C circuit theory with energy playing a role
analogous to electric charge and modal energy taking the
role of electrical potential. A typical SEA model is drawn
in Fig. 1.1 and the analogous electrical circuit that might
be used to represent it is shown in Fig. 1.2. We shall not
make direct use of this electrical analogy in this report,
but rather directly work with the simultaneous equations.

The fundamental element of the SEA model is a group of
"similar" energy storage modes. These modes are usually
modes of the same type (flexural, torsional, etc.) that
exist in some section of the system which we may call a "subsystem" (an acoustic volume, a beam, a bulkhead, etc.). In
selecting the modal group, we are concerned that it meets
the criteria of similarity and significance. Similarity
means that we expect the modes of this group to have nearly
equal excitation by the sources, coupling to modes of other
subsystems, and damping. If these criteria are met, they
will also have nearly equal energy of vibration. Significance
means that they play an important role in the transmission,
dissipation, or storage of energy. Inclusion of an
"insignificant" modal group will not cause errors in the calculations, but may needlessly complicate the analysis.

Input power from the environment, labelled Π_{in}, may
result from a turbulent boundary layer, acoustical noise, or
mechanical excitation. It is usually computed for some frequency band, possibly a one-third or full octave band. In
order to evaluate Π_{in}, we need to know certain input
impedances for the subsystem. The important requirement is
that this input power not be sensitive to the state of
coupling between subsystems. If it is, then the system providing the power (a connecting structure for example) has
important internal dynamics and should be modeled as another
subsystem.

The dissipation of power for each subsystem Π_{diss}
represents energy truly lost to the mechanical vibration and
will depend only on the amount of energy stored in that subsystem. It may be truly dissipated by friction or viscosity,
or it may be merely radiated away into the air or surrounding
structure. The important proviso is that this power cannot
be returned to the system. If it can, then it is part of the
power flow through coupling elements and will require the
addition of another subsystem or coupling path to the diagram.

The transmitted power Π_{12}, represents the rate of energy exchange between subsystems 1 and 2. All energy quantities that we deal with here are time averaged values. There may be very large temporal variations in power flow between the subsystems, even larger than the average flow, but these are ignored. The transmitted power depends on the difference in modal energy of the two subsystems and the strength of the coupling between them.

Parameter Evaluation. Evaluation of the quantities that appear in Fig. 1.1 and that are discussed in the preceding paragraphs require the evaluation of certain parameters, which we may call SEA parameters, that mostly predate SEA. We group them as "energy storage" and "energy transfer" parameters. Energy storage is determined by the number of available modes N_1, N_2...for each subsystem in the chosen frequency band $\Delta\omega$. The ratio of N to $\Delta\omega$ is called the modal density n of the subsystem and is frequently used in SEA calculations instead of the mode count N.

Energy transfer parameters include the input impedance to the system for the determination of input power, the loss factor, which relates subsystem energy to dissipated power, and the coupling loss factor relating transmitted power to subsystem modal energy. In the following paragraphs we provide a brief indication of how these parameters are usually evaluated.

Modal density may be measured by exciting the subsystem with a pure tone and gradually varying the frequency. As a resonance is encountered a maximum in response will occur. If a chart of response amplitude as a function of frequency is drawn, these peaks may be counted. This technique may miss some modes and is, therefore, not perfectly reliable. Experimental methods have been devised to reduce the number of modes missed, but the error cannot be completely eliminated.

The most commonly used way of evaluating modal density is simply to calculate it from theoretical formulas. Most systems have modal densities that may be calculated in terms of relatively simple gross parameters, such as overall dimensions, and the average speed of waves in the system. A few examples of the calculation of modal density are given in Chapter 2.

The dissipation of energy is measured by the loss
factor of the system (defined in Chapter 2). Unless the dissipation mechanism is of a very particular type, it is more
reliable to measure the loss factor than it is to calculate
it. This is done by measuring the rate of energy decay in
the system when the excitation has been removed, or by
measuring the response bandwidth of individual resonance
peaks. The relation of both of these measures to the loss
factor is developed in Chapter 2.

The input power from the environment is sometimes
measured, but more often calculated. It may be measured by
isolating the subsystem of known damping (loss factor) and
observing its response to the environment. By equating input
to dissipated power, Π_{in} is known. The input power may be
calculated by evaluating the load exerted on the subsystem
by the environment and the impedance of the subsystem to this
load. Such calculations may be quite involved, but a simple
example is given in Chapter 2 for flat plates excited by a
point force.

The coupling loss factor is the parameter governing
transmitted power. It is defined in Chapter 3 and has been
measured for some systems, although it is often calculated.
It is also related to quantities that may have been calculated or measured for other reasons - the junction
impedances of mechanical systems, the transmission loss of
walls, the acoustic radiation resistance of a piston, and
other similar measures. Since it in general depends on both
subsystems, and the variety of systems of application for
SEA is very large, the number of coupling loss factors that
we may be concerned with increases as the square of the
number of subsystems on the list. The better strategy would
appear, therefore, to express the coupling loss factor where
possible in terms of subsystem impedances, as we have done in
Chapter 3, and then list the impedances.

Calculation of Response. After the model has been
decided upon and the parameters evaluated, the final step
is to calculate the response. The first part of this calculation is the evaluation of the vibrational energy in the
various energy storage elements or mode groups. As noted
above, this amounts to solving a set of linear algebraic
equations, one equation to each subsystem. A simple
example is worked out in Chapter 3. These average energies
will depend on the various input power values, the modal
densities, and the coupling and dissipative loss factors.

On the basis of the formalism adopted in Chapters 3 and 4, the most direct relation between vibrational energy and another dynamical variable is with the velocity of motion. By formulas developed in the analysis, the velocity may in turn be related to still other variables -- displacement, strain, stress, pressure, etc. These variables are still forms of a spatially averaged response. However, on the basis of system geometry and mode shapes, estimates can also be made of the spatial distribution of response.

The variability just referred to occurs even with the SEA assumption of equal energy for every mode in the subsystem. In addition, however, there is variability because we deal with a particular system in the laboratory, not an ensemble of systems. As noted in the introduction to this chapter, the ensemble average will not fit each member exactly, and we may expect some variation in the various parameters that we have been discussing. The analysis of these variations is an important part of the response prediction process and tells us how much reliance we may place on a response estimate based on the average behavior. The discussion of this topic is developed in Chapter 4.

1.3 Future Developments of SEA

The future development of SEA is, in large part, likely to be a continuation of certain features of its development to date, i.e., its application to a larger group of subsystems. This group will surely include new structures and seismic (ground vibration) systems, and possibly water waves and ship or offshore structures. Such applications will increase the glossary of SEA parameters and make SEA even more useful than it has been until now.

The wider use of SEA is also likely to involve a larger group of professionals within each subject area. This means that research workers and analysts, test engineers, and designers are all likely to use it as one tool among many that are available for dynamical analysis. We may expect to see SEA become more computerized, not only for the purpose of solving the simultaneous equations governing response energy, but also for evaluating coupling loss factors for very complex structures using finite element methods.

Finally, we may expect to see developments in the basic theory of the statistical analysis of systems that will greatly expand the conceptual framework of SEA. To be more specific, let us see how SEA has expanded upon the framework of statistical mechanics, and what may develop in this regard in the future.

Statistical mechanics represents the most disordered state of motion possible for a system. Every mode through-out the system has the same average energy. This includes modes in different subsystems and those that resonate at different frequencies. Thermodynamics is the macroscopic counterpart to this totally equilibrium system of analysis.

The possibility for different temperatures in different parts of a system is allowed in the kinetic theory version of statistical mechanics and in "non-equilibrium" thermo-dynamics. The modes within each subsystem all have the same energy, but the average modal energies of the subsystems may be different.

This latter situation is the one we deal with in SEA, except that we proceed even farther from equilibrium by allowing modes in different frequency bands to have differing energies. Basically, we can get away from this because the systems that we deal with are presumed linear, and modes that resonate in different frequency bands may be considered to be uncoupled from each other. Thus, SEA represents still another step away from the "maximally disordered" state described by statistical mechanics.

At this point, it is logical to ask whether there is another step to be made in this sequence toward a model that is less disordered than the current SEA model that would provide use with useful answers in situations in which SEA has limitations. One answer might be -- forget about this chain of logic, go all the way back to the deterministic system. That is a possible answer, but it may not be the most useful one.

The most glaring deficiency of SEA is its inability to deal with modal coherence effects. In simplest terms, modal coherence may lead to such phenomena as "direct waves," discussed in Chapter 2. It is not clear at present whether this represents the "next logical step" in the chain that we spoke of, but it bears examination.

A second deficiency of SEA is its assumption that all modes that resonate in $\Delta\omega$ are equally probable over that interval, and that their resonance frequencies are unaffected by the values of the resonance frequencies of the other modes. There have already been some developments on improving upon this hypothesis and we may expect more. We are on more solid ground in predicting developments here than we are in our concern for modal coherence effects.

1.4 Organization of Part I

Part I is concerned with presenting the basic theoretical elements of SEA. We may cite these as:

(1) The theory of energy exchange between two resonators. This is a problem in random vibration theory.

(2) The extension of the two resonator interaction to the interaction of two systems having modal behavior. This extension requires that we introduce the basic idea of statistical populations of systems.

(3) The representation of system interactions by the SEA model with the attendant necessity to identify and evaluate SEA parameters.

(4) The calculation of average energy. This is exemplified by applying the formulas to some fairly simple cases.

(5) The calculation of variance in response and the use of variance to generate estimation intervals and their associated confidence coefficients.

Since the emphasis in Part I is the explication of SEA, we have included generous discussion of the ideas of energy storage and transfer, statistical ensembles of systems, isolated vs. coupled systems, etc. Our intention is to provide the most clear cut statement on these matters that can be made at present. The reader will have to decide how well we have succeeded.

Since our attention is on SEA, however, we do not spend time on other subjects that we employ in the various examples. Thus, for example, we use the equations of bending motion of plates and beams, but we do not derive them. We also use certain input impedances to beams, plates and sound fields, but these are not derived. For the most part, such derivations are readily available to the student, and in any case, in the annotated bibliography, guidance is given to where this information may be found.

The reader will notice that no references are given in the text, which is an unusual procedure. In adopting this procedure, we do not wish to deny credit to anyone, rather we have followed regular textbook practice of providing a bibliography, but not breaking the presentation with references. Partly this procedure is a matter of self-discipline; sometimes it is very tempting for an author to try to avoid a difficult matter by slipping in a reference.

Chapter 2 deals with some fundamental ideas in the energy of vibration of single and multi degree of freedom systems. The storage of kinetic and potential energy by modes in free and forced vibration, and the decay or rate of energy removal due to damping are discussed, partly as a review of this important subject and partly to establish some of the basic vocabulary of the work to follow. Also, since both modal and wave descriptions of vibration are employed, energy storage, dissipation, and flow in a wave description are reviewed also.

Chapter 3 is concerned with developing the theory of average power flow (average in both a temporal and ensemble sense) between single and multi degree of freedom systems. First, power flow between two simple resonators is calculated. The important ideas of a blocked system is introduced and power flow is calculated in terms of both blocked and coupled system energies. The idea of averaged modal interaction as a white noise source is also introduced.

The latter sections of Chapter 3 are the central to our discussion. The very important ideas of blocked systems, ensemble averages of system interactions, and the definition and use of the coupling loss factor are introduced here. The use of reciprocity for development of certain useful relations for evaluation of the coupling loss factor is also discussed. The chapter ends with some elementary applications of the SEA relations.

   Chapter 4 completes Part I and is concerned with
the problems of estimating response, based on
the average energy distribution calculations. The first part
of this problem is estimating response variables of more
particular interest than energy -- displacement, stress, etc.
The second is the development of estimation intervals and
their associated confidence coefficients, particularly when
statistical analysis of variance indicates that the
standard deviation is an appreciable fraction of the mean.

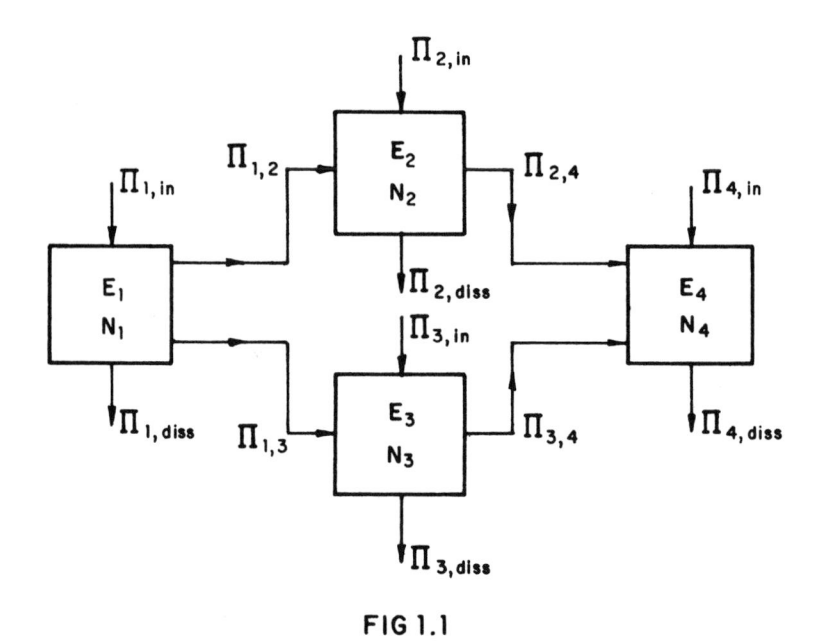

FIG 1.1

TYPICAL FORM OF SEA SYSTEM MODEL

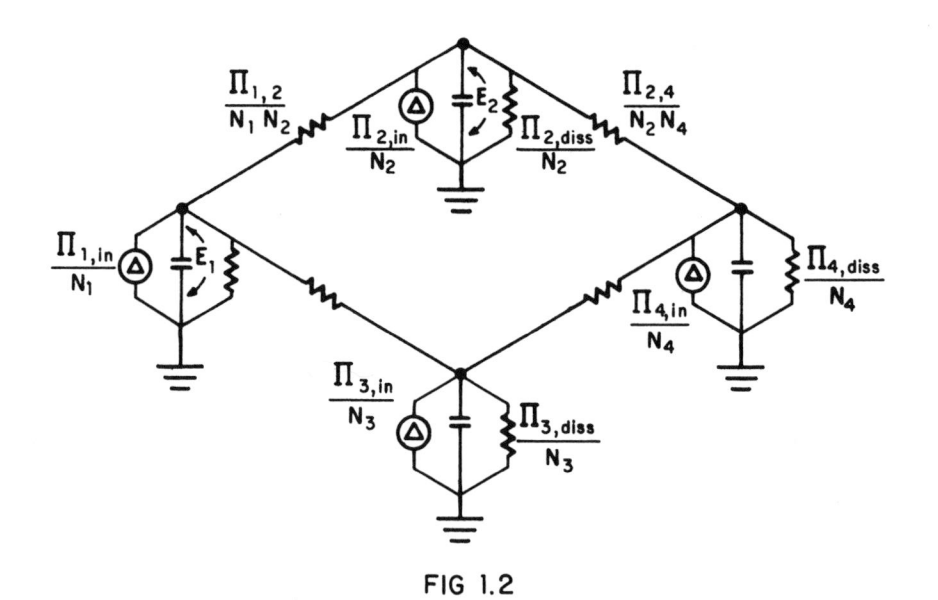

FIG 1.2

AN R·C ELECTRICAL CIRCUIT THAT MAY BE USED TO REPRESENT THE SYSTEM OF FIG 1.1

CHAPTER 2.   ENERGY DESCRIPTION OF VIBRATING SYSTEMS

2.0   Introduction

This book is concerned with the exchange of energy in coupled vibrating systems. This chapter introduces the use of energy variables in describing the vibration of systems. Common ways of analyzing such systems employ modal oscillator or wave descriptions of the motion. A major goal of this chapter is to show how energy analysis applies to both of these models and how certain relations between them are revealed by the use of energy variables.

The chapter considers free vibration and both sinusoidal and random excitation of vibrating systems. This is a very large topic, but we are only interested in certain aspects of it. We specialize our interest to linear systems and to some interesting measures of response that are particularly related to vibrational energy. These measures are mean square response, spatial and temporal coherence and admittance functions.

The chapter begins with a discussion of the energetics of modal resonators in free vibration. The cases of sinusoidal and random excitation are studied next. The role of damping in causing decay of free vibration and bandwidth in forced vibration is of particular interest. Systems having many modes of vibration are studied next, and the notions of modal density and average admittance are introduced. At this point, the very important and pervasive idea of statistical modeling of real systems is introduced for the first time.

Vibrating systems modeled as collections of free and forced waves are introduced next. Certain concepts unique to such a description like energy velocity and wave impedance are introduced. Descriptors that are common to wave and modal descriptions such as mass and damping are also discussed. In paragraph (2.4) of this chapter, we point out some relationships between modal quantities, such as modal density and their wave descriptor counterparts, such as energy velocity.

The goal in this mode-wave interplay is to develop a way of thinking about vibrating systems that allows one shift back and forth between these two viewpoints, exploiting the one that is best suited to the problem at hand.

2.1 Modal Resonators

   To begin our discussion, we examine the energetics of
the simple linear resonator shown in Fig. 2.1. We shall
show later in this chapter that this system is a useful model
of the dynamics of the modal amplitudes of multi-degree-of-
freedom (dof) systems.

   The dashpot or mechanical resistance R in Fig. 2.1
produces a force, $- R\dot{y}$, opposite in direction to the
velocity $\dot{y}$ of the mass M. The spring or stiffness element
K produces a force $- Ky$ opposite in direction to displacement y from the equilibrium position of the mass. These
forces, in combination with the force $\ell(t)$ applied by an
external agent, results in an acceleration of the mass.

$$\ell(t) - R\dot{y} - Ky = M\ddot{y} \quad ,$$

or, more conventionally,

$$\ddot{y} + \omega_0 \eta \dot{y} + \omega_0^2 \, y = \ell(t)/M \qquad\qquad (2.1.1)$$

where $\omega_0 \equiv \sqrt{K/M}$ (natural radian frequency) and $\eta \equiv R/\omega_0 M$
(loss factor).

   Free Vibration - No Damping. To study the case of free
vibration, we set $\ell(t) = 0$ in Eq. (2.1.1). For the moment,
we also neglect damping by setting $\eta = 0$. We then have

$$\ddot{y} + \omega_0^2 y = 0, \qquad\qquad (2.1.2)$$

which has the two solutions $\cos \omega_0 t$ and $\sin \omega_0 t$. The general
solution, therefore, is

$$y = A \cos \omega_0 t + B \sin \omega_0 t = C \sin (\omega_0 t + \phi) \qquad (2.1.3)$$

We say that $\omega_0 = 2\pi f_0$ is the radian frequency of free, un-damped oscillation of the resonator. The amplitudes A and B (or C and ϕ) are real numbers but otherwise arbitrary. They are determined if y and $\dot{y}$ are known at any time.

The kinetic energy of the mass at any time is

$$KE = \frac{1}{2}\, M\, \dot{y}^2 \;=\; \frac{1}{2}\, MC^2\omega_0^2\, \cos^2(\omega_0 t + \phi) \qquad (2.1.4)$$

and the potential energy in the stiffness element is

$$PE = \frac{1}{2}\, K\, y^2 = \frac{1}{2}\, KC^2\, \sin^2(\omega_0 t + \phi)\; . \qquad (2.1.5)$$

The sum of these is

$$E = KE + PE = \frac{1}{2}\, KC^2$$

which is time independent and dependent only on the peak amplitude of vibration. Since we have an isolated system vibrating without damping, it is evident that its vibrational energy should not vary in time.

The displacement and velocity repeat themselves in a period $1/f_0$. If we average the kinetic energy and the potential over such a period, we get

$$\langle KE \rangle_\sim \;=\; \langle PE \rangle_\sim \;=\; \frac{1}{4}\, KC^2 \;=\; \frac{1}{2}\, E \qquad (2.1.6)$$

Thus, the time average kinetic and potential energies are both equal to 1/2 of the energy of vibration. We shall make considerable use of this relation in the work that follows.

Free Vibration with Damping. When damping is present,
then $\eta \neq 0$ and we have

$$\ddot{y} + \omega_0\eta \; \dot{y} + \omega_0^2 \; y = 0 \tag{2.1.7}$$

and if we assume a form of solution $y \sim e^{\alpha t}$, we find that
α must equal one of the two values

$$\alpha = - \; \tfrac{1}{2} \; \omega_0\eta \pm i\omega_d \tag{2.1.8}$$

where $\omega_d \equiv \omega_0\sqrt{1-(\eta^2/4)}.$ The solution for $y(t)$ in this case
is then

$$y(t) = C \; e^{-\frac{1}{2}\omega_0\eta t} \; \sin(\omega_d t + \; \phi). \tag{2.1.9}$$

In this case, the oscillation occurs at radian frequency ω_d
and the amplitude of the oscillations decreases exponentially
in time due to the extraction of energy by damping. If the
loss factor η is 0.5 or smaller, then ω_d is very nearly equal
to ω_0 and the period of damped oscillations is essentially the
same as that for undamped oscillations.

The potential energy of vibration in this case is

$$PE = \; \tfrac{1}{2} \; K \, y^2 = \; \tfrac{1}{2} \; KC^2 \; e^{-\omega_0\eta t}\sin^2(\omega_d t + \phi) \tag{2.1.10}$$

The kinetic energy is a little more complicated. It is

$$KE = \frac{1}{2} \ M \ \dot{y}^2 = \frac{1}{2}c^2 \ M\omega_0^2 \ e^{-\omega_0 \eta t} \left[- \frac{\omega_d}{\omega_0} \ \cos \ (\omega_d t + \phi) \right.$$

$$\left. + \frac{1}{2} \ \eta \ \sin \ (\omega_d t + \phi) \right]^2 \qquad (2.1.11)$$

These expressions simplify considerably if we average over a cycle of oscillation, neglecting the slight change in amplitude in this period due to the exponential multiplier. The result is

$$\langle PE \rangle_\sim \ = \ \frac{1}{4} \ KC^2 e^{-\omega_0 \eta t} \ = \ \langle KE \rangle_\sim \ = \ \frac{1}{2} \ \langle E \rangle_\sim \qquad (2.1.12)$$

In free damped vibration, for which $\eta \leq 0.5$, we have the same relations between kinetic, potential, and total energy that we obtain for undamped vibration. Note that for both damped and undamped vibration

$$\langle y^2 \rangle_\sim \ = \ \langle \dot{y}^2 \rangle_\sim \ / \omega_0^2. \qquad (2.1.13)$$

The loss factor η is simply related to other standard measures of damping. For example, by its definition, $\eta = 1/Q$, where Q is the resonator quality factor, much used in electrical engineering. From Eq. (2.1.9), the amplitude decays as $e^{-\pi \eta t/T}$ and, therefore, $\pi\eta$ is the logarithmic decrement (nepers per cycle). Finally, the oscillations cease when $\eta \rightarrow 2$ so that η is twice the critical damping ratio.

From Eq. (2.1.12), we have

$$\langle E \rangle_\sim \ = \ E_0 e^{-\omega_0 \eta t} \qquad (2.1.14)$$

A measure of damping widely employed in acoustics is the reverberation time T_R, which is the time required for the vibrational energy to decrease by a factor of 10^{-6}. Thus,

$$e^{-\omega_0 \eta} T_R = 10^{-6}$$

which results in

$$T_R = \frac{2.2}{f_0 \eta} \quad . \tag{2.1.15}$$

Sinusoidal Forced Vibration. If the applied force $\ell(t)$ in Fig. 2.1 is sinusoidal at radian frequency ω, then it is convenient to use the exponential form

$$\ell(t) = |L|\cos(\omega t + \psi) = \text{Re}\{|L|\exp[-i\omega t - i\psi]\} \tag{2.1.16}$$

where Re $\{...\}$ means "real part of". The dynamical equations are linear and one can, therefore, consider response to the actual excitation as a combination of (complex) response to the complex excitation and its complex conjugate. This means that we may express the response variables in terms of complex exponentials also.

Accordingly, without loss of generality, we describe the complex force and velocity as

$$\ell(t) = L\, e^{-i\omega t}$$

$$\dot{y}(t) = V\, e^{-i\omega t} \tag{2.1.17}$$

where L and V are complex numbers of the form $|L|e^{-i\psi}$ and $|V|e^{-i\alpha}$, respectively. Differentiation and integration of these functions is particularly simple

$$\frac{d}{dt}\ell(t) = -i\omega \; L \; e^{-i\omega t} = -i\omega\ell$$

$$\int \ell(t)dt = (-i\omega)^{-1}L \; e^{-i\omega t} = \ell/(-i\omega) \qquad (2.1.18)$$

Substituting for ℓ and $\dot{y}$ in Eq. (2.1.1), we obtain

$$L = V(-i\omega_0 M) \; \{(\omega/\omega_0-\omega_0/\omega) + i\eta\} \equiv VZ \qquad (2.1.19)$$

where Z is mechanical impedance of the resonator. This can also be expressed by an admittance $Y = 1/Z$,

$$V/L \equiv Y = \{\omega_0\eta \; M -i(\omega M - K/\omega) \}^{-1} \qquad (2.1.20)$$

Since the magnitudes of L and V are the peak force and velocity, we can graph Eq. (2.1.20) as shown in Fig. 2.2.

The average power supplied to the resonator by the source is $\Pi = \langle \ell\dot{y}\rangle_t$. When variables are described as complex variables, the time average of their product is easily expressed by their complex amplitudes, as follows:

$$\Pi = \langle \ell\dot{y}\rangle_t = \frac{1}{2} \; \text{Re} \; (LV^*) = \frac{1}{2} \; |L|^2 \; \text{Re} \, (Y^*)$$

$$= \frac{1}{2} \; |V|^2 \; \text{Re} \; (Z) \; . \qquad (2.1.21)$$

where ( )* denotes the complex conjugate. Since

$$\text{Re}(Y) = \text{RE}(1/Z) = \text{RE}(Z)/|Z|^2 = \omega_0 \eta \, M \, |Y|^2, \quad (2.1.22)$$

the input power has the frequency dependence shown in Fig. 2.2.
The maximum value is $\Pi = 1/2 |L|^2/\omega_0 \eta \, M = <\ell^2>_t/R$, the dissipation
of a system with a resistance only.

At frequencies $\omega \neq \sqrt{K/M}$, the power diminishes, reaching
1/2 its maximum value when

$$|Y|^2 = \frac{1}{2(\omega_0 \eta M)^2} = \frac{1}{(\omega_0 \eta M)^2 \{1 + (\omega^2 - \omega_0^2)^2/\omega_0^2 \; \omega^2 \eta^2\}} (2.1.23)$$

which occurs when $\omega = \omega_0 \pm \omega_0 \eta/2$ (assuming $\eta < 0.3$), as
shown in Fig. 2.2. These frequencies are both the limits for
half-power and the boundaries for simple forms of dynamical
behavior. When $\omega < \omega_0(1-\eta/2)$, the admittance is adequately
represented by neglecting R and M. and keeping the stiffness term only

$$Y = - i \frac{\omega}{K} \quad [\omega < \omega_0(1-\eta/2)] \qquad\qquad (2.1.24)$$

we denote this region as "stiffness controlled".

For frequencies $\omega < \omega_0(1+\eta/2)$, the admittance is usefully
approximated by the mass term only

$$Y \simeq i/\omega M \quad [\omega > \omega_0(1+\eta/2)] \qquad\qquad (2.1.25)$$

The region is called "mass controlled". The intermediate
region is called "damping controlled". These regions are

noted in Fig. 2.2 also. These simplified limiting forms of
behavior of the resonator are of great importance in our
consideration of systems with many degrees of freedom.

It is possible to discuss sinusoidal excitation very
extensively, and this is done in many textbooks. For our
purposes here, however, we will only note some simple
relations. First note that the relative phase of the
velocity $\dot{y}$ with respect to the force f is the phase of the
admittance function Y, which is the same as the phase of
Z*,

$$\text{Arg}\{Z*\} \ = \ \tan^{-1}|\{\omega - \omega_0^2/\omega\}/\omega_0\eta|$$

$$= \ \tan^{-1}|2(\omega - \omega_0)/\omega_0\eta| \ . \qquad (2.1.26)$$

It is clear from this that the phase is changing rapidly as the
system goes through resonance-the more quickly the smaller the
damping. This is in contrast to the behavior of the amplitude
which has a horizontal slope at $\omega=\omega_0$. For this reason,
resonance can frequently be more accurately determined from
examination of the phase rather than the amplitude of response.

The second point is concerned with mean square response
at resonance:

$$<\dot{y}^2> \ = \ \frac{1}{2} \ |V|^2 \ = \ \frac{1}{2} \ |L|^2/\omega_0^2\eta^2M^2 \qquad (\omega=\omega_0) \quad (2.1.27)$$

Since $y = (V/-i\omega)e^{-i\omega t}$ and $\ddot{y} = -i\omega\,V\,e^{-i\omega t}$, then when
$\omega=\omega_0$,

$$<y^2> \ = \ \frac{1}{2} \ |V|^2/\omega_0^2 \ = \ <\dot{y}^2>/\omega_0^2$$

$$<\ddot{y}^2> \ = \ \frac{1}{2} \ \omega_0^2|V|^2 \ = \ \omega_0^2<\dot{y}^2>, \qquad (2.1.28)$$

the same relations as found for free vibrations. These
relations do not hold outside the damping controlled region,
however, but they do hold to an acceptable degree of accuracy
for $|\omega - \omega_0| < 1/2 \ \omega_0 \eta$.

Random Excitation. Most of the applications of SEA that
concern us are situations in which the excitation by the loading environment is random. We should emphasize here, however,
that the critical feature of SEA is that we assume a
statistical model for the system being excited, not necessarily
for the excitation. Thus, it is perfectly proper to apply SEA
to systems excited by pure tones if a statistical model of the
system and the description of its response using energy
variables are appropriate. Nonetheless, the existence of
random excitation generally means that much less averaging
of system parameters is necessary and consequently less
variability of response from the calculated mean in any
particular situation may be expected.

There is no single satisfactory definition of a stationary
random signal from all viewpoints, but one that has particular
appeal from an experimental viewpoint can be readily developed.
Imagine a filter having the frequency response shown in
Fig. 2.3 and that the load function $\ell(t)$ is applied to the
input of this filter. Now let the bandwidth Δf of the filter
become very small. If the mean square output of the filter
becomes proportional to Δf, then the force $\ell(t)$ is random.
Note that a pure tone does not satisfy this requirement since
the mean square output would be independent of bandwidth as
long as the frequency of the tone were in the pass band of
the filter. A similar statement can be made for any
deterministic, periodic signal.

This statement not only supplies a definition of a
stationary random signal that fits out intuition and is
mathematically respectable, it also provides a direct
indication of how the frequency decomposition of a random
signal is effected. The mean square (m.s.) force corresponding
to the band of frequencies is, therefore:

$$\langle \ell^2 \rangle_{\Delta f} = S_\ell \ \Delta f \tag{2.1.29}$$

S_ℓ being the factor of proportionality. It is usually the
case that this proportionality factor depends on the center
frequency f, so it is written $S_\ell(f)$, and it is called the

power spectral density (psd) of the random variable $\ell(t)$.

Let us suppose that the psd of $\ell(t)$ has been determined for all values of f and the result has been plotted as in Fig. 2.4. We also suppose that $\ell(t)$ is applied to two filters like the one in Fig. 2.3, except they are centered at frequencies f_1 and f_2. We now combine the output of the two filters to yield

$$\langle \ell^2 \rangle = S_\ell(f_1)\ \Delta f + S_\ell(f_2)\ \Delta f\ , \qquad (2.1.30)$$

since the mean squares of two time functions containing different frequency components are simply added together to form the mean square of the sum. Proceeding in this way, the mean square output of a filter that has unity gain from frequency f_1 to f_2 is simply

$$\langle \ell^2 \rangle = \int_{f_1}^{f_2} S_\ell(f)\ df\ , \qquad (2.1.31)$$

or, if the filter has a gain $G(f)$ instead of unity, the dependence is

$$\langle \ell^2 \rangle = \int_{f_1}^{f_2} S_\ell(f)\ G(f)\ df. \qquad (2.1.32)$$

The total unfiltered m.s. value of the loading function is found by setting $f_1 = 0$ and $f_2 = \infty$ in Eq. (2.1.31).

Let us suppose that the loading force in Eq. (2.1.1) is a noise excitation having a psd S_ℓ that is constant. Such a noise signal is called "white noise". The m.s. force produced within a very narrow frequency band df is, therefore:

$$\langle \ell^2 \rangle_{df} = S_\ell df. \qquad (2.1.33)$$

The m.s. velocity response of the resonator to this force at frequency f is given by

$$\langle \dot{y}^2 \rangle_{df} = S_\ell df \, |Y|^2 \qquad\qquad (2.1.34)$$

According to Eq. (2.1.32), therefore, the psd of $\dot{y}$ is given (aside from a constant) by Fig. 2.2, and the total m.s. velocity is found from

$$\langle \dot{y}^2 \rangle = \int_0^\infty S_\ell |Y|^2 df = S_\ell \int_0^\infty |Y|^2 df \qquad (2.1.35)$$

for white noise.

It is clear from the form of Fig. 2.2 that most of the contribution to the integral in Eq. (2.1.35) will come from the region denoted "damping controlled". This observation leads to a useful concept of "equivalent bandwidth", Δ_e. This is the bandwidth of a system with a rectangular pass band (as in Fig. 2.3) that has a constant admittance determined by the damping along, $Y_{eq} = (\omega_0 \eta M)^{-1}$, that has the same response to the white noise excitation that the actual system does. Thus, by this definition,

$$\langle \dot{y}^2 \rangle = S_\ell \Delta_e (\omega_0 \eta M)^{-2}$$

$$\qquad\qquad\qquad\qquad\qquad (2.1.36)$$

$$= \frac{1}{2\pi} S_\ell (\omega_0 \eta M)^{-2} \int_0^\infty \frac{d\omega}{1 + (\omega^2 - \omega_0^2)^2 / \eta^2 \omega^2 \omega_0^2}$$

This integrand is simplified by changing variables to $\xi = 2(\omega - \omega_0)/\eta\omega_0$ and noting that the greatest contribution to the integrand is at $\omega = \omega_0$ or $\xi = 0$, Thus,

$$\int_0^\infty \quad \cdots \rightarrow \frac{\eta\omega_0}{2} \int_{-\infty}^\infty \frac{d\xi}{1+\xi^2} = \frac{\pi\eta\omega_0}{2}$$

and, therefore,

$$\Delta_e = \frac{\pi}{2} \eta f_0 , \qquad\qquad (2.1.37)$$

We note that this bandwidth is greater than the "half power" bandwidth in Fig. 2.2 by the factor of $\pi/2$. The "replacement" of the resonator by this equivalent filter becomes a very useful approximation in many situations.

There is an alternative time domain definition of white noise that is useful. In this case, we consider the force to be made up of a series of impulses of strength $\pm a$, occurring randomly in time as shown in Fig. 2.5. When this excitation is applied to the system in Fig. 2.1, the mass is given a sequence of changes in velocity. Let each change in velocity be $\Delta v = a/M$. These occur randomly in time with random sign. The corresponding change in energy of vibration is

$$\Delta E = \frac{1}{2}M\{(\dot{y}+\Delta v)^2 - \dot{y}^2\} = M \dot{y} \Delta v + \frac{1}{2}M(\Delta v)^2 \qquad (2.1.38)$$

If we average this over a sequence of impulses, the term $\langle \dot{y}\Delta v\rangle$ will vanish since a positive Δv is just as likely as a negative one. The average energy increment is, therefore, $1/2\ M \langle\Delta v\rangle^2 = a^2/2M$.

If the average rate of impulses per second is ν, then the power fed into the resonator is $\nu a^2/2M$. But this must also equal the dissipated power $\langle\dot{y}^2\rangle \ \omega_0\eta M$. We have, therefore, an estimate for the m.s. velocity,

$$\langle\dot{y}^2\rangle = \nu a^2/2\omega_0\eta M^2 = S_\ell \Delta_e (\omega_0\eta M)^{-2} , \qquad (2.1.39)$$

which leads to $S_\ell(f) = 2\nu a^2$, which is a constant. More importantly, however, this relation gives us an additional physical interpretation of the spectral density function for white noise.

If we return to Eq. (2.1.36), the expression for m.s. displacement is simply

$$\langle y^2 \rangle = \frac{1}{2\pi} \, S_\ell \, (\omega_0 \eta M)^{-2} \int_0^\infty \frac{d\omega}{\omega^2 [1 + (\omega^2 - \omega_0^2)^2 / \eta^2 \omega^2 \omega_0^2]}$$

(2.1.40)

and if the same assumptions are made regarding the level of damping are made as previously, we get

$$\langle y^2 \rangle = \langle \dot{y}^2 \rangle / \omega_0^2 \, ,$$

(2.1.41)

If we try to calculate the m.s. acceleration in the same way, we run into difficulty, since the integral

$$\langle \ddot{y}^2 \rangle = \frac{1}{2\pi} \, S_\ell (\omega_0 \eta M)^{-2} \int_0^{f_{max}} \frac{\omega^2 d\omega}{[1 + (\omega^2 - \omega_0^2)^2 / \eta^2 \omega^2 \omega_0^2]}$$

(2.1.42)

will not converge as $f_{max} \to \infty$. For large ω, the integrand simply $\omega_0^2 \eta^2$. We can solve the problem by "subtracting out" this part, and obtain

$$\langle \ddot{y}^2 \rangle = \omega_0^2 \, \langle \dot{y}^2 \rangle + S_\ell \, f_{max}/M^2$$

(2.1.43)

where the first term represents the damping controlled acceleration and second part is the mass controlled acceleration.
In most instances of random excitation, the resonant part will
dominate and we can simply use

$$<\ddot{y}^2> \simeq \omega_0^2 <\dot{y}^2> \qquad\qquad (2.1.44)$$

We have seen that free vibration and the resonant controlled response to sinusoidal and random excitation all produce the
same relations between m.s. displacement, velocity and acceleration. This is very useful to us since it allows us to transform from one response variable to another without concern for
the exact nature of the excitation as long as the general requirements of resonant dominated response for the validity of these
relations are met.

2.2 Modal Analysis of Distributed Systems

The systems of interest to the mechanical designer are
much more complicated than a linear resonator. In real
systems, the stiffness, inertia and dissipation are all distributed over the space occupied by the structure. A displacement of the system that increases the potential energy is
resisted by the elastic restoring forces. A rate of change
in the displacement is resisted by the damping forces, and
these forces, along with the loading excitation cause acceleration of the mass elements.

If we represent the generalized displacement of the
system by y, then an equivalent statement of the above is

$$\rho\ddot{y} + r\dot{y} + \Lambda y = p, \qquad\qquad (2.2.1)$$

where ρ is the mass density, r is a viscous resistance
coefficient and Λ is a linear operator consisting of
differentiations with respect to space. In the case of a
flat plate, for example, ρ is the mass per unit area of the
plate and $\Lambda = B \nabla^4$, where B is the bending rigidity. The
use of simple viscous damping is a valuable simplification and
does not affect the utility of our results as long as the
damping is fairly small.

When this system is bounded, with well defined boundary
conditions, the solution is frequently sought by expansion in
eigenfunctions ψ_n, which are solutions to the equation

$$\frac{1}{\rho} \Lambda \psi_n = \omega_n^2 \psi_n \qquad\qquad (2.2.2)$$

where the functions ψ_n satisfy the same boundary conditions that y does, and the quantities ω_n^2 are determined by the boundary conditions. The response and the excitation are then expanded in these functions

$$y = \sum_n Y_n(t) \psi_n(x)$$

$$p/\rho = \sum_n L_n(t) \psi_n(x) \qquad\qquad (2.2.3)$$

If we multiply Eq. (2.2.2) by ψ_m and integrate over the region of the structure and subtract from this the same equation with the indices reversed, we get

$$\int \{\psi_m \Lambda \psi_n - \psi_n \Lambda \psi_m\} \, dx = (\omega_n^2 - \omega_m^2) \int \psi_m \rho(x) \psi_n \, dx.$$

$$(2.2.4)$$

When $n = m$, this is satisfied in a trivial fashion. When $n \neq m$, there are certain specific (but nevertheless very useful) conditions under which the differential operator has the property that

$$\int (\psi_m \Lambda \psi_n - \psi_n \Lambda \psi_m) \, dx$$

will vanish. This means that

$$\int \psi_m \rho \, \psi_n \, dx$$

must vanish, a condition that is referred to as orthogonality of the eigenfunctions ψ_m with a weighting function ρ.

A convenient normalization of the amplitude of the eigenfunctions is simply

$$\frac{1}{M} \int \psi_m \rho \, \psi_n \, dx \quad \equiv \, <\psi_m \psi_n>_\rho \, = \, \delta_{m,n} \qquad (2.2.5)$$

which we may think of as a mass density weighted average of the product $\psi_m \psi_n$. For systems in which the mass density is uniform, this becomes a simple average over the spatial coordinates.

If we now place the expansions of Eq. (2.2.3) into Eq. (2.2.1), we obtain

$$\rho \sum_n \, (\ddot{Y}_n + \frac{r}{\rho} \dot{Y}_n + \omega_n^2 Y_n) \, \psi_n = \frac{\rho}{m} \sum_m L_m \, \psi_m \qquad (2.2.6)$$

We can simplify this immediately if we also assume that $r(x)$ is proportional to $\rho(x)$; $r = \Delta\rho$. We can do this on two counts. To take a somewhat cynical view, since we have introduced the damping in a rather ad hoc fashion, we can feel free to configure it any way we like. The other basis for the assumption is that research studies of this problem have shown that the consequence of this assumption is to ignore a degree of inter-modal coupling that is less significant than other forms of coupling that we will be concerned about.

We now multiply Eq. (2.2.6) by $\psi_n(x)$ and integrate over the system domain. Using Eq. (2.2.5), we obtain

$$M\{\ddot{Y}_m + \Delta\dot{Y}_m + \omega_m^2 Y_m\} = L_m(t) \qquad (2.2.7)$$

Thus, each modal response amplitude obeys the equation of a linear resonator of the sort discussed in paragraph 2.1. This result, in conjunction with the spatial orthogonality of the mode shapes according to Eq. (2.2.5), leads us to the concept of a complex dynamical system as a group of independent resonators of mass M, stiffness $\omega_m^2 M$ and mechanical resistance $M\Delta$.

Since there is a one-to-one correspondence between the modes and the ω_n's, we can use the latter to keep track of the modes. An example should be helpful here. For the two-dimensional, simply supported, isotropic and homogeneous rectangular flat plate of dimensions ℓ_1 by ℓ_2, drawn in Fig. 2.6, the mode shapes are

$$\psi_{n_1,n_2} = 2 \sin \frac{n_1 \pi x_1}{\ell_1} \sin \frac{n_2 \pi x_2}{\ell_2} \qquad (2.2.8)$$

and ω_n^2 is

$$\omega_{n_1,n_2}^2 = \left[\left(\frac{n_1 \pi}{\ell_1} \right)^2 + \left(\frac{n_2 \pi}{\ell_2} \right)^2 \right]^2 \kappa^2 c_\ell^2$$

$$\equiv (k_1^2 + k_2^2)^2 \kappa^2 c_\ell^2 \equiv k^4 \kappa^2 c_\ell^2 \qquad (2.2.9)$$

where κ is the radius of gyration of the plate cross-section, $c_\ell = \sqrt{Y_p/\rho_m}$ is the longitudinal wavespeed in the plate material, where Y_p is the plate Young's modulus and ρ_m is the material density and n_1 and n_2 are integers.

Eq. (2.2.9) suggests a very convenient ordering of the modes, which is shown in Fig. 2.7. By inspection, we can see that each point in this "wave-number lattice" corresponds to a mode. Further, the distance from the origin to that

point will determine the value the resonance frequency of
the mode ω_n. The main value of this ordering is that it
enables us for example to count the modes that will resonate
in some frequency interval. It will not always be as convenient as it is in this case, but the ordering by some
parameter is a necessary condition to allow us to do the
counting at all.

When the ordering indices form a lattice as shown in
Fig. 2.7, then each lattice point corresponds to an area
$\Delta A_k = \pi^2/A_p$, where $A_p = \ell_1 \ell_2$ is the area of the plate. As we
increase the wavenumber from k to $k + \Delta k$, we include a new
area $1/2\, \pi k \Delta k$. On the average, this will include $1/2\, \pi k \Delta k / \Delta A_k$
new modes. Thus, the average number of modes per unit
increment of wavenumber is

$$n(k) = \frac{\pi k \Delta k}{2 \Delta A_k \Delta k} = \frac{\pi k}{2 \Delta A_k} \qquad (2.2.10)$$

which we may call the modal density in wavenumber.

To find the modal density in frequency $n(\omega)$, which tells
us how many modes on the average are encountered when we
increase frequency by 1 unit, we use the relation
$n(\omega) \Delta \omega = n(k)\, \Delta k$, or

$$n(\omega) = \frac{\pi k}{2 \Delta A_k} \cdot \frac{\Delta k}{\Delta \omega} = \frac{\pi k}{2 c_g \Delta A_k} \qquad (2.2.11)$$

where we have used the result from elementary physics that
$\Delta \omega / \Delta k$ is the group velocity for waves in a system that has a
phase velocity ω/k.

For a flat plate, the group velocity c_g is twice the
phase velocity $c_b = \omega/k = \sqrt{\omega \kappa c_\ell}$, so that the modal density
in cycles per second (hertz) is

$$n(f) = n(\omega) \frac{d\omega}{df} = \frac{2\pi^2 \omega A_p}{4 c_b^2 \pi^2} = \frac{A_p}{2 \kappa c_\ell} = \frac{\sqrt{3} A_p}{h c_\ell} \qquad (2.2.12)$$

where we have used $\kappa = h/2\sqrt{3}$, the radius of gyration for a homogeneous plate of thickness h. Note that the modal density in this case is independent of frequency. As an example, consider a plate that has an area of 10 ft.^{2} and a thickness of 1/8 in. (approximately 10^{-2} ft.). Then, since $c_\ell \simeq 17{,}000$ ft./sec. (for steel or aluminum), we have

$$n(f) = \frac{(1.7) \cdot 10}{(.01)\ (17{,}000)} \simeq 0.1 \text{ mode/Hz} \qquad (2.2.13)$$

For this plate, one mode is encountered on average whenever the excitation frequency is increased by 10 Hz. A plate with a larger area or smaller thickness will have a modal density that is greater than this.

The kinetic energy of vibration is

$$\frac{1}{2} \int dx\ \rho \left(\frac{\partial y}{\partial t}\right)^2 = \frac{1}{2} \int dx \quad \sum_{m,n} \dot{Y}_m(t)\dot{Y}_n(t)\rho\psi_m(x)\psi_n(x)$$

$$= \frac{1}{2} M \sum_m \dot{Y}_m^2(t)\ , \qquad (2.2.14)$$

where we have used the orthogonality relation. Thus, the kinetic energies of the modes add separately to give the KE of the system. Since the kinetic and potential energies of each resonator are equal at resonance

$$\langle \dot{Y}_n^2 \rangle = \omega_n^2\ \langle Y_n^2 \rangle \qquad ,$$

the total energies of the modes simply add to form the total system energy.

Response of System to Point Force Excitation. We now imagine that a point force of amplitude L_0 is applied at a location x_s on the structure. If we solve for the modal amplitudes from Eq. (2.2.3), we get

$$L_m = \int p\psi_m \, dx = L_0 \, \psi_m(x_s) \quad , \qquad (2.2.15)$$

where M is the system mass. If the excitation and response are proportional to $e^{-i\omega t}$, Eq. (2.2.7) becomes

$$M(\omega_n^2 - i\omega \, \omega_n \eta - \omega^2) \, Y_n = L_0 \psi_n(x_s) \quad , \qquad (2.2.16)$$

so that the formal expression for the response is

$$y(x,t) = \frac{L_0 e^{-i\omega t}}{M} \sum_n \frac{\psi_n(x_s)\psi_n(x)}{\omega_n^2 - \omega^2 - i \, \omega_n \omega \eta}. \qquad (2.2.17)$$

We shall look at some ways of simplifying this complicated result.

The velocity at the excitation point x_s is $-i\omega y(x_s)$. The ratio of this velocity to the applied force is the input conductance of the system.

$$\frac{-i\omega Y(x_s, \omega)}{L_0} = \frac{-i\omega}{M} \sum_n \frac{\psi_n^2(x_s)}{\omega_n^2 - \omega^2 - i\omega_n \eta}$$

$$\equiv G - iB \quad , \qquad (2.2.18)$$

where G, the real part of the sum is the conductance and
B, the imaginary part, is the susceptance. By rationalizing
the complex in Eq. (2.2.18), we get

$$G = \sum_n a_n(x_s)g_n(\omega) \quad ,$$
(2.2.19)

where $a_n = \psi_n^2(x_s)/\omega M\eta$ and $g_n = (\xi^2+1)^{-1}$, where $\xi = 2(\omega_n - \omega)/\omega\eta$. The
susceptance is

$$B = \sum_n a_n(x_s)b_n(\omega)$$
(2.2.20)

where $b_n = \xi(\xi^2+1)^{-1}$. In deriving these relations, we have
assumed that the damping is small enough so that the individual
modal admittance $a_n(g_n - ib_n)$ is quite sharp in frequency.

The total admittance Y is a rapidly fluctuating function
of frequency. We can simplify this result however, if we consider averages of Y with respect to the variable ξ. Such
an average is appropriate if we are exciting the system with
a band of noise, since the response to noise having a uniform
spectral density from ω_1 to ω_2 is the same as the average of
the m.s. response to a pure tone as may be seen by referring
to Eqs. (2.1.32) and (2.1.35).

Another possibility is to assume that the system itself
is "random". That is, that the exact mode shapes and resonance
frequencies are not known, either because of random irregularities
in their construction, or because the detailed calculation
procedures are not accurate enough to calculate them. In this
case, we assume that the resonance frequencies ω_n are uniformly distributed over some frequency interval. Such an
approach is one example of statistical modeling, a central
theme in Statistical Energy Analysis. In the work that
follows we take this latter approach, assuming that ω is
fixed and that ω_n is the random variable.

If the interval of resonance frequency uncertainty is $\Delta\omega$, then if $\Delta\omega \gg \frac{\pi}{2}\omega\eta$, the average over ω_n will give

$$\langle g_n \rangle_{\omega_n} = \frac{\omega\eta}{2\Delta\omega} \int_{-\infty}^{\infty} \frac{d\xi}{\xi^2+1} = \frac{\pi\omega\eta}{2\Delta\omega} \qquad (2.2.21)$$

which, interestingly enough, is simply the ratio of the modal bandwidth to the averaging bandwidth. In the interval $\Delta\omega$, the number of modes that can contribute to the average is $n(\omega)\Delta\omega$. Thus, from Eq. (2.2.19), the average conductance is

$$\langle G \rangle_{\omega_n, Y_s} = \frac{n\Delta\omega}{\omega\eta\ M} \cdot \frac{\pi}{2\Delta\omega}\ \omega\eta\ \langle\psi_n^2\rangle = \frac{\pi}{2}\frac{n(\omega)}{M} \qquad (2.2.22)$$

where the average on ψ^2 is over the mass distribution of the system, which is of course the same as a spatial average for a uniform mass density.

Eq. (2.2.22) is a useful and general result for multi-modal systems. In the case of a flat plate, $n(\omega)=A_p/4\pi\ \kappa c_\ell$, and $M=\rho_s A_p$, where ρ_s is the surface density of the plate. Then,

$$\langle G \rangle = (8\rho_s\ \kappa c_\ell)^{-1} \qquad (2.2.23)$$

which is also the admittance of an infinite plate. It very often happens that average impedance functions of finite systems are the same as those for the same system infinitely extended.

The average susceptance is of less interest, but we note that when the modal density is constant, then the average susceptance will vanish because the integral of $b_n(\xi)$ vanishes. When the modal density is not constant, the calculation is more complicated, and has been dealt with in the references.

Before leaving the discussion of impedance, let us consider the same problem from the point of view of noise excitation. If the force in Eq. (2.2.15) has a flat spectrum S_ℓ over the band $\Delta\omega$, then the power fed into any one mode can be found from the dissipation $\omega_0 \eta M \langle y^2 \rangle$ is, according to Eq. (2.1.39), just $S_\ell \psi_n^2(x_s)/4M$. If the bandwidth of the noise is $\Delta\omega$ (radians/sec) or $\Delta\omega/2\pi$ Hz then the number of modes randomly excited is $n\Delta\omega$ and the power input to the system, averaged over source location is

$$\langle \Pi \rangle = \frac{\pi}{2} \; \frac{S_\ell \Delta\omega}{2\pi} \; \cdot \; \frac{n}{M} \; = \; \langle \ell^2 \rangle \quad \langle G \rangle \qquad\qquad (2.2.24)$$

where again $\langle G \rangle = \pi n/2M$. Thus, we can view the conductance as a measure of the number of modes that are available to absorb energy from a noise source. On this basis, it is apparent that there should be a close tie between the average conductance and modal density. This relation can in fact be exploited as a way of measuring modal density when the modes are so closely packed or the damping is so large that counting resonance peaks is not feasible.

2.3 Dynamics of Infinite Systems

When the system is infinitely extended, then an alternative formulation is needed. The differential equation governing the motion is still given by Eq. (2.2.1). We now assume, however, that the mass and damping distributions are uniform (ρ and r = constant) and that the linear differential operator Λ is a simple polynomial in the spatial derivatives, $\Lambda(\partial/\partial x_i)$ with constant coefficients. With these assumptions, we may assume a "wave" solution to the equation for unforced motion in the form

$$y \sim e^{i(\vec{k}.\vec{x} - \omega t)} \; . \qquad\qquad (2.3.1)$$

With this substitution, the equation of motion becomes the "dispersion relation" between frequency and wave number:

$$-\rho\omega^2 - i\omega r + \Lambda(ik_i) = 0 \qquad (2.3.2)$$

Let us consider some simple examples of Eq. (2.3.2). For the undamped string, ρ = lineal density, $r = 0$, and $\Lambda = -T(\partial/\partial x)^2$. Thus, Eq. (2.3.2) becomes

$$Tk^2 = \rho\omega^2, \text{ or } k = \pm\omega/c , \qquad (2.3.3)$$

where $c = \sqrt{T/\rho}$ is the speed of free waves on the string. In the case of undamped bending motions of a thin beam, we would have

$$\Lambda(\partial/\partial x) = B(\partial/\partial x)^4 \qquad (2.3.4)$$

where $B = \rho c_\ell^2 \kappa^2$ is the bending rigidity of the beam. In this case the dispersion relation is

$$k^4 = \omega^2/\kappa^2 c_\ell^2 \equiv (\omega/c_b)^4 \qquad (2.3.5)$$

the same as Eq. (2.2.9). The parameter c_b is the phase velocity for bending waves on the beam. In the case of a two dimensional plate, the wave vector $\vec{k}$ in Eq. (2.3.1) has the two components (k_1, k_2). For either the beam or the plate, the phase velocity c_b is a function of frequency

$$c_b = \sqrt{\omega \kappa c_\ell} \qquad (2.3.6)$$

and the system is said to be dispersive. When damping is included, the propagation constant k is complex and an attenuated wave results.

Energy variables are of great importance in infinite, free wave systems just as they are for finite, modal descriptions. In the infinite system, we are interested in energy density, the energy of vibration per unit length, area, or volume depending on the dimensionality of the system. It is shown in advanced textbooks that for the system that we are considering, the kinetic and potential energy densities of free waves are equal, so that the total energy density is just twice the kinetic energy density, which is $\rho(\partial y/\partial t)^2$.

The intensity I of a free wave is equal to the power flowing through a unit width (or area) of the system due to that wave as it propagates. If this power flows for 1 second, then the amount of structure that has filled with energy is numerically equal to $c_g = d\omega/dk$, the energy velocity. If $\mathcal{E}$ is the energy density, then the energy that passed the reference location is $\mathcal{E}c_g=I$. From the dispersion relation, Eq. (2.3.2) with r=0, the energy velocity is

$$c_g = \frac{d\omega}{dk} = \frac{i}{2\rho\omega} \quad \Lambda'(ik) \ .\qquad\qquad (2.3.7)$$

Since $E = \rho(\partial y/\partial t)^2$,

$$I = -\frac{i\omega}{2} \quad \langle y^2 \rangle_t \qquad \Lambda'(ik) \ .\qquad\qquad (2.3.8)$$

Let us consider some examples of the use of Eq. (2.3.8) for some familiar systems. In the case of the string, $\Lambda(ik) = - T(ik)^2$ and $\Lambda'(ik) = - 2T(ik)$. Thus,

$$I_{string} = -\frac{i\omega}{2}\langle y^2\rangle_t \ (-\rho c^2) \ 2ik = \rho c \langle(\partial y/\partial t)^2\rangle_t$$

$$\qquad\qquad\qquad\qquad\qquad\qquad\qquad\qquad (2.3.9)$$

In the case of the beam, $\Lambda = \rho \kappa^2 c_\ell^2 (ik)^4$ and $\Lambda' = 4\rho \kappa^2 c_\ell^2 (ik)^3$. Thus

$$I_{beam} = \frac{i}{2\omega} < \left(\frac{\partial y}{\partial t}\right)^2 >_t \; 4\rho \; \kappa^2 c_\ell^2 (ik)^3 \; = \; 2\rho c_b < \left(\frac{\partial y}{\partial t}\right)^2 >_t$$

$$(2.3.10)$$

In both of these cases, the intensity is a mean square velocity of motion of the system times an impedance term of the form ρc, where ρ is the density and c is a wavespeed.

When damping is included, its most important effect is to cause the propagation constant to become complex. The new dispersion relation is found by the substitution,

$$k(\omega) \; \rightarrow \; k\left[\omega \left(1 + \frac{i\eta}{2}\right)\right] \qquad , \qquad\qquad (2.3.11)$$

Thus, for the string,

$$e^{ik_s x} = e^{i\frac{\omega}{c}[1+(i\eta/2)]x} = e^{i\frac{\omega x}{c}} e^{-\frac{\omega \eta}{2c} x} \qquad (2.3.12a)$$
$$(\text{string})$$

and for flexural motion of the beam $(k_b \equiv \omega/c_b)$

$$e^{ik_b x} = e^{i\sqrt{\dfrac{\omega(1+(i\eta/2))}{\kappa c_\ell}}\, x}$$

$$= e^{i\frac{\omega}{c_b} x} \; e^{-\frac{\omega\eta}{4c_b}} \qquad (\text{beam}) \qquad\qquad (2.3.12b)$$

From this, it is clear that the form of the dispersion relation
will affect the rate at which the wave decays in space as
damping is added. For a non-dispersive system, the attenuation
is $2\pi\eta$ nepers or $27.3\ \eta$ dB per wavelength. It is clear that
if we used the energy velocity, the attenuation factors in
Eqs. (2.3.12a) and (2.3.12b) would be formally the same since
$c_g = 2\ c_b$ for the plate.

Most of the structures of interest to structural designers
consist of segments of beams and plates, so that we shall place
most of our emphasis on such systems. In paragraph 2.2, we
studied the impedance looking into a finite plate, and found that
the average impedance was real and related to the modal density
and the mass in a particularly simple fashion. We now want to
carry out this same calculation for the infinite plate.

The two dimensional flat plate has the equation of motion
Eq. (2.2.1) with

$$\Lambda = \rho\kappa^2\ c_\ell^2\ \left[(\frac{\partial}{\partial x_1})^2 + (\frac{\partial}{\partial x_2})^2\right]^2.$$

The point force of strength $L_0 e^{-i\omega t}$ is assumed to act at x = 0.
The problem is to calculate the motion y at x = 0 and form the
ratio of $(\partial y/\partial t)$ to the force to find the admittance of the
plate. We solve the problem by using two-dimensional Fourier
transforms

$$p(\vec{x}) = \frac{1}{(2\pi)^2}\iint dk_1\ dk_2\ e^{i\vec{k}.\vec{x}}\ P(\vec{k})$$

$$P(\vec{k}) = \iint dx_1\ dx_2\ e^{-i\vec{k}.\vec{x}}\ p(\vec{x}) \qquad (2.3.13)$$

with similar relations between y(x) and its transform Y(k).
Since p(x) acts only at x = 0 with strength L_0, the second
integral above is simple $P(k) = L_0$. Placing the transform into
the equation of motion

$$[-\omega^2(1+i\eta)\rho + \Lambda(ik_1,ik_2)]\; Y(\vec{k}) = L_0 \qquad (2.3.14)$$

which gives

$$y(0) = \frac{L_0}{(2\pi)^2}\iint dk_1\, dk_2\; \frac{1}{\Lambda(ik) - \omega^2(1+i\eta)\rho} \qquad (2.3.15)$$

Since there is no azimuthal dependence in the (k_1,k_2) integration, we replace the area element by $dk_1\, dk_2 \to 2\pi k\, dk$ and have a simple integration over the magnitude k.

$$y(0) = \frac{L_0}{2\pi}\int_0^\infty k\, dk\; \frac{1}{\rho\kappa^2 c_\ell^2[k^4 - k_b^4(1+i\eta)]}$$

$$= \frac{L_0}{4\pi\rho\kappa^2 c_\ell^2}\int_0^\infty d\xi\; \frac{1}{\xi^2 - \xi_b^2(1+i\eta)}$$

where $\xi = k^2$ and $\xi_b = k_b^2$. Thus,

$$y(0) = \frac{L_0}{8\pi\rho\kappa^2 c_\ell^2}$$

$$\times \int_{-\infty}^{\infty} \frac{d\xi}{[\xi - \xi_b(1+i\eta/2)]\,[\xi + \xi_b(1+i\eta/2)]} \qquad (2.3.16)$$

or,

$$y(0) = \frac{L_0/(-i\omega)}{8\rho\kappa c_\ell}$$

allowing η to vanish, and the path of integration is taken as in Fig. 2.8. Thus, the ratio of velocity $-i\omega\, y(0)$ to the force F_0 is the infinite plate admittance

$$Y_\infty = (8\ \rho\kappa c_\ell)^{-1} \qquad\qquad (2.3.17)$$

which is the impedance that we found for the finite plate when an average was taken over modal response and source location.

A mean square force $\langle \ell^2 \rangle$ applied at a point on an infinite plate will, therefore, inject an amount of power

$$\Pi_{in} = \langle \ell^2 \rangle\, Y_\infty \qquad\qquad (2.3.18)$$

into the plate. This power will propagate at the energy velocity outward from the point of excitation as a circular wave. When a boundary is encountered, a reflection occurs and the energy propagates unimpeded until another boundary is encountered. The average distance traversed between reflections is called the "mean free path", d, and is given by $d = \pi A_p/P$, where P is the perimeter of the plate and A_p is the area of the plate. The attenuation rate (in space), according to Eq. (2.3.12b) is $\omega\eta/2c_g$ nepers per unit length, or $4.34\omega\eta$ dB per second (c_g is the energy velocity defined by Eq. (2.3.7)).

For systems in which the energy is contained in the propagation of free waves, the damping is frequently expressed in terms of the time (referred to as the reverberation time T_R) required for the energy to decay by 60 dB. Thus, $4.34\ \omega\eta\ T_R = 60$ or,

$$T_R = \frac{2.2}{f\ \eta},$$
(2.3.19)

a result that we found earlier for single dof and modal resonators. Measurement of decay rate or reverberation time is a commonly used procedure for determining loss factor. The decay rate procedure applies equally well whether our model of the system is one of a collection of modes or a group of waves rebounding within the system boundaries.

In Chapter 3, we will be concerned with the dynamics of interacting systems. As a preview, let us consider here the interaction between the single dof system and a finite plate shown in Fig. 2.9. We wish to calculate the mean square velocity of the resonator as a result of its attachment to a plate which is vibrating randomly.

The resonator consists of a mass M, stiffness K, and dashpot resistance R, configured as shown. A diffuse reverberant vibrational field (equal wave intensity in all directions) is assumed to exist on the plate, resulting in a transverse velocity v. At the point of resonator attachment x_s, the plate velocity is v_s. The velocity of the resonator mass is v_M. The reaction force ℓ resulting from compression of the spring K is due to a difference in velocities v_s and v_M:

$$\ell = K \int (v_s - v_M)\ dt = M\ \frac{dv_M}{dt} + R\ v_M$$
(2.3.20)

The velocity v_s is equal to the velocity that would exist if there were no resonator v, less the "reaction" or "induced" velocity v_r that is induced by the force; $v_r = \ell <G>$, where $<G>$ is the plate admittance $(8\rho_s \kappa c \ell)^{-1}$. In using $<G>$, we assume that the modal density of the plate η_p is sufficiently high so that several plate modes resonate within the combined equivalent bandwidth of plate and resonator modes:

$$\frac{\pi}{2}\ \omega(\eta_p + \eta_0)\ n_p \gg 1.$$

We now differentiate Eq. (2.3.20) with respect to time to obtain

$$M \frac{d^2 v_M}{dt^2} + R \frac{dv_M}{dt} + K v_M = K v - K \ell <G> \qquad (2.3.21)$$

Substituting for ℓ from Eq. (2.3.20), we get

$$\frac{d^2 v_M}{dt^2} + \omega_0 (\eta_0 + \eta_{coup}) \frac{dv_M}{dt} + \omega_0^2 (1 + \eta_0 \eta_{coup}) v_M = \omega_0^2 v \qquad (2.3.22)$$

where $\omega_0 \equiv \sqrt{K/M}$, $\eta_0 = R/\omega_0 M$ and η_{coup} is the combination of of parameters $\omega_0 M <G>$.   Eq. (2.3.22) is the equation for the response of a resonator to a random base excitation velocity v.   The effects of the plate are expressed by modified damping and resonance frequency of the resonator. If v is assumed to have a flat spectrum $<v^2>/\Delta\omega$ over a bandwidth $\Delta\omega$, then from the discussion following Eq. (2.1.35), the response of the resonator is

$$<v_M^2> = \frac{\pi}{2} \frac{\omega_0}{\eta_0 + \eta_{coup}} \frac{<v^2>}{\Delta\omega} \qquad (2.3.23)$$

or,

$$. M <v_M^2> = \frac{\eta_{coup}}{\eta_0 + \eta_{coup}} \frac{M_p <v^2>}{n_p \Delta\omega} \qquad (2.3.24)$$

The term $M_p <v^2>/n_p \Delta\omega$ is simply the vibrational energy of the plate divided by the number of modes that resonate in the band $\Delta\omega$. The ratio represents, therefore, the average energy

per mode of vibration within that band. Since the ratio
$\eta_{coup}/(\eta_{coup} + \eta_0)$ is always less than or equal to one,
Eq. (2.3.24) says that the average energy of vibration of the
resonator cannot exceed the average modal (resonator) energy
of the plate. These resonator energies will in fact be equal
when the loss factor due to the coupling between the resonator
and the plate η_{coup} (also called the coupling loss factor) is
large compared to the internal damping of the resonator. This
tendency toward energy equality is an example of "equipartition
of energy", a principle that is well known to students of
statistical mechanics.

2.4 Modal-Wave Duality

In paragraphs 2.2 and 2.3 of this chapter, we have discussed modal and wave descriptions of finite systems. These
two ways of describing the motion are in most cases fully
equivalent to each other, but that does not mean that the
choice of description is arbitrary. Certain aspects of
structural vibration are much more readily interpreted in one
description than in the other. The effect of damping at the
boundaries of a plate is better described by wave reflection
processes. The spatial variations in vibration amplitude are
better described by the modal analysis. But we must emphasize,
that it is always possible, at least in principle, to arrive
at the same conclusions by either approach.

We have already demonstrated an important example of
modal-wave duality, the average point conductance of finite
plates. We have found that the conductance computed by
averaging over modes of a finite plate is the same as that
found by considering only waves radiated outward by driving
an infinite plate at one point. The fact that these waves
reflect from the boundaries and may return to the drive point
is presumed to be unimportant in affecting the drive point
impedance because they will have random phases, or be incoherent with the excitation, particularly if the drive point
is assumed to be randomly located over the surface of the
structure.

Another aspect of this duality is in the description of
damping. We found in discussing resonators in paragraph (2.1)
that the damping was closely associated with resonator (or
modal) bandwidth and decay of vibrations in time. In discussing waves, in paragraph (2.2) we found that the damping

was related to the spatial decay of energy. However, if this
decay in space were related to the time required for the
energy to propagate from one point to the other, the time decays
in both descriptions were related to the loss factor in the
same way. Thus, decay rate of energy forms a link between the
two descriptions even though modal bandwidth is a very difficult concept to explain by a wave analysis and spatial decay
of vibration is an equally difficult concept to explain using
a modal description.

In the chapters that follow, particularly in examples
dealing with applications of SEA, we find that certain
quantities that enter the modal description for example, are
equivalent to other quantities in the wave description. For
example, modal energy is usually equivalent to a spatial
energy density. For example, in a plate, the energy is $\rho_s<v^2>$,
whereas the average modal energy for the same plate is
$(4\pi \ \kappa c_\ell) \ \rho_s \ <v^2>$. In a sound field, the energy density is
$<p^2>/\rho c^2$, and the average modal energy for the same sound
field is $(2\pi^2 c^3/\omega^2)<p^2> \ \rho \ c^2$, where c is the speed of sound
and p is the pressure fluctuation.

Other equivalences exist between coupling loss factors,
appropriate for modal systems, and junction impedances,
appropriate for wave descriptions. An example of this
equivalence was used in the example of the combined plate-
resonator systems of Fig. 2.9, where we found $\eta_{coup} \ \omega_0 M \ <G>$.
We shall encounter more examples of this in the chapters to
come.

In the remainder of this section, we shall discuss a very
important aspect of wave-mode duality. This is the problem of
the "coherent" and "incoherent" wave fields in a plate. In
paragraph 2.3 we began this discussion, but we wish to delve
further into the matter at this point. Let us begin by considering the wave description first.

Let us suppose that a point force located at position x_s
on a very large plate excites the plate in a bandwidth $\Delta\omega$.
If the rms force is L, then the power injected into the plate
at x_s is simply

$$\Pi_{in} = L^2 Y_\infty = L^2 <G> \qquad\qquad (2.4.1)$$

where <G> is given by Eq. (2.3.17). This energy flows
radially outward from this point, so that at a distance r

from x_s, the mean square velocity is found from the intensity relation

$$\rho \, c_g \, \langle v_D^2 \rangle \, 2\pi r = \Pi_{in}. \qquad (2.4.2)$$

This is the velocity that would exist on an infinite plate without damping. If damping is present, the energy is reduced by the factor $\exp[-\eta \omega r/c_g]$, so that

$$\langle v_D^2 \rangle = \frac{\Pi_{in}}{2\pi \, \rho \, c_g} \, \frac{1}{r} \, e^{-\omega \eta r/c_g} \qquad (2.4.3)$$

This part of the plate motion is called the "direct field" of the source and has a "geometric attenuation" of 3 dB per double distance and an attenuation due to damping that increases linearly with r. The direct field is the only contribution to plate vibration if the plate is infinite in extent or if η is so large that the vibration has nearly ceased by the time the direct wave reaches the boundary of the plate.

When the energy in the direct field reflects from the boundary, then if the boundary is not perfectly regular, it is frequently assumed that coherence with the direct field is lost, particularly as the number of reflections mounts. The motion v_R corresponding to this reflected energy is called the "reverberant" field and is determined by the power dissipated by it:

$$\Pi_R = \Pi_{in} \, \exp[-\omega \eta d/2c_g] = M \, \langle v_R^2 \rangle \, \omega \eta \qquad (2.4.4)$$

where d is the mean free path. This leads to an expression for the reverberant velocity

$$\langle v_R^2 \rangle = \Pi_{in} \, \exp[-\omega \eta d/2c_g]/\omega \eta M. \qquad (2.4.5)$$

Since the direct and reverberant velocity components are
assumed to be incoherent, the total mean square velocity of
vibration is obtained by simple addition,

$$<v^2> = <v_D^2> + <v_R^2>. \qquad (2.4.6)$$

Near the point of excitation, the direct field will dominate,
but at large distances, the reverberant field will dominate.
The "boundary" between the direct and reverberant fields is
defined as the distance r_D for which these fields are equal

$$r_D = \frac{\omega\eta \ M}{2\pi \ \rho \ c_g} \qquad (2.4.7)$$

The total mean square velocity pattern is shown in Fig. 2.10.

We can also analyze this situation by a modal description.
From paragraph 2.2, the response of a plate to a point force is

$$V = \frac{i \ \omega \ L_0}{M} \ \sum_m \ \frac{\psi_m(x) \ \psi_m(x_s)}{\omega^2 (1+i\eta) - \omega_m^2} \qquad (2.4.8)$$

Let us assume that the plate is simply supported so that the
mode shapes are

$$\psi_m(x) \ \psi_m(x_s) = \frac{1}{4} \prod_{i=1}^{2} \ (e^{ik_i x_i} - e^{-ik_i x_i}) \ (e^{ik_i x_i^s} - e^{-ik_i x_i^s})$$

$$(2.4.9)$$

There are 16 terms in this product, each representing a plane
wave having one of the four wave vectors shown in Fig. 2.11.

The phase factors are the "dot products" of these wave vectors with the four position vectors shown in Fig. 2.12. These position vectors are the difference vectors between the point source and the observation point and its three images in the coordinate axes.

If the drive point is excited at frequency ω, then because of the modal response bandwidth, we will assume that all modes within a band $\Delta\omega = \pi\omega\eta/2$ will be excited, and that modes outside this bandwidth are not excited. As we sum over the circle of excited modes in k-space (the quarter circle of Fig. 2.7 becomes a full circle because of the addition of the images of $\hat{k}$ in Fig. 2.11), some of the terms will fluctuate wildly in phase as we sum over the indices m_i, while others will combine because the phase variation between them is small. The smallest phase variation will occur for the smallest vector, $\vec{x} - \vec{x}_s$, as seen in Fig. 2.12. We shall assume that only these terms contribute to the "coherent" part of the summation in Eq. (2.4.8).

We, therefore, replace Eq. (2.4.9) by its coherent part

$$\psi_m(x)\ \psi_m(x_s) \simeq \frac{1}{4}\, e^{i\vec{k}\cdot\vec{R}} \qquad\qquad (2.4.10)$$

where $\vec{R} \equiv \vec{x} - \vec{x}_s$, and $\vec{k}$ now varies over the complete circle in Fig. 2.11. If the phase variation in going from one lattice point (allowable wave) to another when summing over m is less than $\pi/2$, then the summation in Eq. (2.4.8) may be replaced by an integration over angle in k-space.

$$v(x) = \frac{-i\omega L_0}{4M\kappa^2 c_\ell^2} \sum_m \frac{e^{ikr\cos\theta}}{k^4 - k_b^4(1+i\eta)} \qquad\qquad (2.4.11)$$

where θ is the angle between $\vec{k}$ and $\vec{R}$ and $k_b^2 = \omega/\kappa c_\ell$.

Let us note that $M = \rho\, A_p = \rho\pi^2/\Delta A_k$. We can now write $\Delta A_k = k\, \Delta k\, \Delta\theta$, and the summation in Eq. (2.4.11) becomes

$$v(x) \approx \frac{-i\omega L_0}{4\pi^2 \rho \kappa^2 c_\ell^2} \int_0^\infty \frac{k \, dk}{k^4 - k_b^4(1+i\eta)} \int_0^{2\pi} e^{ikr \cos\theta} d\theta \qquad (2.4.12)$$

The second integral is simply $2\pi J_0(kr)$. The integral to be evaluated then is

$$v(x) = \frac{-i\omega L_0}{\pi \rho \kappa^2 c_\ell^2} \int_0^\infty \frac{k \, dk J_0(kr)}{k^4 - k_b^4(1+i\eta)} . \qquad (2.4.13)$$

Unfortunately, this integral is quite complicated to evaluate, but is a standard form in tables of Hankel transforms. The result of the integration is (as $\eta \to 0$),

$$v(x) = \frac{L_0}{8\rho\kappa c_\ell} \{H_0^{(1)}(k_b r) - H_0^{(1)}(ik_b r)\}$$

$$\xrightarrow[k_b r > 1]{} \frac{L_0}{8\rho\kappa c_\ell} \sqrt{\frac{2}{\pi k_b r}} \{e^{ik_b r} - e^{-i\pi/2} e^{-k_b r}\} \qquad (2.4.14)$$

The first term on the rhs of Eq. (2.4.14) represents a cylindrically spreading wave, diminishing in amplitude as $1/\sqrt{r}$. The second is a pure exponential decay, characteristic of the bending vibrations of a plate. Vibrational energy is carried by the first term in the equation, so that

$$\langle v_0^2 \rangle = \frac{\langle L_0^2 \rangle}{8\rho\kappa c_\ell} \frac{1}{4\pi\rho\kappa c_\ell k_b r} = \Pi_{in}/2\pi\rho c_g r \qquad (2.4.15)$$

as found in Eq. (2.4.3). Thus, the direct calculation of the coherent field from the modal summation gives the proper result for the amplitude of this component of the total motion, but also gives us the phase of response and the near field non propagated component as well.

The incoherent part of the velocity field is found from Eq. (2.4.8) by adding the mean square values of modal response incoherently. The mean square value of each term is

$$\frac{\omega^2 L_0^2}{2M^2} \cdot \frac{\pi}{2} \, \omega\eta n_s \cdot \frac{1}{\omega^2\eta^2} = \frac{L_0^2}{2} \cdot \frac{1}{\omega\eta M} \cdot \frac{\pi}{2} \frac{n_s}{M}$$

$$= \frac{1}{2} L_0^2 \, G/\omega\eta \quad M, \qquad (2.4.16)$$

which is the same as the result in Eq. (2.4.5) if we assume that very little dissipation of the direct field occurs before the first reflection from a boundary.

We have shown here that the direct and reverberant fields that arise naturally in the wave analysis have their direct counterparts in the "coherent" and "incoherent" components of the summations of the modal description. Such correspondences are very useful in that they allow us to interpret some of the phenomena described by a wave picture in terms of their effects on a model analysis. For example, the transmission of vibrational energy through a connection between two plates will impose a degree of coherence between the modes of the receiving

plate. But if the bulk of the vibrational energy may be
judged to be in the reverberant field, we may treat the modes
of the "source" and "receiving" structures to be incoherent
with each other.

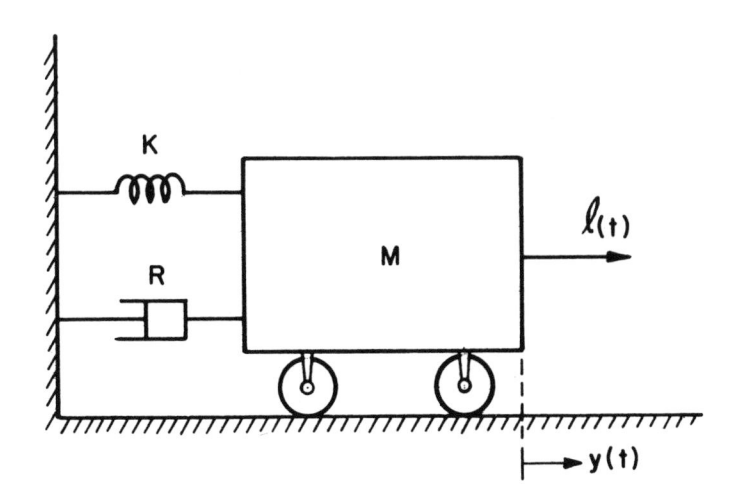

FIG. 2.1

LINEAR RESONATOR

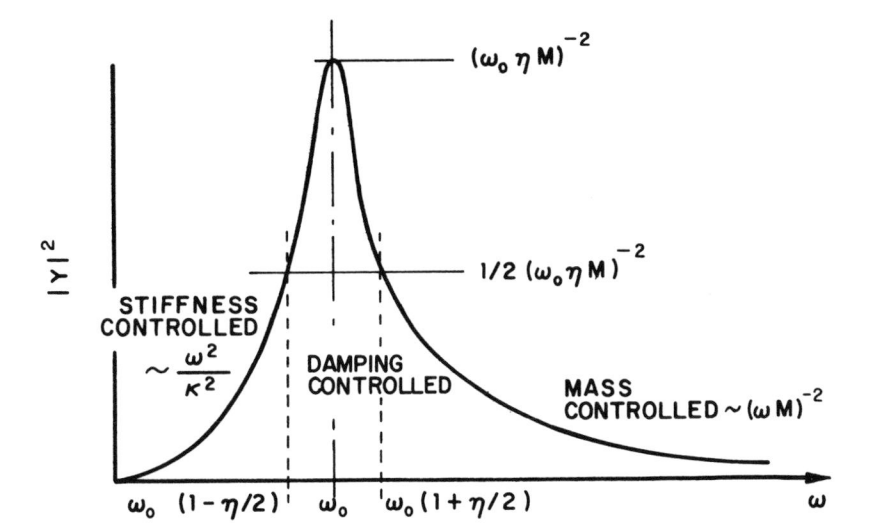

FIG. 2.2

ADMITTANCE OF LINEAR RESONATOR

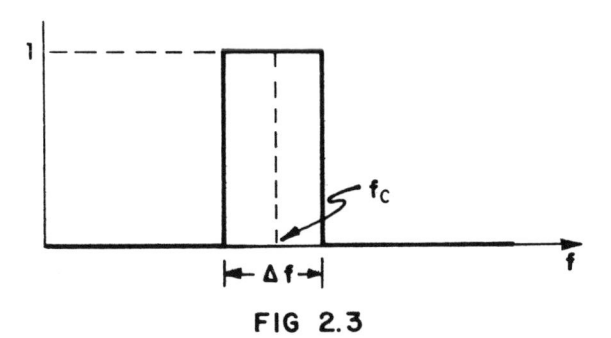

FIG 2.3

FREQUENCY RESPONSE OF IDEAL RECTANGULAR FILTER

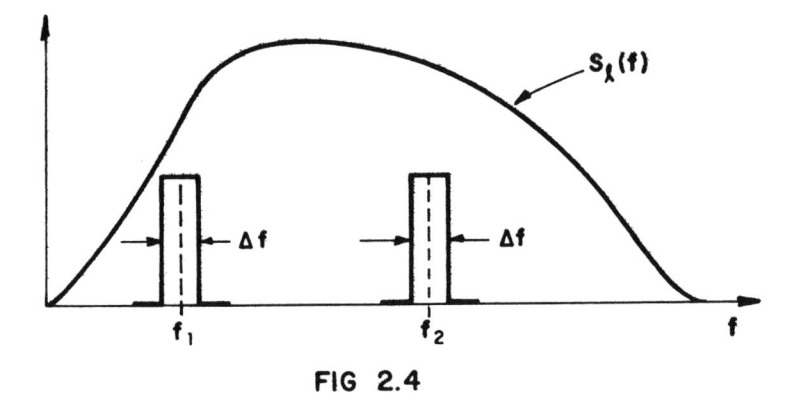

FIG 2.4

SAMPLING OF LOADING SPECTRUM BY TWO NARROW BAND FILTERS

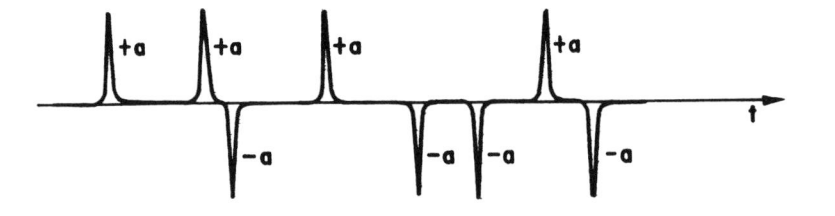

FIG 2.5

TEMPORAL REPRESENTATION OF WHITE NOISE

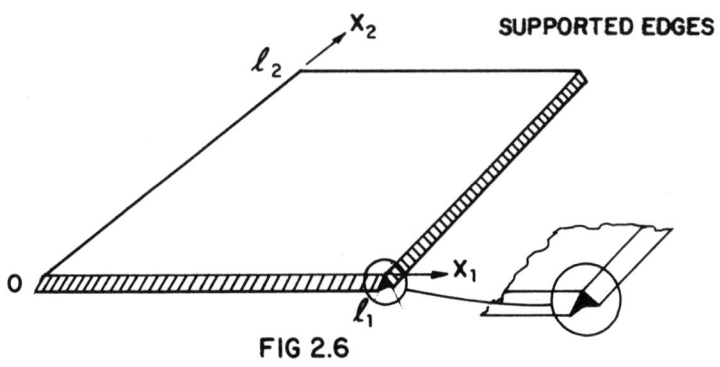

FIG 2.6

COORDINATES OF RECTANGULAR PLATE

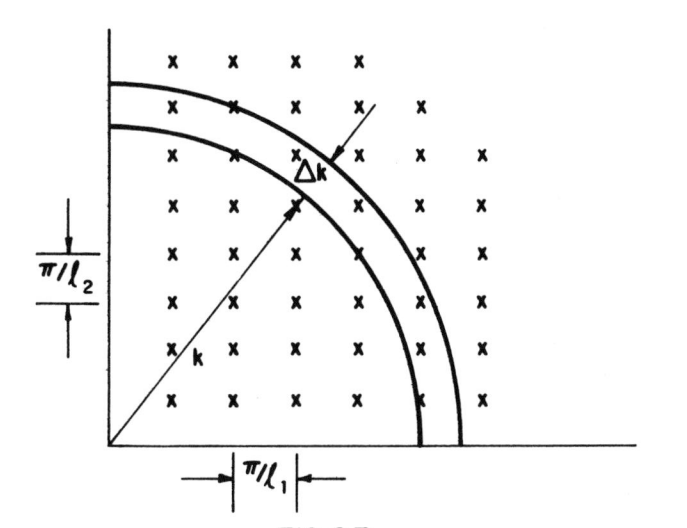

FIG 2.7

WAVE NUMBER LATTICE FOR RECTANGULAR SUPPORTED PLATE

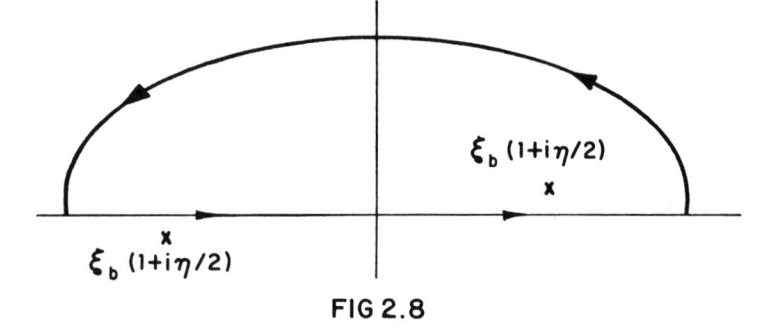

FIG 2.8

CONTOUR FOR EVALUATION OF INTEGRAL IN EQUATION 2.3.16

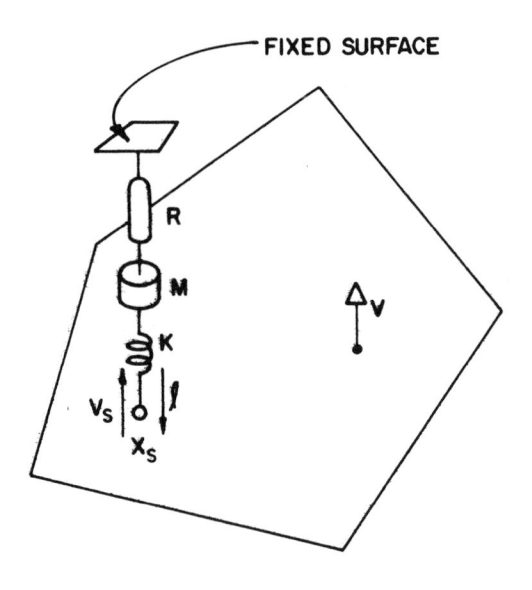

FIG 2.9

INTERACTION OF SINGLE RESONATOR AND FINITE PLATE

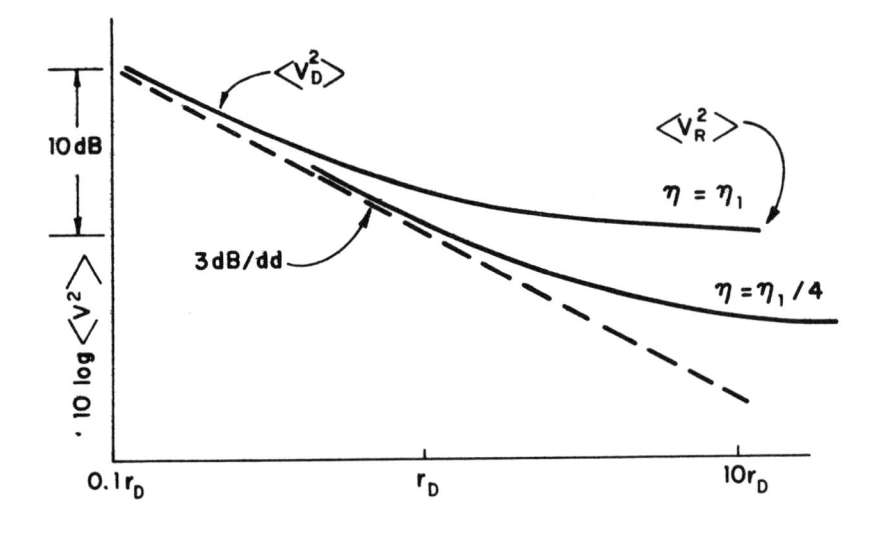

FIG 2.10

DIRECT AND REVERBERANT FIELDS ON ELASTIC PLATE

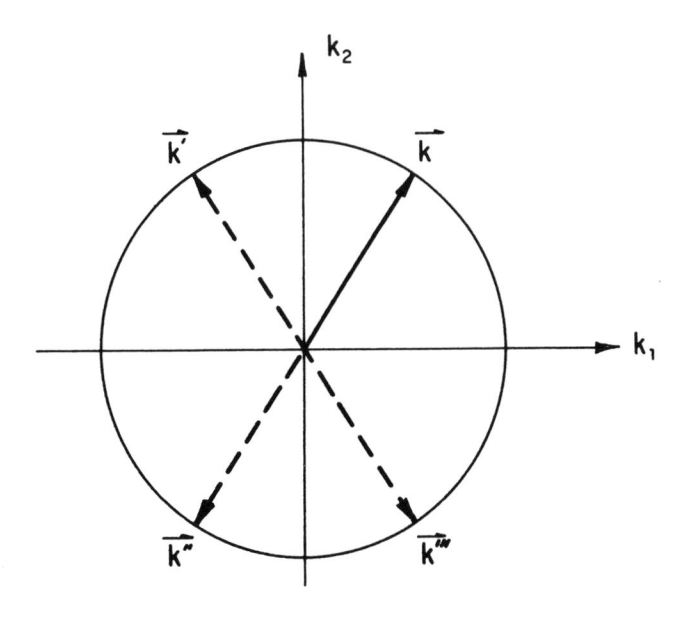

FIG 2.11

**CONSTRUCTION OF 4 WAVE VECTORS
CORRESPONDING TO A SINGLE MODE**

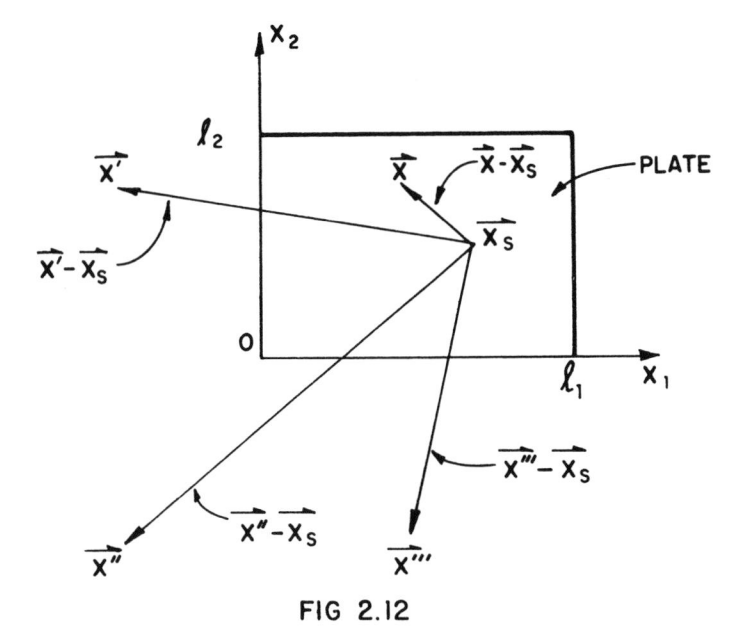

FIG 2.12

**CONFIGURATION OF SOURCE POINT AND THE
OBSERVATION POINT AND ITS IMAGES**

CHAPTER 3. ENERGY SHARING BY COUPLED SYSTEMS

3.0 Introduction

In SEA, we are principally concerned with systems that may be subdivided into subsystems that are directly excited, and other subsystems that are indirectly excited. For example, a vehicle excited by a turbulent boundary layer may be thought of as a single system excited by random noise, but we are more likely inclined to treat it as a collection of subsystems. One of these will be the exterior shell of the vehicle, which is directly in contact with the turbulence. A second system will include interior structures, such as bulkheads, shelves, etc. which are excited through their connections to the exterior shell. In this case, such a breakdown is immediate because of the nature of these structural elements and their configuration.

When the subsystems have been defined, we find that the effect of their connections is to provide coupling between them. This coupling is the mechanism by which vibrational energy is transferred from directly excited to indirectly excited systems. If a modal description has been applied to the subsystems, then intermodal coupling forces result from the connection. If a wave description has been used, the coupling is more naturally represented by impedances at the junction between the structure. In certain cases, one finds it convenient to use a mixed description, one of the subsystems being treated as a wave system and the other as a modal or resonator system. It is, therefore, quite important that we be able to shift our viewpoint between these descriptions as the need arises.

An important key to use of SEA is the definition of subsystems. This is a part of the modeling process that we shall deal with later on. Even for "obvious" choices of subsystems, the breakdown has an arbitrary element to it. In many cases, structural elements are the subsystems, but in others, modal (or wave) types may be the key. The shell of a vehicle has modes that represent mainly transverse or flexural vibration and modes of longitudinal or torsional in-plane motion. It may be necessary to treat transverse and in-plane modes as two separate subsystems, particularly if their excitation by the environment or their interaction with the rest of the structure is markedly different.

We begin the chapter by discussing the interaction of (sub) systems described in modal terms. Single resonator systems are dealt with first, and some important basic theorems of SEA are derived. Then, multi-modal systems are studied. Various expressions for the energy flow are presented in this section.

In paragraph 3.3 of this chapter, we study interactions of systems using a wave description. Impedances and energy densities become the important interaction and dynamical quantities here. The definition of transmissibility coefficients is also made, a parameter that is widely used in acoustical analysis of interacting systems. Equivalences between the parameters of modal and wave descriptions are then drawn. We also show how reciprocity arguments may be used to develop response estimates which result in the same results as previously found. Reciprocity methods are very powerful in obtaining equivalences between response parameters in mechanical systems and one of the goals of this chapter is to show how they fit into the general SEA framework.

We complete the chapter with the analysis of some fairly simple systems that have practical importance. This serves to illustrate the methods developed in the chapter, and also to show how we begin to develop a catalog of SEA parameters as a result of the applications that have been made of SEA.

3.1   Energy Sharing Among Resonators

We began our discussion of vibration in Chapter 2 with a single dof system. Although quite artificial, we were able to develop basic ideas from the 1 dof system that were applicable to much more complex systems. This approach will be followed in the present chapter also. We shall study the energy interactions of the system shown in Fig. 3.1, consisting of two simple linear resonators coupled by conservative elements. We shall find that much more complex systems can be represented by this system in the following sections.

The equation of motion of mass M_1 is given by calculating all forces on it, including those arising from the motion of mass M_2.

In this instance, this is best done by using the Lagrangian. The kinetic energy of the system is

$$KE = \frac{1}{2} M_1 \dot{y}_1^2 + \frac{1}{2} M_2 \dot{y}_2^2 + \frac{1}{8} M_c (\dot{y}_1 + \dot{y}_2)^2$$

The potential energy of the system is:

$$PE = \frac{1}{2} K_1 y_1^2 + \frac{1}{2} K_2 y_2^2 + \frac{1}{2} K_c (y_1 - y_2)^2 \qquad (3.1.1)$$

The equations of motion in the absence of velocity dependent forces are given by

$$\frac{d}{dt} \frac{\partial KE}{\partial \dot{y}_i} - \frac{\partial PE}{\partial y_i} = L_i \qquad (3.1.2)$$

Using Eq. (3.1.1) in Eq. (3.1.2) results in velocity independent equations of motion. In addition, we include the damping force $R\dot{y}$ and the gyroscopic coupling force $G\dot{y}$, to obtain

$$(M_1 + \frac{1}{4} M_c) \ddot{y}_1 + R_1 \dot{y}_1 + (K_1 + K_c) y_1$$

$$= L_1 + K_c y_2 + G\dot{y}_2 - \frac{1}{4} M_c \ddot{y}_2 \qquad (3.1.3a)$$

$$(M_2 + \frac{1}{4} M_c) \ddot{y}_2 + R_2 \dot{y}_2 + (K_2 + K_c) y_2$$

$$= L_2 + K_c y_1 - G\dot{y}_1 - \frac{1}{4} M_c \ddot{y}_1 \qquad (3.1.3b)$$

These equations clearly display the effect of motion in one system causing forces (and power input to or absorption from) the other.

In application to real systems, the question invariably arises: "What are the uncoupled structures?" From a mathematical viewpoint, we may obtain uncoupled equations by "clamping" system (2) [setting $y_2 = \dot{y}_2 = \ddot{y}_2 = 0$ in Eq. (3.1.3a)] and by clamping system (1) in Eq. (3.1.3b). An alternative is to reduce the coupling parameter to zero, i.e., setting $K_c = M_c = G = 0$ in both equations. Either is possible, but note that the effective mass and stiffness of each resonator will change in the process of decoupling if we use the latter approach. Since we do not want to change the resonator we deal with in the process of removing the coupling, we prefer the "clamping" definition of the uncoupled system.

The clamping or blocking concept is also appealing from an experimental viewpoint. A welded junction between a pipe and a plenum chamber for example is not readily identified in terms of the coupling parameters of Eq. (3.1.3). Nevertheless, we can imagine reducing the displacements of one element to zero by suitable clamping arrangement, even though it might be a very difficult thing to do from a practical viewpoint. If stress variables are used then the systems are decoupled by being cut apart. Such a thought experiment is of particular relevance in the event that one has a finite element computer model of each subsystem since the decoupling can be done by some simple instructions. In this monograph we will consistently define the decoupled forms of Eq. (3.1.3) or its counterparts to be found by setting the response variables of the other systems to be zero.

Eq. (3.1.3) displays more symmetry when rewritten as

$$\ddot{y}_1 + \Delta_1 \dot{y}_1 + \omega_1^2 y_1 + \frac{1}{\lambda} \left[\mu \ddot{y}_2 - \gamma \dot{y}_2 - \kappa y_2 \right] = \ell_1 \qquad (3.1.4a)$$

$$\ddot{y}_2 + \Delta_2 \dot{y}_2 + \omega_2^2 y_2 + \lambda \left[\mu \ddot{y}_1 + \gamma \dot{y}_1 - \kappa y_1 \right] = \ell_2 \qquad (3.1.4b)$$

where

$$\Delta_i = R_i / \omega_i \, (M_i + M_c/4) \; , \; \omega_1^2 = (K_i + K_c) / (M_i + M_c/4) \; ,$$

$$\mu = (M_c/4) \, (M_1 + M_c/4)^{-\frac{1}{2}} \, (M_2 + M_c/4)^{-\frac{1}{2}}$$

$$\gamma = G / (M_1 + M_c/4) \quad (M_2 + M_c/4)^{\frac{1}{2}} \; ,$$

$$\kappa = K_c / (M_1 + M_c/4)^{\frac{1}{2}} \, (M_2 + M_c/4)^{\frac{1}{2}}$$

and $\qquad \lambda \; = \; (M_1 + M_c/4)^{\frac{1}{2}} \, (M_2 + M_c/4)^{-\frac{1}{2}}.$

Note that the equations could be made completely symmetric (that is, λ would vanish) by further defining variables $y_i' = y_i (M_i + M_c/4)^{\frac{1}{2}}$. Also note $\ell_i = L_i / (M_i + M_c/4)$.

In Chapter 2, we were interested in the energy of vibration of a single resonator excited by "white" or broad band noise. In this Chapter we are concerned with the response of two resonators when both ℓ_1 and ℓ_2 are independent white noise sources. Since Eqs. (3.1.4) are linear, such an analysis is fairly straightforward, but it is instructive to go through it in some detail. Since we want to apply these results to energy flow problems, we shall be particularly interested in calculating the power flow between the resonators.

Before we treat Eqs. (3.1.4) is all of their detail, however, let us consider a particularly simple case: that for $\Delta_1 = \Delta_2 = \Delta$, $\omega_1 = \omega_2 = \omega_0$, $\lambda = 1$, and $\mu = \gamma = 0$. Thus, we have two identical resonators with stiffness coupling only. If we add and subtract Eqs. (3.1.4), we obtain,

$$\ddot{z}_1 + \Delta \dot{z}_1 + \Omega_1^2 \, z_1 = g_1 \tag{3.1.5a}$$

$$\ddot{z}_2 + \Delta \dot{z}_2 + \Omega_2^2 \, z_2 = g_2 \tag{3.1.5b}$$

when

$$z_1 = y_1 + y_2, \quad z_2 = y_1 - y_2, \quad g_1 = \ell_1 + \ell_2, \quad g_2 = \ell_1 - \ell_2,$$

$$\Omega_1^2 = \omega_0^2 - \kappa, \quad \Omega_2^2 = \omega_0^2 + \kappa.$$

From elementary dynamics, we know that if we start this system vibrating without damping or other excitation present, that the general free vibration response of either mass is of the form

$$y_1(t) = A \sin(\Omega_1 t + \phi_2) + B \sin(\Omega_2 t + \phi_2), \qquad (3.1.6)$$

If A and B are nearly equal in magnitude (this will occur if one mass is held fixed while the other is displaced and then released) then the motion of either mass is of the form

$$y_1(t) \simeq C \sin\left(\frac{\kappa}{\omega_0} t + \psi_a\right) \sin(\omega_0 t + \psi_b), \qquad (3.1.7)$$

which represents an oscillation that is modulated by an envelope that varies at radian frequency κ/ω_0. Indeed, this beating phenomenon is one of the best known and most remarkable features of coupled oscillators.

If the resonators have damping Δ, however, then they will have a decaying motion represented by an additional "modulation" of Eq. (3.1.7) by the term $e^{-\Delta t/2}$ [see Eq. (2.1.9)]. If the damping is very light, so that $\kappa/\omega_0 > \Delta$, then the beating phenomenon will still be apparent but if the damping is very strong compared to the coupling, so that $\Delta >> \kappa/\omega_0$, then the energy of vibration will be dissipated before the beating oscillations can take place. We shall find that quite parallel conclusions may be drawn when the resonators are excited by a white noise source.

Let us now return to Eqs. (3.1.5) and set $\ell_2 = 0$. Then by examining $\langle \dot{y}_2^2 \rangle$, we can determine the power that is flowing into the indirectly excited resonator. When $\ell_2 = 0$, then $g_1 = g_2$ and the responses y_1, y_2 will be identical, except that their response curves will be shifted as shown in

Fig. 3.2. When there is no coupling, then $z_1 = z_2$
and $\dot{y}_2 = 1/2\,(\dot{z}_1 - \dot{z}_2) = 0$. When the coupling is such that
$\kappa/\omega_0 << \Delta$, then the spectrum of $\dot{y}_2$ is very nearly the difference between the spectra $S_{\dot{z}_1}$ and $S_{\dot{z}_2}$, as shown in
Fig. 3.2. The response y_2 is weak in this case, very broad
band but with a slight double peak in its spectrum that
would show a slight beating effect. The response
$\dot{y}_2 = 1/2\,(\dot{z}_1 + \dot{z}_2) = \dot{z}_1$ has the spectrum $S_{\dot{y}_1} \approx S_{\dot{z}_1}$.

When the coupling is increased, the two spectra shift
farther apart, and the difference spectrum is increased, as
shown. This means that $\langle \dot{y}_2^2 \rangle$ is increasing with the increase
in coupling. The spectrum of $\dot{y}_2$ is even more strongly
double peaked, and fluctuations at the difference frequency
κ/ω_0 (beats) will become more pronounced. Also, since
$\dot{z}_1$ and $\dot{z}_2$ are no longer identical, $\langle \dot{y}_1^2 \rangle$ will be diminished.

Finally, when the coupling is strong enough so that the
resonances have shifted beyond the damping bandwidth
$(\kappa/\omega_0 > \Delta)$, the response $\dot{y}_1$ and $\dot{y}_2$ are statistically independent
since they do not share common frequency components.
Accordingly, $\langle \dot{y}_1^2 \rangle = \langle \dot{y}_2^2 \rangle = \langle \dot{z}_1^2 \rangle /2$. Thus, each resonator dis-
sipates 1/2 of the power supplied by the white noise source
ℓ_1. This power depends only on the mass of resonator "1"
and not on the coupling [see discussion in connection with
Eq. (2.1.39)]. Now the "beating" phenomenon will be quite
strong implying an oscillation of energy in the system.
This flow is in addition to the steady power flow from
resonator "1" to resonator "2" required to supply the power
dissipated by resonator $\Delta \langle \dot{y}_2^2 \rangle$.

Let us now return to the more complex system represented
by Eq. (3.1.3) and (3.1.4). We assume that both L_1 and L_2
are independent white noise sources. The power supplied by
the source L_1 is $\langle L_1 \dot{y}_1 \rangle$, or

$$\langle L_1 \dot{y}_1 \rangle = (M_1 + \tfrac{1}{4} M_c)\ \langle \ddot{y}_1 \dot{y}_1 \rangle + R_1 \langle \dot{y}_1^2 \rangle + (K_1 + K_c) \langle y_1 \dot{y}_1 \rangle$$

$$- K_c\ \langle y_2 \dot{y}_1 \rangle - G\ \langle \dot{y}_2 \dot{y}_1 \rangle + \tfrac{1}{4} M_c \langle \ddot{y}_2 \dot{y}_1 \rangle \qquad (3.1.8)$$

Since $\langle \tfrac{d}{dt}\,\dot{y}_1^2 \rangle = \langle \tfrac{d}{dt}\,y_1^2 \rangle = 0$ for stationary processes, the first and third terms on the right hand side of Eq. (3.1.8) vanish. The term $R\langle\dot{y}_1^2\rangle$ represents dissipation by the damper R_1 and the other terms represent power flow into the coupling elements.

In a similar fashion, the power supplied by source 2 is

$$\langle L_2\dot{y}_2\rangle = R \langle\dot{y}_2^2\rangle + K_c \langle y_1\dot{y}_2\rangle + G \langle\dot{y}_1\dot{y}_2\rangle + \tfrac{1}{4}M_c \langle\ddot{y}_1\dot{y}_2\rangle$$

(3.1.9)

Using the fact that

$$\langle\tfrac{d}{dt} \dot{y}_1\dot{y}_2\rangle = \langle\ddot{y}_1\dot{y}_2\rangle + \langle\dot{y}_2\dot{y}_1\rangle = 0$$

and

$$\langle\tfrac{d}{dt} \dot{y}_1\dot{y}_2\rangle = \langle\ddot{y}_1\dot{y}_2\rangle + \langle\ddot{y}_2\dot{y}_1\rangle = 0,$$

we can add Eqs. (3.1.8) and (3.1.9) to obtain

$$\langle L_1\dot{y}_1\rangle + \langle L_2\dot{y}_2\rangle = R_1 \langle\dot{y}_1^2\rangle + R_2 \langle\dot{y}_2^2\rangle \qquad (3.1.10)$$

which states that the total power supplied is dissipated in the two damping elements; the coupling elements in the form presented are non-dissipative.

We also note from Eq. (3.1.8) or (3.1.9) that the power flow from resonator 1 to resonator 2 is simply

$$\Pi_{12} = - K_c \langle y_2\dot{y}_1\rangle - G \langle\dot{y}_2\dot{y}_1\rangle - \tfrac{1}{4}M_c \langle\ddot{y}_2\dot{y}_1\rangle. \qquad (3.1.11)$$

Our next task is to evaluate these averages in terms of the spectral densities of the sources and the system parameters. This evaluation may be done in the time or frequency domain.

In the frequency domain calculation, we introduce the complex system frequency response $H_{pq}(\omega)$ defined as the response y_p to a force $L_q = e^{-i\omega t}$. Thus, all time derivatives in Eq. (3.1.3) are replaced by $\frac{d}{dt} \to -i\omega$, and we have

$$H_{11}(\omega) = (M_1 + \tfrac{1}{4}M_c)^{-1} (-\omega^2 - i\omega\Delta_2 + \omega_2^2)/D \qquad (3.1.12a)$$

$$H_{21}(\omega) = \lambda(M_1 + \tfrac{1}{4}M_c)^{-1} (\omega^2\mu + i\omega\gamma + \kappa)/D \qquad (3.1.12b)$$

$$H_{12}(\omega) = (M_2 + \tfrac{1}{4}M_c)^{-1}\lambda^{-1} (\omega^2\mu - i\omega\gamma + \kappa)/D \qquad (3.1.12c)$$

$$H_{22}(\omega) = (M_2 + \tfrac{1}{4}M_c)^{-1} (-\omega^2 - i\omega\Delta_1 + \omega_1^2)/D \qquad (3.1.12d)$$

where D is the system determinant given by

$$D = \omega^4(1-\mu^2) + i\omega^3(\Delta_1+\Delta_2) - \omega^2(\omega_1^2+\omega_2^2+\Delta_1\Delta_2+\gamma^2+2\mu\kappa)$$

$$- i\omega(\Delta_1\omega_2^2+\Delta_2\omega_1^2) + (\omega_1^2\omega_2^2-\kappa^2). \qquad (3.1.13)$$

Using these transfer functions, the various second moments of the dynamical variables can be generated by integration over frequency from $-\infty$ to $+\infty$:

$$\langle y_1^2 \rangle = S_{L_1} \int |H_{11}|^2 \, d\omega + S_{L_2} \int |H_{12}|^2 \, d\omega \qquad (3.1.14a)$$

$$\langle y_2^2 \rangle = S_{L_2} \int |H_{22}|^2 \, d\omega + S_{L_1} \int |H_{21}|^2 \, d\omega \qquad (3.1.14b)$$

$$\langle \dot{y}_1^2 \rangle = S_{L_1} \int \omega^2 |H_{11}|^2 \, d\omega + S_{L_2} \int \omega^2 |H_{12}|^2 \, d\omega \quad (3.1.14c)$$

$$\langle \dot{y}_2^2 \rangle = S_{L_2} \int \omega^2 |H_{22}|^2 \, d\omega + S_{L_1} \int \omega^2 |H_{21}|^2 \, d\omega \quad (3.1.14d)$$

$$\langle y_2 \dot{y}_1 \rangle = S_{L_1} \int -i\omega^3 H_{11} \, H_{21}^* \, d\omega + S_{L_2} \int -i\omega \, H_{12} \, H_{22}^* d\omega$$

$$(3.1.14e)$$

$$\langle \dot{y}_2 \dot{y}_1 \rangle = S_{L_1} \int \omega^2 \, H_{11} \, H_{21}^* \, d\omega + S_{L_2} \int \omega^2 \, H_{12}^* \, H_{22} \, d\omega$$

$$' \qquad (3.1.14f)$$

$$\langle \ddot{y}_2 \dot{y}_1 \rangle = S_{L_1} \int -i\omega^3 \, H_{11}^* \, H_{21} \, d\omega + S_{L_2} \int -i\omega^3 \, H_{12}^* \, H_{22} \, d\omega$$

$$(3.1.14g)$$

The final three relations in Eqs. (3.1.14 e, f, g) are the terms we need to determine power flow, the first four, Eqs. (3.1.14a, b, c, d) are needed in evaluating the energy of vibration. In these equations we have used the "two-sided" spectrum $S_L(\omega)$ which has a range from $\omega = -\infty$ $\omega = +\infty$ rather than the more "physical" one-sided spectral density in cyclic frequency f which has a range from $f = 0$ to $f = \infty$. Since the integral over the range of each must produce the same mean square value, i.e.,

$$\langle y^2 \rangle = \int_{-\infty}^{\infty} S_y(\omega) \, d\omega = \int_{0}^{\infty} S_y(f) \, df, \qquad (3.1.15)$$

the relation between them must be

$$S_y(f) = 2 \, S_y(\omega) \, \frac{d\omega}{df} = 4\pi \, S_y(\omega) \quad . \qquad (3.1.16)$$

We will not bother to evaluate the integral of Eqs. (3.1.14) here, since they are of standard form. Even so, the form of the relations is quite complicated. They are found to be:

$$\langle y_2 \dot{y}_1 \rangle = - \left\{ \frac{\pi S_{L_1}}{\Delta_1 (M_1 + \frac{1}{4}M_c)} - \frac{\pi S_{L_2}}{\Delta_2 (M_2 + \frac{1}{4}M_c)} \right\} \qquad (3.1.17a)$$

$$\times \; \frac{\Delta_1 \Delta_2 [\mu(\Delta_1 \omega_2^2 + \Delta_2 \omega_1^2) + \gamma(\omega_2^2 - \omega_1^2) + \kappa(\Delta_1 + \Delta_2)]}{(M_1 + \frac{1}{4}M_c)^{\frac{1}{2}} (M_c + \frac{1}{4}M_c)^{\frac{1}{2}} d}$$

$$\langle \dot{y}_2 \dot{y}_1 \rangle = - \left\{ \frac{\pi S_{L_1}}{\Delta_1 (M_1 + \frac{1}{4}M_c)} - \frac{\pi S_{L_2}}{\Delta_2 (M_2 + \frac{1}{4}M_c)} \right\}$$

$$\times \; \frac{\Delta_1 \Delta_2 [\gamma(\Delta_1 \omega_2^2 + \Delta_2 \omega_1^2) - \kappa(\omega_2^2 - \omega_1^2)]}{(M_1 + \frac{1}{4}M_c)^{\frac{1}{2}} (M_2 + \frac{1}{4}M_c)^{\frac{1}{2}} d}$$

$$- \left\{ \frac{\pi S_{L_1}}{M_1 + \frac{1}{4}M_c} + \frac{\pi S_{L_2}}{M_2 + \frac{1}{4}M_c} \right\}$$

$$\times \frac{\mu \left[(\Delta_1 \omega_2^2 + \Delta_2 \omega_1^2)(\omega_1^2 + \omega_2^2 + \Delta_1 \Delta_2 + \gamma^2 + 2\mu\kappa) - (\Delta_1 + \Delta_2)(\omega_1^2 \omega_2^2 - \kappa^2) \right]}{(M_1 + \frac{1}{4}M_c)^{\frac{1}{2}} (M_2 + \frac{1}{4}M_c)^{\frac{1}{2}} (1 - \mu^2) \ d}$$

$$+ \left\{ \frac{\omega_2^2 \pi S_{L_1}}{M_1 + \frac{1}{4}M_c} + \frac{\omega_2^2 \pi S_{L_2}}{M_2 + \frac{1}{4}M_c} \right\} \frac{\mu (\Delta_1 \omega_2^2 + \Delta_2 \omega_1^2)}{(M_1 + \frac{1}{4}M_c)^{\frac{1}{2}} (M_2 + \frac{1}{4}M_c)^{\frac{1}{2}} d}$$

$$\text{(3.1.17b)}$$

$$\langle \ddot{y}_2 \dot{y}_1 \rangle = \left\{ \frac{\pi S_{L_1}}{\Delta_1 (M_1 + \frac{1}{4}M_c)} - \frac{\pi S_{L_2}}{\Delta_2 (M_2 + \frac{1}{4}M_c)} \right\} \frac{\Delta_1 \Delta_2}{(M_1 + \frac{1}{4}M_c)^{\frac{1}{2}} (M_2 + \frac{1}{4}M_c)^{\frac{1}{2}} d}$$

$$\times \left[\frac{\mu}{1 - \mu^2} (\Delta_1 \omega_2^2 + \Delta_2 \omega_1^2)(\omega_1^2 + \omega_2^2 + \Delta_1 \Delta_2 + \gamma^2 + 2\mu\kappa) \right.$$

$$\left. - (\Delta_1 + \Delta_2)(\omega_1^2 \omega_2^2 - \kappa^2) + \kappa (\Delta_1 \omega_2^2 + \Delta_2 \omega_1^2) \right]$$

$$- \left\{ \frac{\pi S_{L_1}}{M_1 + \frac{1}{4}M_c} + \frac{\pi S_{L_2}}{M_2 + \frac{1}{4}M_c} \right\}$$

$$X \quad \frac{\gamma[(\Delta_1 \omega_2^2 + \Delta_2 \omega_1^2)(\omega_1^2 + \omega_2^2 + \Delta_1 \Delta_2 + \gamma^2 + 2\mu\kappa) - (\Delta_1 + \Delta_2)(\omega_1^2 \omega_2^2 - \kappa^2)]}{(M_1 + \frac{1}{4}M_c)^{\frac{1}{2}} (M_2 + \frac{1}{4}M_c)^{\frac{1}{2}} (1 - \mu^2) \, d}$$

$$+ \left\{ \frac{\omega_2^2 \pi S_{L_1}}{M_1 + \frac{1}{4}M_c} + \frac{\omega_2^2 \pi S_{L_1}}{M_2 + \frac{1}{4}M_c} \right\} \quad \frac{\gamma[\Delta_1 \omega_2^2 + \Delta_2 \omega_1^2]}{(M_1 + \frac{1}{4}M_c)^{\frac{1}{2}} (M_2 + \frac{1}{4}M_c)^{\frac{1}{2}} \, d}$$

$$(3.1.17c)$$

The quantity d that appears in these equations is given by:

$$d = \Delta_1 \Delta_2 [(\omega_1^2 - \omega_2^2)^2 + (\Delta_1 + \Delta_2)(\Delta_1 \omega_2^2 + \Delta_2 \omega_1^2)] + \mu^2 (\Delta_1 \omega_2^2 + \Delta_2 \omega_1^2)^2$$

$$+ (\gamma^2 + 2\mu\kappa)(\Delta_1 + \Delta_2)(\Delta_1 \omega_2^2 + \Delta_2 \omega_1^2) + \kappa^2 (\Delta_1 + \Delta_2)^2 .$$

$$(3.1.18)$$

These expressions are now placed into Eq. (3.1.11) to evaluate the power flow. Incredibly, they simplify enormously (although they are still relatively complicated). The result of the substitution is

$$\Pi_{12} = A \left[\frac{\pi S_{L_1}}{\Delta_1 (M_1 + \frac{1}{4}M_c)} - \frac{\pi S_{L_2}}{\Delta_2 (M_2 + \frac{1}{4}M_c)} \right] \qquad (3.1.19)$$

where

$$A = \frac{\Delta_1 \Delta_2}{(1-\mu^2)d} \left\{ \mu^2 [\Delta_1 \omega_2^4 + \Delta_2 \omega_1^4 + \Delta_1 \Delta_2 (\Delta_1 \omega_2^2 + \Delta_2 \omega_1^2)] \right.$$

$$\left. + (\gamma^2 + 2\mu\kappa)[\Delta_1 \omega_2^2 + \Delta_2 \omega_1^2] + \kappa^2 (\Delta_1 + \Delta_2) \right\}$$

$$(3.1.20)$$

and d is given by Eq. (3.1.18).

Comparing Eqs. (3.1.4) to Eq. (2.1.1), we see that $\Delta_1 = \omega_1 \eta_1$ and $\Delta_2 = \omega_2 \eta_2$, where the η's are the loss factors of each uncoupled resonator. Thus, when $x_2 = x_2 = 0$, the term is

$$\frac{\pi S_{L_1}(\omega)}{\Delta_1 (M_1 + \frac{1}{4}M_c)} = \frac{S_{L_1}(f)}{4\omega_1 \eta_1 (M_1 + \frac{1}{4}M_c)} = \langle \dot{y}_1^2 \rangle \ (M_1 + \frac{1}{4}M_c) \qquad (3.1.21)$$

according to Eqs. (2.1.39) and (2.1.40), the energy of

resonator "1". Similarly

$$\frac{\pi S_{L_2}(\omega)}{\Delta_2 (M_2+\frac{1}{4}M_c)} \quad = \quad (M_2+\frac{1}{4}M_c) \quad \langle \dot{y}_2^2 \rangle \qquad\qquad (3.1.22)$$

is the energy of resonator "2" when $y_1=\dot{y}_2=0$. Thus, Eq.
(3.1.19) states that the power flow from resonator "1" to
resonator "2" is

(a)   directly proportional to the difference in decoupled
      energy of the resonators, where decoupling is de-
      fined by constraining the resonator, the energy of
      which is not being evaluated

(b)   since the quantity A is positive definite, the
      average power flow is from the resonator of greater
      to lesser energy

(c)   the quantity A is also symmetric in the system
      parameters so that an equal difference of resonator
      energies in either direction (1→2 or 2→1) will
      result in an equal power flow, we may say that power
      flow is reciprocal.

The concept of power flow between resonators being
proportional to the difference in *uncoupled* energies is
useful. When we are exciting a structure with noise so
that the input power is known, then since S_L/M is proportional to input power, this interpretation of Eq. (3.1.19)
is the appropriate one. However, we often do not know the
input power, we only know the m.s. response values $\langle y^2 \rangle$ and
$\langle y^2 \rangle$ by measurements taken on the system as it vibrates in
its coupled state. We need to relate Eq. (3.1.19) to these
values also. To do this, we require calculations of $\langle y_1^2 \rangle$
$\langle y_2^2 \rangle$, $\langle \dot{y}_1^2 \rangle$ and $\langle \dot{y}_2^2 \rangle$ from Eqs. (3.1.14). The results of the
calculations are as follows:

$$\langle y_1^2 \rangle = \frac{\pi S_{L_1}}{(M_1 + \frac{1}{4}M_c)^2 (\omega_1^2 \omega_2^2 - \kappa^2) d} \left\{ \Delta_2 \omega_2^2 [(\omega_1^2 - \omega_2^2)^2 \right.$$

$$+ (\Delta_1 + \Delta_2)(\Delta_1 \omega_2^2 + \Delta_2 \omega_1^2)] + \mu^2 [\omega_2^4 (\Delta_1 \omega_2^2 + \Delta_2 \omega_1^2)]$$

$$+ (\gamma^2 + 2\bar{\mu}\kappa)[\omega_2^2 (\Delta_1 + \Delta_2)] - \kappa^2 [(\Delta_1 + \Delta_2)(\Delta_2^2 - 2\omega_2^2)$$

$$\left. + (\Delta_1 \omega_2^2 + \Delta_2 \omega_1^2)] \right\} + \frac{\pi S_{L_2}}{(M_1 + \frac{1}{4}M_c)(M_2 + \frac{1}{4}M_c)(\omega_1^2 \omega_2^2 - \kappa^2) d}$$

$$\times \left\{ \mu^2 [\omega_1^2 \omega_2^2 (\Delta_1 \omega_2^2 + \Delta_2 \omega_1^2)] + (\gamma^2 + 2\mu\kappa)[\omega_1^2 \omega_2^2 (\Delta_1 + \Delta_2)] \right.$$

$$\left. + \kappa^2 [\Delta_1 \Delta_2 (\Delta_1 + \Delta_2) + (\Delta_1 \omega_1^2 + \Delta_2 \omega_2^2)] \right\} \qquad (3.1.23a)$$

$$\langle \dot{y}_1^2 \rangle = \frac{\pi S_{L_1}}{(M_1 + \frac{1}{4}M_c)^2 (1-\mu^2) d} \left\{ \Delta_2 [(\omega_1^2 - \omega_2^2)^2 + (\Delta_1 + \Delta_2)(\Delta_1 \omega_2^2 + \Delta_2 \omega_1^2)] \right.$$

$$= \mu^2 [\omega_2^4 (\Delta_1 + \Delta_2) + (\Delta_2^2 - 2\omega_2^2)(\Delta_1 \omega_2^2 + \Delta_2 \omega_1^2)]$$

$$+ (\gamma^2 + 2\mu\kappa) \; [\Delta_1\omega_2^2 + \Delta_2\omega_1^2] + \kappa^2 [\Delta_1 + \Delta_2] \Big\}$$

$$+ \frac{\pi S_{L_2}}{(M_1 + \frac{1}{4}M_c)(M_2 + \frac{1}{4}M_c)(1-\mu^2)d} \Big\{ \mu^2 \; [\Delta_1\omega_2^4 + \Delta_2\omega_1^4]$$

$$+ \Delta_1\Delta_2 \; (\Delta_1\omega_2^2 + \Delta_2\omega_1^2)] + (\gamma^2 + 2\mu\kappa)[\Delta_1\omega_2^2 + \Delta_2\omega_1^2 + \kappa^2(\Delta_1 + \Delta_2)] \Big\}$$

$$(3.1.23b)$$

The expressions for $\langle y_2^2 \rangle$ and $\langle \dot{y}_2^2 \rangle$ are found by revising subscripts in these two equations.

We note from Eqs. (3.1.23) that in coupled vibration $\langle \dot{y}^2 \rangle \neq \omega_1^2 \langle y_1^2 \rangle$ so that a detailed equality of potential and kinetic energy has been lost. Nevertheless, by direct substitution.

$$(M_1 + \frac{1}{4}M_c) \; \langle \dot{y}_1^2 \rangle - (M_2 + \frac{1}{4}M_c) \; \langle \dot{y}_2^2 \rangle = (K_1 + K_c)\langle y_1^2 \rangle$$

$$-(K_2 + K_c) \; \langle y_2^2 \rangle = \frac{\Delta_1\Delta_2}{d} \; [(\omega_1^2 - \omega_2^2)^2 + (\Delta_1 + \Delta_2)(\Delta_1\omega_2^2 + \Delta_2\omega_1^2)]$$

$$\left\{ \frac{\pi S_{L_1}}{\Delta_1(M_1 + \frac{1}{4}M_c)} - \frac{\pi S_{L_2}}{\Delta_2(M_2 + \frac{1}{4}M_c)} \right\}$$

$$(3.1.24)$$

Thus, we arrive at the remarkable conclusion that the power flow is proportional to the difference in the <u>actual</u> kinetic energy, or potential energy, or total energy. This is a very useful result because it allows us to calculate power flow from input power to the system or from measured vibration. We are assured that we will get the correct answer whether we base our calculations on the uncoupled systems or on the actual vibrational energy of the coupled systems.

If the average energy of resonator "1" is E_1 and the average energy of resonator "2" is E_2, then the power flow between them is, according to Eqs. (3.1.19) and (3.1.24)

$$\Pi_{12} = B(E_1 - E_2) \qquad (3.1.25)$$

where

$$B = \frac{\mu^2 [\Delta_1 \omega_2^4 + \Delta_2 \omega_1^4 + \Delta_1 \Delta_2 (\Lambda_1 \omega_2^2 + \Delta_2 \omega_1^2)] + (\gamma^2 + 2\mu\kappa)(\Delta_1 \omega_2^2 + \Delta_2 \omega_1^2) + \kappa^2 (\Delta_1 + \Delta_2)}{(1 - \mu^2)[(\omega_1^2 - \omega_2^2)^2 + (\Delta_1 + \Delta_2)(\Delta_1 \omega_2^2 + \Delta_2 \omega_1^2)]}$$

$$(3.1.26)$$

Let us suppose that resonator "2" has no direct excitation, so that $S_{L_2} = 0$. Then, the power dissipated by the resonator must equal the power transferred from "1" to "2"; $\Pi_{2,diss} = \Pi_{12} = \Delta_2 E_2 = B(E_1 - E_2)$ or,

$$\frac{E_2}{E_1} = \frac{B}{\Delta_2 + B} \qquad (3.1.27)$$

This relation shows that the largest value E_2 can have is E_1, which occurs when the coupling (determined by B) is strong compared to the damping Δ_2.

When the two resonators are identical and have stiffness coupling only, then from Eq. (3.1.26)

$$B = \frac{2\Delta\kappa^2}{4\Delta^2\omega_0^2} = \frac{\kappa^2}{2\Delta\omega_0^2} \qquad (3.1.28)$$

and,

$$\frac{E_2}{E_1} = \frac{\kappa^2/2\Delta^2\omega_0^2}{1+\kappa^2/2\Delta^2\omega_0^2}$$

which shows that $E_2 \to E_1$ as $\omega_0\Delta/\kappa \to 0$ and $E_2 \to 0$ as $\kappa \to 0$, which is consistent with our earlier discussion of this system.

We can now make some additional points to those already made in the discussion following Eq. (3.1.22). They are:

(d)   the power flow is also proportional to the actual vibrational energies of the system, the constant of proportionality being B, defined by Eq. (3.1.26)

(e)   the parameter B is positive definite and symmetric in system parameters; the system is reciprocal and power flows from the more energetic resonator to the less energetic one

(f)   if only one resonator is directly excited, the greatest possible value of energy for the second resonator is that of the first resonator.

These six points form the basis for our studies of energy
flow in mechanical and acoustical systems, since the calculation of power flow between resonators according to the
schemes indicated in Eqs. (3.1.19) and (3.1.25) will be
fundamental to our study of more complex systems. We will
have to make some additional assumptions as we proceed, but
we could not begin without these (conceptually) simple
relationships.

     Before we leave this subject, let us develop an
additional result of modal interaction that will be useful
to us in dealing with multi-dof systems. We derive the
average power into resonator 2 when the only active source is
L_1 and ω_1 is allowed to vary randomly over an interval $\Delta\omega$
that includes ω_2. To do this calculation, we see from Eq.
(3.1.19) that for any particular value of ω_1, the power flow
is $\Pi_{12} = A \ \pi S_{L_1} /\Delta_1 (M_1 + \tfrac{1}{4} M_c)$ and that an average over ω_1
requires that we evaluate $<A>_{\omega_1}$. As long as the interval
$\Delta\omega$ is not too large compared to the values of ω_1 and ω_2,
we can replace both by ω, the center frequency of the
averaging band, everywhere except in the $(\omega_1^2 - \omega_2^2)^2$ term in d,
since this is the term that is sensitive to the relative
values of ω_1 and ω_2 . We then form the average

$$\left<\frac{1}{d}\right>_{\omega_1} = \frac{1}{\Delta_1\Delta_2} \ \left<\frac{1}{(\omega_1^2-\omega_2^2)^2+\omega^2\overline{\Delta}^2}\right>_{\omega_1} = \frac{1}{\Delta_1\Delta_2} \ \frac{1}{\Delta\omega} \ \frac{\pi}{2} \ \frac{1}{\omega^2\overline{\Delta}}$$

$$(3.1.29)$$

where we have used the result leading to Eq. (2.1.37), and
have defined

$$\overline{\Delta}= (\Delta_1 +\Delta_2) \ [1+(\mu^2\omega^2+\gamma^2+2\mu\kappa+\kappa^2/\omega^2)/\Delta_1\Delta_2]^{\frac{1}{2}}. \quad (3.1.30)$$

We see that $\overline{\Delta}$ is very nearly the total damping of the two
resonators when the coupling is very weak but is more complicated for stronger coupling. The average power flow then,
is

$$\langle \Pi_{12} \rangle_\omega = \frac{\pi S'_{L_2}}{M_2 + \frac{1}{4}M_c} \tag{3.1.31}$$

and

$$S'_{L_2} = S_{L_1} \frac{M_2 + \frac{1}{4}M_c}{M_1 + \frac{1}{4}M_c} \frac{\pi \Delta_2}{2 \Delta \omega} \frac{\lambda}{(1 + \lambda)^{\frac{1}{2}}} \tag{3.1.32}$$

with

$$\lambda \equiv [\mu^2 \omega^2 + (\gamma^2 + 2\mu\kappa) + \kappa^2/\omega^2]/\Delta_1 \Delta_2 . \tag{3.1.33}$$

and we have assumed $\mu^2 \ll 1$ and $\Delta/\omega \ll 1$.

The average interaction, therefore, is equivalent to
a white noise source acting on resonator 2 of spectral
density S'_{L_2}. This effective spectral density S'_{L_2} is proportional to S_{L_1}, to the ratio of the effective bandwidth
$\pi \Delta_2/2$ of resonator 2, to the averaging bandwidth $\Delta \omega$, to the
mass ratio of the systems, and to the strength of the system
coupling as measured by $\lambda(1+\lambda)^{-\frac{1}{2}}$. The replacement of modal
interactions averaged over system parameters with white noise
sources is an important device in the development of SEA
methods.

3.2 Energy Exchange in Multi-Degree-of-Freedom Systems

We now consider two subsystems that are connected together. We combine the analysis of distributed systems in paragraph 2.2 with that for two single resonators in paragraph 3.1 to develop a theory of multi-modal interactions. The actual system that we are interested in is the situation in Fig. 3.3a, in which the two subsystems are responding to their own excitation and the interaction "forces". Each of the sub-systems vibrates "on its own" when the other system is blocked or clamped as shown in Figs. 3.3b,c, and can be analyzed according to the methods of paragraph 2.2. Thus, the equations for the blocked subsystems are:

$$\ddot{y}_i^{(b)} + (r_i/\rho_i)\dot{y}_i^{(b)} + \Lambda_i y_i^{(b)} = p_i/\rho_i \quad i=1,2 \quad (3.2.1)$$

As in Paragraph 2.2, the operators have eigenfunctions and eigenvalues

$$\Lambda_i \psi_{i\alpha} = \omega_{i\alpha}^2 \psi_{i\alpha} \qquad (3.2.2)$$

with, of course

$$<\psi_{i\alpha}\psi_{i\beta}>_{\rho_i} = \delta_{\alpha\beta} \quad . \qquad (3.2.3)$$

The boundary conditions satisfied for $\psi_{i\alpha}$ include the clamped condition on subsystem $j \neq i$.

We now suppose that the spectral densities of the forces

$$L_{i\alpha}(t) = \int dx_i p \psi_{i\alpha}$$

is flat (white) over a finite range of frequency $\Delta\omega$, and that within this band there are $N_i = n_i \Delta\omega$ modes of each subsystem, where $n_i(\omega)$ is the modal density of the ith subsystem. The modes for these subsystems may be illustrated as shown in Fig. 3.4a. Each mode group represents a model of the subsystem. In the applications of SEA, this model has some very particular properties, which we now list and discuss:

1. Each mode is assumed to have a natural frequency $\omega_{i\alpha}$ that is uniformly probable over the frequency interval $\Delta\omega$. This means that each subsystem is a member of a population of systems that are generally physically similar, but differ enough to have randomly distributed parameters. The assumption is based on the fact that nominally identical structures or acoustical spaces will have uncertainties in modal parameters, particularly at higher frequencies.

2. We assume that every mode in a subsystem is equally energetic, and that its amplitudes

$$Y_{i\alpha}(t) = \int \rho_i Y_i \psi_{i\alpha} dx_i / M$$

    are incoherent, that is,

$$\langle Y_{i\alpha} Y_{i\beta} \rangle_t = \delta_{\alpha\beta} \langle Y_{i\alpha}^2 \rangle.$$

    This assumption requires that we select mode groups for which this should be approximately correct, at least, and is an important guide to proper SEA modeling. It also implies that the excitation functions L_i are drawn from random populations of functions that have certain similarities (such as equal frequency and wavenumber spectra) but are individually incoherent.

3. As a matter of convenience, we may also assume that the damping of each mode in a subsystem is

the same. This is not essential, but it greatly simplifies the formalism and tends to be nearly true for reasonably complex subsystems.

The conditions just described are the basis for the word "statistical" in SEA. The concept of systems derived by selection of individual modes and modal excitations from random populations is of greater importance than the use of random (or noise) excitation. Indeed, we can use SEA to good accuracy in situations in which the excitation is a pure tone if there are a sufficient number of modes in inter-action to provide "good statistics". We shall see an example of this in Chapter 4.

We now "unblock" the system and consider the new equations of motion, which are

$$\ddot{y}_i + (r_i/\rho_i)\dot{y}_i + \Lambda_i y_i = [p_i + \mu_{ij}(x_i,x_j)\ddot{y}_j$$

$$+ (-)^j \gamma_{ij}(x_i,x_j)\dot{y}_j + \kappa_{ij} y_j]/\rho_i$$

$$[i \neq j; \; i,j = 1,2] \qquad\qquad (3.2.4)$$

One now expands these 2 equations in the eigenfunctions $\psi_{i\alpha}(x_i)$ to obtain

$$M_1[\ddot{Y}_\alpha + \Delta_1 \dot{Y}_\alpha + \omega_\alpha^2 Y_\alpha] = L_\alpha + \sum_\sigma [\mu_{\alpha\sigma}\ddot{Y}_\sigma + \gamma_{\alpha\sigma}\dot{Y}_\alpha + \kappa_{\alpha\sigma}Y_\sigma] \qquad (3.2.5a)$$

$$M_2[\ddot{Y}_\sigma + \Delta_2 \dot{Y}_\sigma + \omega_\sigma^2 Y_\sigma] = L_\sigma + \sum_\alpha [\mu_{\sigma\alpha}\ddot{Y}_\alpha - \gamma_{\sigma\alpha}\dot{Y}_\alpha + \kappa_{\sigma\alpha}Y_\alpha] \qquad (3.2.5b)$$

where we now reserve the indices $\alpha, \beta, \ldots$ for subsystem 1 and $\sigma, \tau, \ldots$ for subsystem 2, and we have also set $\Delta_i^\bullet = r_i/\rho_i$. The mass of subsystem i is M_i in these equations. The coupling parameters are

$$\mu_{\alpha\sigma} = \int_{\text{boundary}} \mu_{12}[x_1, x_2(x_1)] \psi_\alpha(x_1) \psi_\sigma[x_2(x_1)] \; dx_1, \text{etc.}$$

(3.2.6a)

$$\mu_{\sigma\alpha} = \int_{\text{boundary}} \mu_{21}[x_1(x_2), x_2] \psi_\alpha[x_1(x_2)] \psi_\sigma(x_2) dx_2, \text{ etc.}$$

(3.2.6b)

where the integrations are taken along the boundary between the subsystems and, therefore, over the same range of x_1, x_2 in both integrals. Conservative coupling requirements are met by $\mu_{\sigma\alpha} = \mu_{\alpha\sigma}, \gamma_{\alpha\sigma} = \gamma_{\sigma\alpha}, \kappa_{\alpha\sigma} = \kappa_{\sigma\alpha}$ or $\mu_{12} = \mu_{21}, \gamma_{12} = \gamma_{21}$, and $\kappa_{12} = \kappa_{21}$.

The coupled systems may now be represented as shown in Fig. 3.4b, with the interaction lines showing the energy flows that result from the coupling. Suppose we are interested in the energy flow between resonator (mode) α of subsystem 1 and mode σ of subsystem 2. We showed at the end of paragraph 3.1 that energy modal pair interaction, when averaged over the ensemble of systems will act like a white noise generator. Thus, modes α and σ have energies E_α and E_σ (yet to be determined) as a result of these noise generators. Further, the modal energies of subsystem 1 modes are all equal, so that $E_\alpha = E_1 = $ const. and $E_\sigma = E_2 = $ constant. (Note that E_1 and E_2 are <u>modal</u> energies). Under these circumstances we have for the inter-modal power flow

$$\Pi_{\alpha\sigma} = \langle B_{\alpha\sigma} \rangle_{\omega_\alpha, \omega_\sigma} (E_1 - E_2) \quad , \qquad (3.2.7)$$

according to Eq. (3.1.25). The average value of $B_{\alpha\sigma}$ is found in a manner completely analogous to the result in Eq. (3.1.29) to be

$$\langle B_{\alpha\sigma} \rangle_{\omega_\alpha, \omega_\sigma} = \frac{\pi}{2} \frac{\Delta_1 \Delta_2}{\Delta\omega} \langle \lambda \rangle_{\alpha, \sigma} \qquad (3.2.8)$$

where λ is defined by Eq. (3.1.33) and the average is taken with respect to frequencies $\omega_\alpha, \omega_\sigma$.

The total power flow from all N_1 modes of subsystem 1 to mode σ of subsystem 2 is, therefore,

$$\Pi_{1,\sigma} = \langle B_{\alpha\sigma} \rangle \; N_1 (E_1 - E_2) \quad .$$

Finally, the total power from subsystem 1 to subsystem 2 is found by summing over the N_2 modes of subsystem 2;

$$\Pi_{12} = \langle B_{\alpha\sigma} \rangle \; N_1 N_2 \; (E_1 - E_2)$$

$$= \frac{\pi}{2} \; \Delta\omega \; \frac{\mu^2\omega^2 + (\gamma^2 + 2\mu\kappa) + \kappa^2/\omega^2}{\delta\omega_1 \delta\omega_2} (E_1 - E_2) \qquad (3.2.9)$$

where $\delta\omega_i = \Delta\omega/N_i$ is the average frequency separation between modes.

Several interesting features may be seen in this result. The first is that the power flow is proportional to the bandwidth $\Delta\omega$, a result that agrees with our intuition. Secondly, the power flow is proportional to the difference in average actual modal energies of the two subsystems, which by now we also expect. We recall from Fig. 3.2 and Eq. (3.1.7) that κ/ω is the frequency shift produced by stiffness coupling alone. We may suspect (and more detailed analysis confirms) that the numerator of the fraction in Eq. (3.2.9) is the square of the frequency deviation produced by the coupling. The ratio of this to the square of the geometric mean of the average modal spacings of the two systems is a measure of the strength of the power flow.

Let us define the total energy of the subsystems by $E_{1,tot}$ and $E_{2,tot}$. Then, clearly since $E_1 = E_{1,tot}/N_1$ and $E_2 = E_{2,tot}/N_2$,

$$\Pi_{12} = <B_{\alpha\sigma}> N_1 N_2 \left[\frac{E_{1,tot}}{N_1} - \frac{E_{2,tot}}{N_2} \right]$$

$$\equiv \omega \eta_{12} \left[E_{1,tot} - \frac{N_1}{N_2} E_{2,tot} \right] \qquad (3.2.10)$$

where $\eta_{12} \equiv <B_{\alpha\sigma}> N_2/\omega$. If we define $\eta_{21} = N_1 \eta_{12}/N_2$, then

$$\Pi_{12} = \omega \left(\eta_{12} E_{1,tot} - \eta_{21} E_{2,tot} \right). \qquad (3.2.11)$$

The quantities η_{12} and η_{21} are called the <u>coupling loss factors</u> for the systems 1, 2. By introducing them, we have lost the symmetry of the energy flow coefficients of the preceding equations, but the coupling loss factor has strong physical appeal. Notice that $\omega\eta_{12} E_{1,tot}$ represents the power lost by subsystem 1 due to coupling, just as the quantity $\omega\eta_1 E_{1,tot}$ represents the power lost to subsystems 1 by damping, as measured by the damping loss factor $\eta = \Delta/\omega$. Of course, if system 1 is connected to another system with energy $E_{2,tot}$, it will receive an amount of power $E_{2,tot}\omega\eta_{21}$. The basic relationship

$$\eta_1\eta_{12} = \eta_2\eta_{21} \qquad (3.2.12)$$

is very useful in practical situations of power flow calculations.

We may now use these results to calculate the still unknown system energies. Consider the system of Fig. (3.4b) to be more simply represented as shown in Fig. 3.5. Then the power flow equations for systems 1 and 2 respectively are

$$\Pi_{1,in}=\Pi_{1,diss}+\Pi_{12}=\omega\left[\eta_1 E_{1,tot}+\eta_{12}E_{1,tot}-\eta_{21}E_{2,tot}\right] \qquad (3.2.13a)$$

$$\Pi_{2,in}=\Pi_{2,diss}+\Pi_{21}=\omega\left[\eta_2 E_{2,tot}+\eta_{21}E_{2,tot}-\eta_{12}E_{1,tot}\right] \qquad (3.2.13b)$$

First, consider the case where only one of the systems is directly excited by an external source. Set $\Pi_{2,in}=0$, and accordingly, from Eq. (3.2.13b) we get

$$\frac{E_{2,tot}}{E_{1,tot}} = \frac{\eta_{12}}{\eta_2+\eta_{21}} = \frac{\eta_2}{\eta_1}\frac{\eta_{21}}{\eta_2+\eta_{21}} \qquad (3.2.14)$$

Which is the multi-modal equivalent of Eq. (3.1.27). A solution of Eq. (3.2.13) in terms of both power sources is

$$E_1 = \{\Pi_{i,in}(\eta_2+\eta_{21}) + \Pi_{2,in}\eta_{21}\}/\omega D \qquad (3.2.15a)$$

$$E_2 = \{\Pi_{2,in}(\eta_1+\eta_{12}) + \Pi_{1,in}\eta_{12}\}/\omega D \qquad (3.2.15b)$$

$$D = (\eta_1+\eta_{12})(\eta_2+\eta_{21}) -\eta_{12}\eta_{21} . \qquad (3.2.15c)$$

The coupling loss factor or its various equivalent expressions is a measure of inter-modal forces at the system junction, averaged over frequency and over the modes of the interacting systems. Sometimes this calculation may be carried out directly, but more often it is found useful to adopt another view of the interaction - that developed by the impingement of waves on the boundary. Consequently, we now proceed to a discussion of the coupling of wave bearing systems.

3.3   Reciprocity and Energy Exchange in Wave Bearing Systems

In Paragraph 3.2, we described the vibrations of the systems that we are examining in terms of modes. An alternative formulation of the problem in terms of waves exists that sometimes offers both conceptual and practical computational advantages over a modal analysis. This approach was touched on in paragraph 2.3, wherein we found the dispersion relation between wavenumber and frequency in the form of Eq. (2.3.2). We examined the admittance function for a flat plate, both finite and infinite, to a point force transverse to the plane of the plate.

In paragraph 2.3, we also discussed the vibration of a simple resonator, shown in Fig. 2.9 as a result of its attachment to a plate undergoing random vibration. We found that the energy of the resonator was given by Eq. (2.3.24), which is identical in form to Eq. (3.2.14), where (the resonator is subsystem 2 and the plate is subsystem 1), one has $n_2\Delta\omega =1$ (one resonator mode), and $E_{1,tot}=M_p<v_p^2>$ and $\eta_{coup} = \eta_{21}$. Thus, the quantity $\dot{\eta}_{coup}$ that we calculated on the basis of an interaction of the resonator with many modes of the plate is just the coupling loss factor of Paragraph 3.2. But, it is also the same as

would be calculated on the basis of the resonator interacting with an infinite plate [See Eq. (2.2.23)]. Consequently,
it is often advantageous to develop the equations for coupling loss factors on the basis of interaction with an infinite
system, since such calculations may be much simpler than they
would be on a modal basis.

We should also make another point about our use of the
term "blocked" as it applies to both interacting modal and
wave systems. We define the blocked system to be that
system that results when the other system has a vanishing
response for all time. Thus, for example, when a sound
field described by its pressure response interacts with a
structure described by its displacement, the "blocked"
sound field is that field that occurs when the structure is
absolutely rigid. The "blocked" structural vibration is
that vibration which occurs when the pressure vanishes, i.e.,
the in vacuo condition of the structure. We would ordinarily
think of such a vibration as "free", but we must distinguish
between the "blocked" or "uncoupled" interaction, and our
ordinary ideas of free or constrained boundary conditions.
The interpretation depends on the variables that we use to
define "response" in the adjoining system.

We saw in Eq. (2.4.9) and in Fig. 2.11 that we could
form four simple waves out of an eigenfunction of the
rectangular plate. Although such a replacement is not
universal, it is useful to think of it as a typical example
of wave-mode duality. The wave number lattice for normal
modes such as shown in Fig. 2.7 may be re-interpreted as
a distribution of waves by reflecting the lattice points
in the various coordinate axes (or planes) of the wave
number space. Incoherence of modes means incoherence of
waves in different directions. Equality of modal energy
is translated into equality of intensity for wave groups
contained by equal angles of spheres or circles in k-space
as shown in Fig. 3.6. A cylinder has a more complicated
locus for bending modes than that shown in Fig. 3.6 but in this
case also, the equivalence between waves and modes is
essentially as shown here.

The arguments for coherence and power are based upon
the decomposition of modes into waves as developed in
Paragraph 2.4. The relation between modal energy and
power is obtained by simply noting that equal modal response
amplitudes will result in equal amplitudes of each wave in

the decomposition of Eq. (2.4.9). Secondly, if the modal responses are incoherent, the wave motions for adjacent points on the wave number circle in Fig. 3.6 will be incoherent. There is still the possibility that there will be appreciable coherence between waves in different directions at any point, since there is functional relationship between the phase of any one wave component of a mode and the 3 (for a two-dimensional mode) other components. However, as noted in the discussion preceding Eq. (2.4.10), this phase coherence is quite small unless the source and observation points are close together. If the source may be assumed to be randomly located over the surface (or volume) of the system, even this coherence will vanish.

We imagine, therefore, a diffuse wave field (in the SEA sense) possessing energy distributed uniformly over the frequency interval $\Delta\omega$. The energy density (Energy per unit "area") of this field is $\Delta\mathcal{E}$. The various parcels of energy are carried by the waves with an energy velocity c_g as defined by Eq. (2.3.7), and the power is distributed among the various directions according to an intensity

$$d(\Delta I) = \Delta\mathcal{E} \, c_g \; D(\Omega) \; d\Omega/\Omega_t \qquad (3.3.1)$$

where $d(\Delta I)$ is the intensity of wave energy in the interval of directions $d\Omega$, $D(\Omega)$ is a weighting function that essentially measures the distribution of area in the interval between k, $k + \Delta k$ as a function of Ω in Fig. 3.6(b), and Ω_t is the total range of the angle Ω. One must have $<D(\Omega)>_\Omega = 1$, and it often occurs that $D(\Omega) \equiv 1$ (this is called "isotropy").

Reciprocity. There is a general principle that applies to systems composed of linear, passive and bilateral elements, called reciprocity, that is quite useful in our discussions of wave interactions. If we imagine for the moment that the system is cut up into a very large number of tiny masses, springs and dashpots (dampers) these descriptors mean the following:

(a) linear: the mechanical response of each element is directly proportional to the force (including its sense) causing the response

(b) passive: the only sources operative are those
     explicitly set aside as sources in the
     equations of motion; no element of the system
     can generate energy.

(c) bilateral: in transferring forces from one
     neighbor to the next, a reversal of roles
     between the neighbors as to force interactions
     will result in an exact reversal of the
     relative motions. Notice that gyrators are
     excluded from the system at this level, even
     though the interaction of system modes may end
     up as a gyroscopic interaction.

The statement of reciprocity may be made with
reference to Fig. 3.7. If at any two terminal pairs of
a reciprocal system we generate a drop ℓ at terminal pair
1 and measure a flow U through a "wire" connecting the
terminal pair 2, then if a drop p' is applied to terminal
pair 2, a short circuit flow v' will be developed at
terminal pair 1. The statement of reciprocity is

$$\frac{v'}{p'} = \frac{U}{\ell} \qquad\qquad\qquad (3.3.2a)$$

The only restriction is that p', U and ℓ,v' be conjugate variables, that is, their time average product
equals the power flow appropriate to that set of terminals.
Clearly

$$\left|\frac{v'}{p'}\right| = \left|\frac{U}{\ell}\right| \qquad\qquad\qquad (3.3.2b)$$

and if p' and ℓ are noise signals with identical spectral
shapes, then

$$\frac{\langle v'^2 \rangle}{\langle p'^2 \rangle} = \frac{\langle U^2 \rangle}{\langle \ell^2 \rangle} \qquad\qquad\qquad (3.3.2c)$$

Similar statements are easily derived for the systems shown in Figs. 3.8 and 3.9. They are

$$\frac{<p^2>}{<v^2>} = \frac{<\ell'^2>}{<U'^2>}$$

(3.3.3)

for system B, and

$$\frac{<U^2>}{<v^2>} = \frac{<\ell'^2>}{<p'^2>}$$

(3.3.4)

for system C. These various statements have their applications depending on whether it is more natural for us to apply forces or velocities at one location, and measure forces or velocities at the second location.

Systems Connected at a Point. As a simple example of reciprocity consider the force exerted on a point clamp at the edge of a plate, as shown in Fig. 3.10, due to vibration of the plate. Let $\ell(t)$ be a band limited noise load applied to an arbitrary point x_s on the surface of the plate, and let ℓ_B be the load that results on the clamp. The load ℓ will inject an amount of power $<\ell^2>G_p$ into the plate (G_p is the mechanical conductance for the plate) and this will result in a m.s. plate velocity $<v_p^2>$ given by

$$<v_p^2> = <\ell^2>G_p/\omega\eta_p M_p$$

(3.3.5)

where η_p is the loss factor of the plate modes and M_p is the plate mass. Now we say that

$$<\ell_B^2> = \Gamma <v_p^2>$$

(3.3.6)

where Γ is the parameter that we seek.

In the reciprocal situation, we drive the clamp with a prescribed noise velocity $v'_B(t)$ that has the same band limited spectrum that $\ell(t)$ has. This velocity source generates an amount of power $\langle v'^2_B \rangle R_1$, ($R_1$ is the input resistance at the edge of the plate) which results in a plate m.s. velocity $\langle v'^2_p \rangle$ given by

$$\langle v'^2_p \rangle = \langle v'^2_B \rangle R_1/\omega \eta_p M_p \qquad (3.3.7)$$

If we now apply the reciprocity statement of Eq. (3.3.4), or Fig. 3.9, we obtain $\langle \ell^2_B \rangle / \langle \ell^2 \rangle = \langle v'^2_p \rangle / \langle v'^2_B \rangle$, or

$$\Gamma = R_1/G_{1,\infty}. \qquad (3.3.8)$$

Let us now extend this example by connecting a second plate to the first at the edge point under discussion, requiring that they move together in a transverse way but (to keep the matter simple) no moment is transmitted. Again, we let $\ell_1(t)$ be the band limited noise that generates a m.s. velocity on plate 1 given by $\langle v^2_1 \rangle$. We are interested in finding the velocity $\langle v^2_2 \rangle = \Gamma' \langle v^2_1 \rangle$ that results from this motion. Clearly

$$\langle v^2_2 \rangle = \Gamma' \langle v^2_1 \rangle = \Gamma' \beta \langle \ell^2_1 \rangle G_{1,\infty} \quad . \qquad (3.3.9)$$

If we now apply the load $\ell'_2(t)$ (same spectrum as ℓ) at the position where v_2 was measured, we will inject an amount of power $\langle \ell'^2_2 \rangle G_{2,\infty}$ into plate 1 through the junction (call it Π'_{21}) and a certain amount will be dissipated. From our earlier definitions of loss factors and coupling loss factors, this fraction is

$$\Pi'_{21} = \langle \ell'^2_2 \rangle G_{2,\infty} \eta_{21}/(\eta_{21} + \eta_2). \qquad (3.3.10)$$

This power, transmitted into plate 1 will produce a m.s. velocity

$$\langle v_1'^2 \rangle = \Pi_{21}'/\omega n_1 M_1 = \Pi_{21}' \beta \; . \tag{3.3.11}$$

Applying the reciprocity condition,

$$\Gamma' = \frac{G_{2,\infty}}{G_{1,\infty}} \; \frac{n_{21}}{n_2 + n_{21}} = \frac{\langle v_2^2 \rangle}{\langle v_1^2 \rangle} \; . \tag{3.3.12}$$

If the conditions are such that $G_{2,\infty} = (\pi/2)(n_2/M_2)$ and $G_{1,\infty} = (\pi/2)(n_1/M_1)$. Then Eq. (3.3.11) can be rewritten as

$$\frac{M_2 \langle v_2^2 \rangle}{n_2} = \frac{M_1 \langle v_1^2 \rangle}{n_1} \; \frac{n_{21}}{n_2 + n_{21}} \; , \tag{3.3.13}$$

which, of course, is the same result as Eq. (3.2.14). In a sense, this offers an alternative derivation of the basic power flow relations of SEA and it avoids discussion of either modes or waves. Of course, to evaluate $G_{1,\infty}$ and $G_{2,\infty}$, one must have either a modal or wave model. Also, to evaluate n_{21}, one must either do a modal analysis, following the chain from Eqs. (3.2.4, 5, 6), through Eq. (3.2.9) and (3.2.10), or proceed by wave analysis.

We can relate the power flow from plate 1 to plate 2 to the blocked force given by Eqs. (3.3.6) and (3.3.8), by subdividing the situation shown in Fig. 3.12a with the two problems shown in Fig. 3.12b. The actual force ℓ that the junction applies to plate 1 is equal to the blocked force ℓ_B, less the force induced by the velocity, $\ell_i = v Z_1$. Since our convention is that upward forces and velocities are positive, the upward velocity v imposed at the boundary will result in upward force $v Z_1$ on the edge of plate 1. The motion v will also produce an upward force $v Z_2$ on plate 2

and a consequent downward reaction force on the edge of
plate 1. Consequently,

$$\ell = -vZ_2 = \ell_B + vZ_1$$

or

$$\langle \ell_B^2 \rangle = \langle v^2 \rangle \ |Z_1 + Z_2|^2 \qquad (3.3.14)$$

We now relate $\langle \ell_B^2 \rangle$ to $\langle v_1^2 \rangle$ by Eq. (3.3.6). We can
relate $\langle v^2 \rangle$ to $\langle v_2^2 \rangle$ by a relation similar to that of
Eq. (3.3.7);

$$\langle v_2^2 \rangle = \langle v^2 \rangle \ R_2 / \ \omega M_2 \eta_2 \ . \qquad (3.3.15)$$

We may then combine Eqs. (3.3.6, 8, 14, 15) to obtain

$$\frac{M_2 \langle v_2^2 \rangle}{n_2} = \frac{M_1 \langle v_1^2 \rangle}{n_1} \left\{ \frac{2}{\pi} \ \frac{R_1 R_2}{|Z_1 + Z_2|^2} \right) \frac{1}{\omega n_2} \qquad (3.3.16)$$

In many ways, Eq. (3.3.16) is the multi-dof equivalent
of Eq. (3.1.19) since $M \langle v_1^2 \rangle$ is determined in Eq. (3.3.16) by
the power injected into system 1, and the power flow coefficient in the relation

$$a_{12} = \frac{2}{\pi \omega n_1} \ \frac{R_1 R_2}{|Z_1 + Z_2|^2} \qquad (3.3.17)$$

is the "coupling loss factor" for the uncoupled system
energies. It satisfies the same reciprocity relation that
n_{12} does - namely

$$n_1 a_{12} = n_2 a_{21} \; .$$

(3.3.18)

    The quantity a_{12} will take on various forms depending
on the nature of the interacting systems, and its relation
to n_{12} will depend on the system. We shall come back to
this in paragraph 3.4.

    The result of Eq. (3.3.17) for the system in
Eq. 3.11 is typical. Since n_{12} must ultimately be determined by a_{12}, the coupling loss factor is usually
representable in terms of certain junction impedances or
averages over such impedances. If we had allowed moment
constraints between the systems, then our impedances would
have become matrices involving both force and moment terms,
but the structure of the result would remain the same. The
impedances that enter Eq. (3.3.17), $Z_1 = R_1 - iX_1$ and
$Z_2 = R_2 - iX_2$ may be evaluated on either a modal or wave basis.
As we saw in Chapter 2, there are circumstances in which a
junction impedance can be quite simply calculated on a wave
basis if resonance frequencies are averaged over a band of
frequencies and if the drive point is also randomly located.
Thus, one important use of wave analysis in SEA is in the
calculation of junction impedances.

    Systems connected along a Line. The coupling loss
factor is fairly easy to derive when the systems are connected at one or more points. Another relatively tractable
situation is when the connection is along a line (for two
dimensional systems) or along a surface (for three dimensional systems), if the dimensions of the line (or surface)
are large compared to the length of a free wave in the
system. In this case, a set of waves in the interval $d\Omega$,
shown in Fig. 3.13 will be partially transmitted and partially
reflected at the junction. In acoustics, the ratio of the
transmitted to incident power is called the transmission
coefficient $\tau(\Omega)$ and will in general depend on the direction
Ω that the incident waves are travelling with respect to
the bounding line or surface.

By our hypothesis, therefore, the total transmitted power will be

$$\Pi_{12} = \int_{\Omega_{inc}} \tau(\Omega) \; d(\Delta I) \; L_p(\Omega) \tag{3.3.19}$$

where $d(\Delta I)$ is given by Eq. (3.3.1), $\Omega_{inc.}$ is the range of angles of propagation that impinge upon the boundary, and $L_p(\Omega)$ is the projection of the boundary length (or area) presented to the waves travelling in direction Ω. Since Π_{12} is the transmitted power due to energy $A_1 \Delta \mathcal{E}_1 = E_1$, where A_1 is the area (or volume) of system 1, we have

$$\eta_{12} = \frac{\Pi_{12}}{\omega A_1 \Delta \mathcal{E}_1} = \frac{c_g}{\omega A_1 \Omega_t} \int_{\Omega_i} \tau(\Omega) \; L_p(\Omega) D(\Omega) d\Omega. \tag{3.3.20}$$

For example, in a 2-dimensional isotropic system, let the transmitting boundary be a straight line of length L and Ω be the angle of wave incidence, as shown in Fig. 3.14. Then $D(\Omega) = 1$, $\Omega_i = \pi$, $\Omega_t = 2\pi$.

$$\eta_{12} = \frac{c_g L}{2\pi\omega A_1} \int_{-\pi/2}^{\pi/2} \tau(\Omega) \; \cos\Omega \; d\Omega = \frac{c_g L}{2\omega A_1} \langle \tau(\Omega)\cos\Omega \rangle_{\Omega_i}$$

$$[\text{2-dim'l}] \qquad (3.3.21)$$

Of course, the transmissibility must be known, but this information is usually available from existing sources, or is calculable. In the 3-dimensional acoustic case, $c = c_g =$ speed of sound; $A_1 \to V_1$, the room volume and $L \to A_w$, the area of the adjoining partition. Also $\Omega_t = 4\pi$ (steradians) and $\Omega_i = 2\pi$ steradians. Then,

$$\eta_{12} = \frac{cA_w}{2\omega V_1} \langle \tau(\Omega)\cos\phi \rangle_{\Omega_i} \quad [\text{3-dim'l}] \tag{3.3.22}$$

where ϕ is the angle between the wave vector and the
normal to the panel, and $-10 \log \langle \tau(\Omega)\cos\phi \rangle_{\Omega}$ is the
ordinary transmission loss (TL) of building acoustics.
In this situation, therefore, we would be advised to evaluate
η_{12} from existing data sources on wall transmission loss.

3.4 Some Sample Applications of SEA

Here, we present some applications of SEA that demonstrate its usefulness in relatively simple situations that
are of practical interest. The emphasis here is in the
use of relations developed in paragraph 3.3, rather than in
development of the theory. Of course, every new application
will generate formulas or data for the parameters that we
are interested in, such as modal densities, coupling loss
factor, and damping loss factor.

Resonator Excited by a Reverberant Sound Field. As
the first example we consider a resonator formed by the
small piston of mass M_0 in the wall of a room as shown in
Fig. 3.15. The piston is supported by a spring of stiffness
$M\omega_0^2$ and a dashpot having a mechanical resistance $\omega_0\eta_0 M_0$.
We shall call the room system "R" and the resonator system
"0". Then, according to Eq. (3.2.14)

$$\frac{E_0}{E_R} = \frac{1}{n_R \Delta\omega} \frac{\eta_{OR}}{\eta_0 + \eta_{OR}} \qquad (3.4.1)$$

where η_{OR} is the coupling loss factor from the resonator to
the room. We evaluate η_{OR} by using Eq. (3.3.17). The
impedance Z_R is the radiation impedance of the piston
"looking into" the room

$$Z_R = R_{rad} - i\omega M_{rad} \quad . \qquad (3.4.2)$$

These quantities are well known for the piston radiator. The impedance "looking into" the resonator was given in Eq. (2.1.19),

$$Z_0 = \omega_0 \eta_0 M_0 - i\omega M_0 (1-\omega_0^2/\omega^2) \tag{3.4.3}$$

Thus, averaging a_{OR} (defined by Eq. (3.3.17)) over the band $\Delta\omega$ gives

$$\omega_0 <a_{OR}>_{\Delta\omega} = \frac{2\Delta\omega}{\pi}\left\{\frac{R_{rad}\omega_0\eta_0 M_0}{|\omega_0\eta_0 M_0 + R_{rad} - i\omega(M_0 + M_{rad}) + i\omega_0^2 M_0/\omega|^2}\right\}_{\Delta\omega}$$

$$= \frac{\omega_0 \eta_0 R_{rad}}{\omega_0\eta_0 M_0 + R_{rad}} \tag{3.4.4}$$

and substituting $\omega_0 <a>$ into Eq. (3.3.16),

$$M_0 <v_0^2> = \frac{<p_R^2> V_R}{\rho c^2}\frac{1}{n_R\Delta\omega}\frac{R_{rad}}{\omega_0\eta_0 M_0 + R_{rad}} \tag{3.4.5}$$

where we have used $\eta_2 = \eta_0$. Comparing Eqs. (3.4.5) and (3.4.1), we see that

$$\eta_{OR} = R_{rad}/\omega M_0 . \tag{3.4.6}$$

The coupling loss factor for the piston resonator, therefore, is determined by the radiation resistance of the piston "looking into" the room. When $R_{rad} >> \omega_0\eta_0 M_0$, the energy of the resonator will equal the average model energy of the sound level.

   Two Beams in Longitudinal Vibration. As a second
example, consider the problem of two long beams connected
by a fairly weak spring, as shown in Fig. 3.16. If beam 1 is
directly excited to energy M_1 $<v_1^2>$ and beam 2 is only
excited through the connection, then by our assumptions,
the energy flow will be fairly weak into beam 2. We
may use the result expressed by Eq. (3.3.16) directly,
where we define system 1 to the beam 1 and the spring,
and system 2 to be beam 2 alone. Then, the average
mechanical impedance looking into beam 2 is

$$R_{2,\infty} = (\rho_2 c_2 A_2)^2 \; \frac{\pi}{2} \; \frac{n_2}{M_2} = \frac{1}{2} \; \rho_2 c_2 A_2 = Z_2 = R_2 \qquad (3.4.7)$$

The impedance looking into system 1 is complex

$$z_1 = \frac{R_{1,\infty}(iK/\omega)}{R_{1,\infty} + i \; K/\omega} \; \simeq \; \frac{i \, K}{\omega} \; \frac{(K/\omega)^2}{R_{1,\infty}} \qquad (3.4.8)$$

Using $R_1 = R_e(Z_1)$ and $K/\omega \ll R_{1,\infty}, R_{2,\infty}$, we get

$$\frac{M_2<v_2^2>}{n_2} = \frac{M_1<v_1^2>}{n_1} \; \frac{2}{\pi\omega n_n} \; \frac{4K^2/\omega^2}{\rho_1 c_1 A_1 \cdot \rho_2 c_2 A_2} \qquad (3.4.9)$$

In this case, $a_{12} \simeq n_{12}$ (modal energy of system 2 << modal
energy of system 1) and we can say

$$\eta_{12} = \frac{8}{\pi} \; \frac{K^2/\omega^2}{\rho_1 c_1 A_1 \cdot \rho_2 c_2 A_2} \qquad . \qquad (3.4.10)$$

Note the similarity between this result and that of Eq.
(3.1.28).

   Two rooms coupled by a limp curtain. As a final
example in this section, we consider an application of
Eq. (3.3.22). We want to evaluate η_{12} for the system shown
in Fig. 3.17. Two rooms, of volume V_1 and V_2 are separated
by a "wall" that has a mass density m and area A_w. The
average transmission coefficient for this situation is known
to be approximately

$$. \; <\tau(\Omega) \; \cos\phi>_{\Omega_i} = 3 \; \left(\frac{\rho c}{\omega m}\right)^2 \; .$$ (3.4.11)

Thus

$$\eta_{12} = \frac{3\rho^2 c^3 A_w}{2\omega^3 m^2 V_1} \; .$$ (3.4.12a)

The energy expression Eq. (3.2.14) in this case becomes simply

$$<p_2^2> V_2 = <p_1^2> V_1 \; \frac{\eta_{12}}{\eta_2 + \eta_{21}} \; .$$ (3.4.13)

Where $\eta_2 = 2.2/fT_R^{(2)}$, $T_R^{(2)}$ is the reverberation time of room
2, and

$$\eta_{21} = \frac{3\rho^2 c^3 A_w}{2\omega^3 m^2 V_2} \; .$$ (3.4.12b)

Equation (3.4.13) predicts that $p_1^2 = p_2^2$ when losses in V_2
due to transmission of sound through the wall exceed the
losses due to sound absorption in room 2.

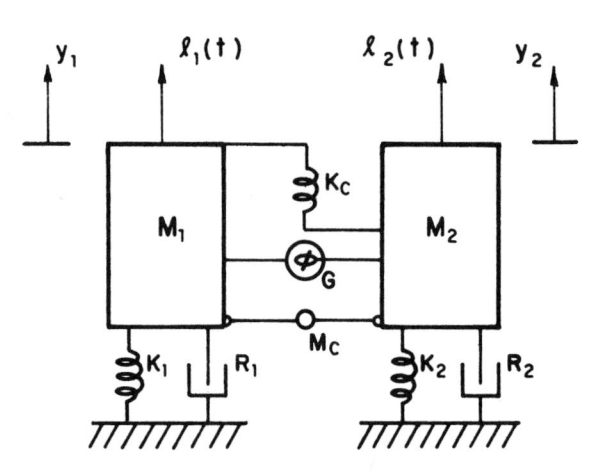

FIG 3.1

TWO LINEAR RESONATORS COUPLED BY SPRING, MASS, AND GYROSCOPIC ELEMENTS

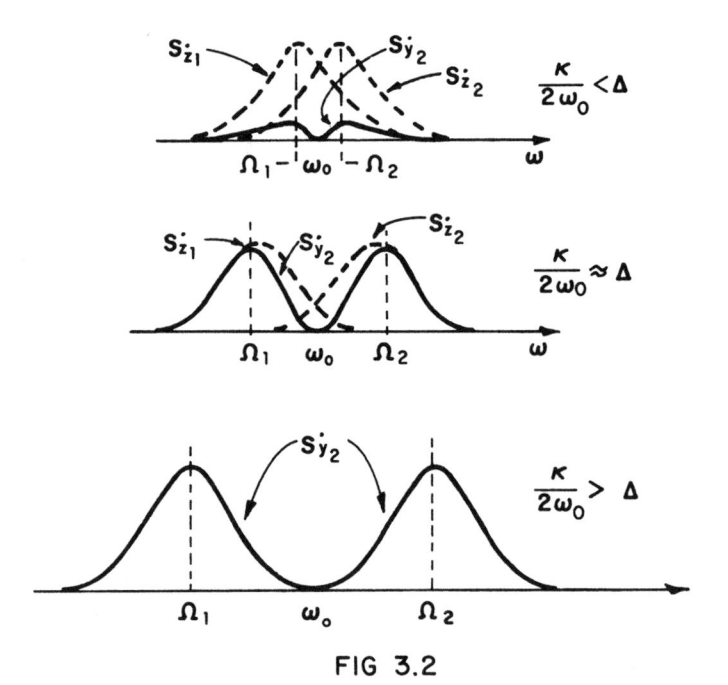

FIG 3.2

SPECTRAL DENSITY OF INDIRECTLY EXCITED RESONATOR VELOCITY AS A FUNCTION OF THE DEGREE OF COUPLING

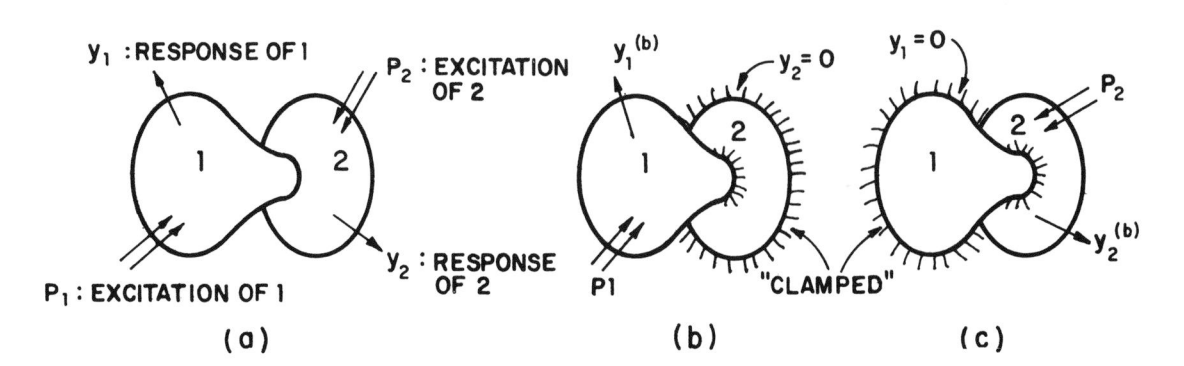

FIG 3.3

(a) SYSTEM OF INTEREST; TWO SUBSYSTEMS INDEPENDENTLY EXCITED
AND RESPONDING BOTH TO EXCITATION AND COUPLING FORCES
(b) SYSTEM 1 VIBRATING IN RESPONSE TO ITS OWN EXCITATION,
SYSTEM 2 IS BLOCKED, (c) SYSTEM 2 IS VIBRATING IN RESPONSE
TO ITS EXCITATION, SYSTEM 1 IS BLOCKED.

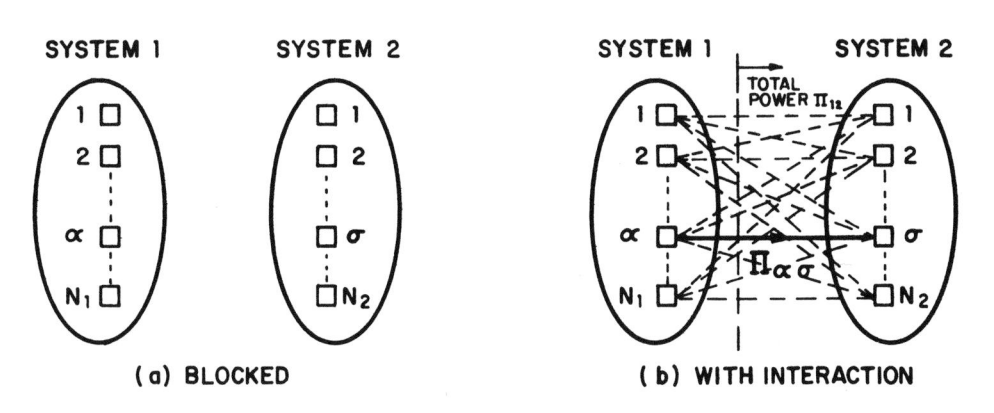

FIG 3.4

(a) SHOWS MODAL SETS WHEN THERE IS NO INTERACTION, AS
ACHIEVED BY CONDITIONS SHOWN IN FIG 3.3 (b,c).
(b) SHOWS MODE PAIR INTERACTIONS THAT OCCUR WHEN ACTUAL
CONDITIONS OBTAIN AT THE JUNCTION

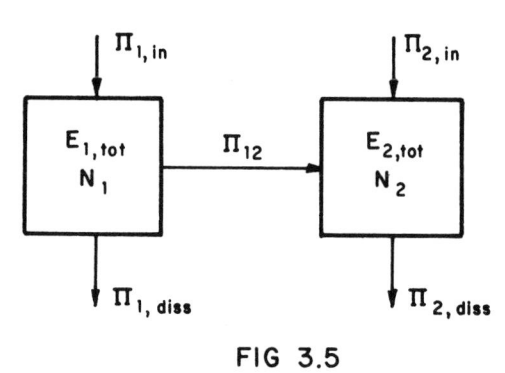

FIG 3.5

AN ENERGY TRANSFER AND STORAGE MODEL THAT CAN REPRESENT THE SYSTEM SHOWN IN FIG 3.4 (b) WHEN THE SEA MODEL IS USED THE DEFINING PARAMETERS ARE THE MASSES M_1, M_2; THE LOSS FACTORS η_1, η_2; THE MODAL DENSITIES $n_1 = N_1/\Delta\omega$ $n_2 = N_2/\Delta\omega$ AND THE COUPLING LOSS FACTOR η_{12}

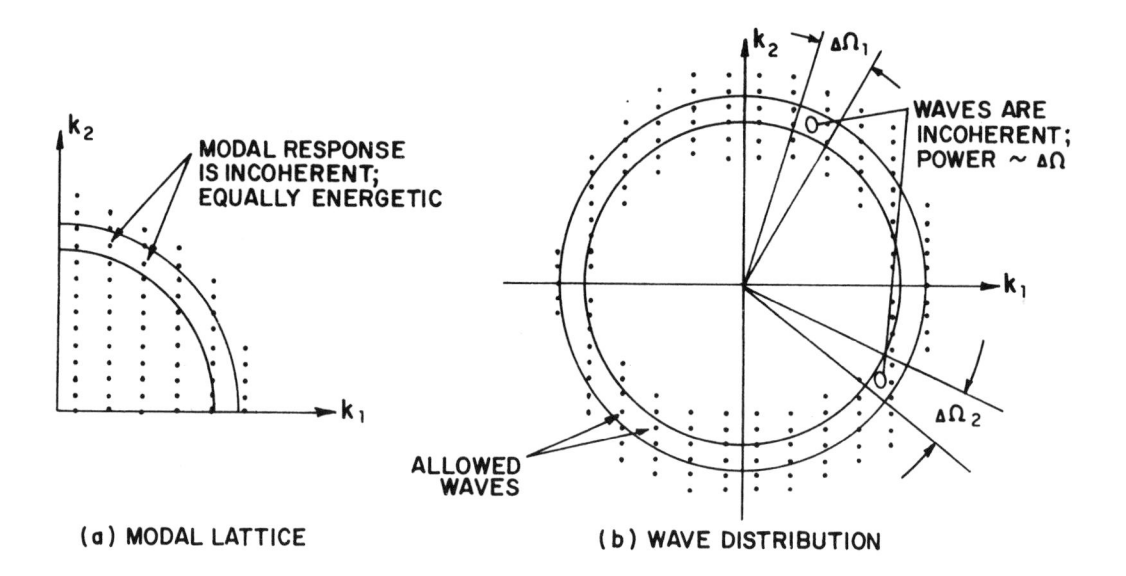

(a) MODAL LATTICE   (b) WAVE DISTRIBUTION

FIG 3.6

EQUIVALENCE OF MODAL AND WAVE COHERENCE ASSUMPTIONS IN SEA

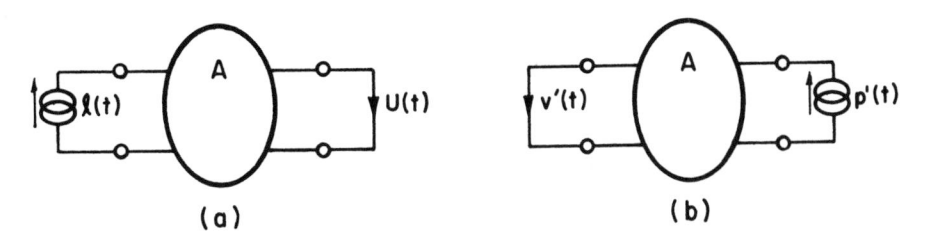

FIG 3.7

IF THE SYSTEM A IS RECIPROCAL, THEN IF ℓ AND p′ ARE "PRESCRIBED
DROP" SOURCES AT FREQUENCY ω, THEN ONE HAS U/ℓ = v′/p′

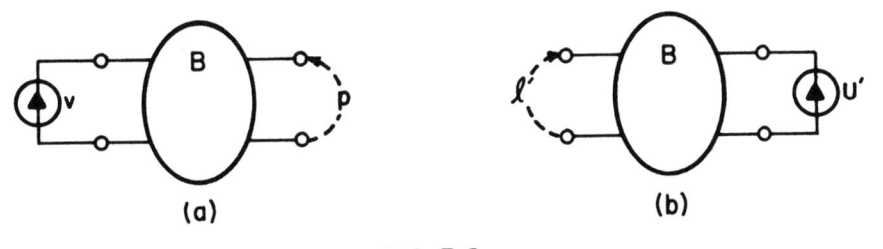

FIG 3.8

SYSTEM B IS THE SAME AS SYSTEM A WITH THE TWO TERMINAL
PAIRS "OPEN" THE RECIPROCITY STATEMENT IN THIS CASE IS p/v = ℓ′/U′

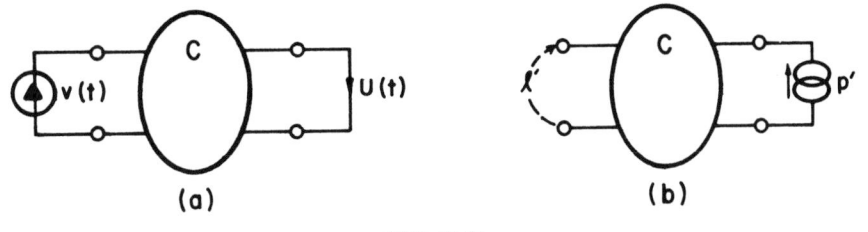

FIG 3.9

SYSTEM C IS THE SAME AS SYSTEM A WITH TERMINAL PAIR 1
OPEN – CIRCUITED AND TERMINAL PAIR 2 SHORT-CIRCUITED. THE
RECIPROCITY STATEMENT IS NOW U/v = ℓ′/p′.

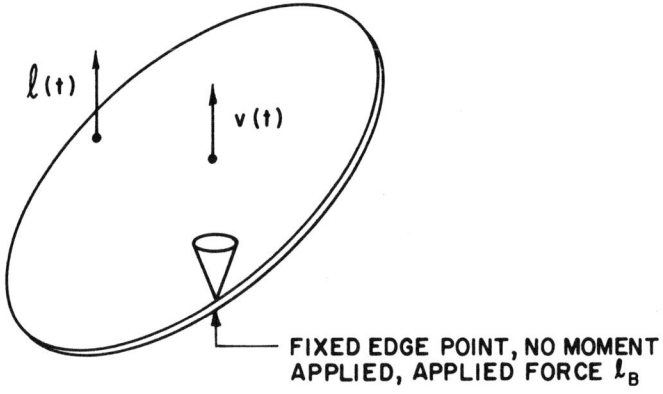

FIG. 3.10

**PLATE DRIVEN BY POINT LOAD NOISE SOURCE WITH
SECOND POINT ON PLATE EDGE FIXED WITHOUT MOMENT RESTRAINT**

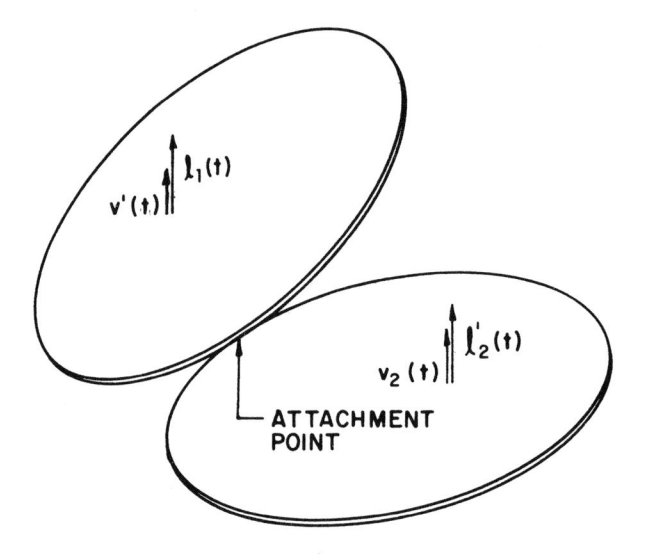

FIG. 3.11

**THE PLATE IN FIG. 3.10 IS NOW JOINED TO
ANOTHER PLATE AT ONE POINT ALONG THE EDGE
WITH NO MOMENT COUPLING OR RESTRAINT**

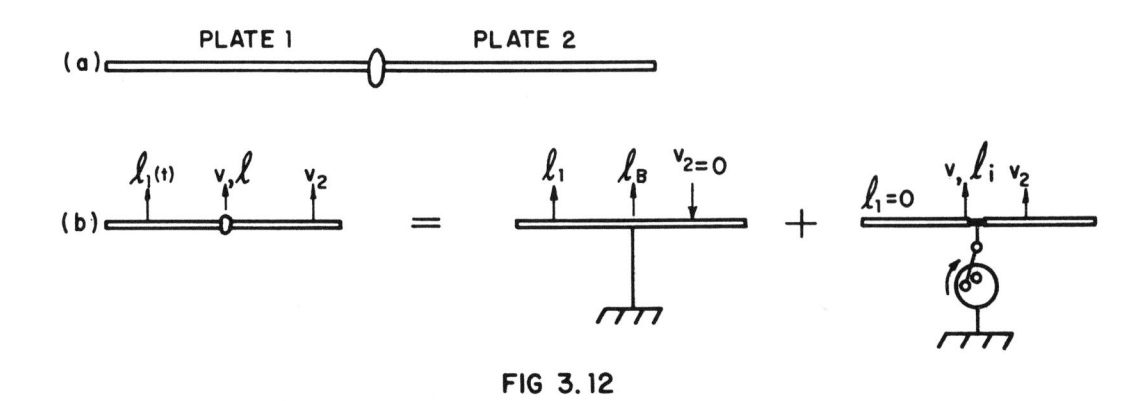

FIG 3.12

**SKETCH OF 2-PLATE SYSTEM OF FIG 3.11 SHOWING HOW THE
ACTUAL INTERACTION IS DEVELOPED AS A COMBINATION
OF 2 IDEALIZED PROBLEMS**

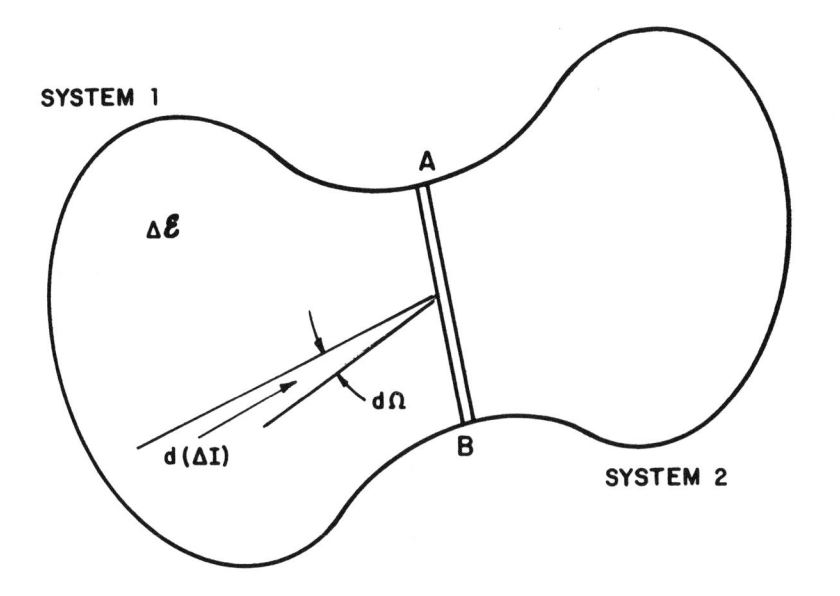

FIG 3.13

ENERGY FLOW TOWARD SYSTEM JUNCTION IN THE ANGULAR INTERVAL $d\Omega$

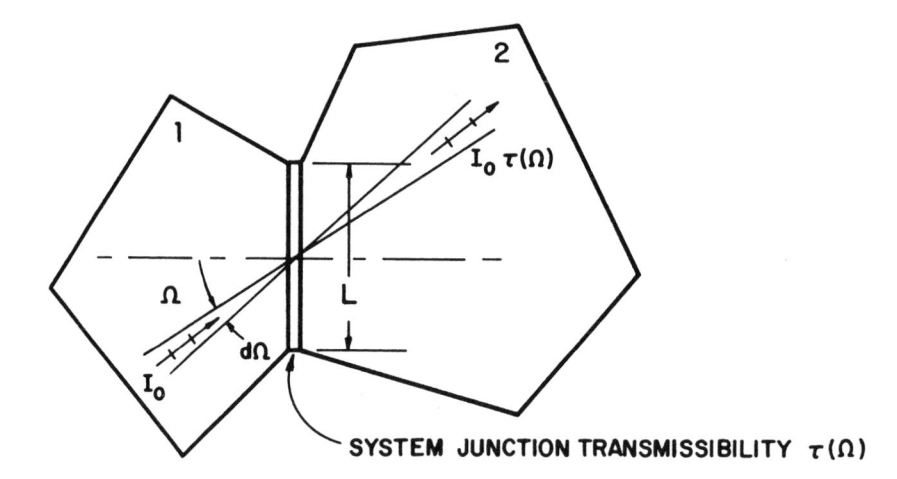

FIG 3.14

**SKETCH SHOWING TRANSMISSION OF POWER BETWEEN TWO
SYSTEMS THROUGH A LINEAR JUNCTION**

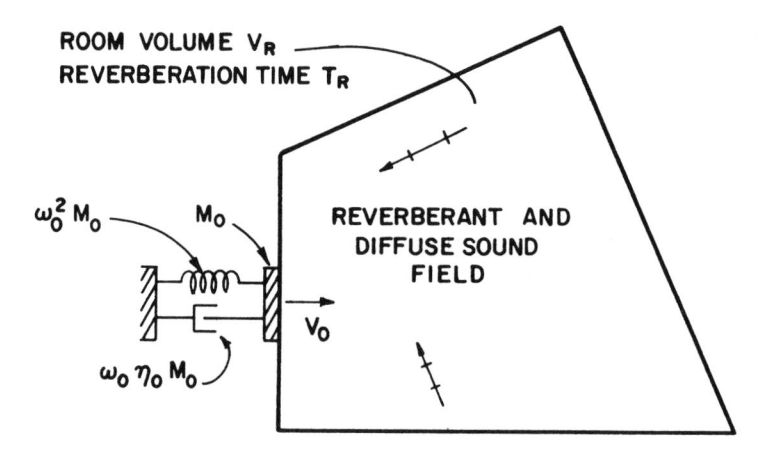

FIG 3.15

**A PISTON RESONATOR IN THE WALL OF A REVERBERATION
CHAMBER INTERACTING WITH THE CONTAINED SOUND FIELD**

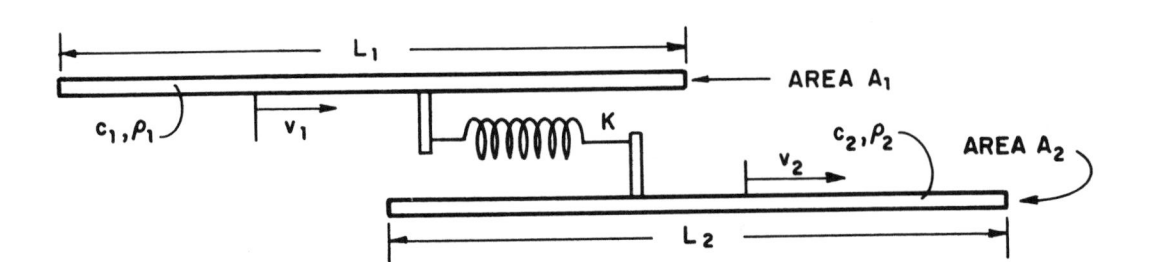

FIG 3.16

**TWO BEAMS IN LONGITUDINAL VIBRATION
COUPLED BY A SPRING OF STIFFNESS K**

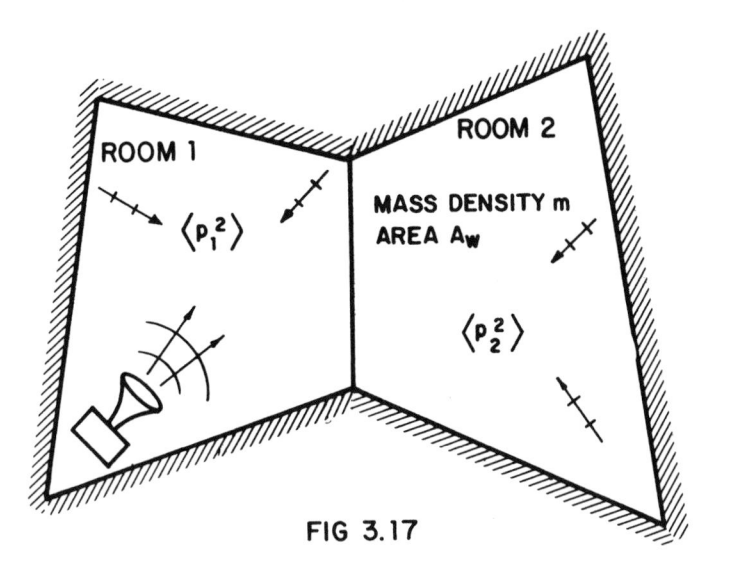

FIG 3.17

TWO ROOMS COUPLED BY LIMP WALL OF SURFACE DENSITY m

CHAPTER 4.   THE ESTIMATION OF RESPONSE IN STATISTICAL ENERGY ANALYSIS

4.0 Introduction

In the preceding chapters, we have shown how the energy of vibration may be estimated in an average sense based on system parameters that are fundamentally descriptive of energy storage and transfer capabilities of the systems. The use of energy quantities has the great advantage that sound, vibration and other resonant systems all are described by the same variables. Consequently, our viewpoint has great generality which is useful when we must deal with systems of such complexity.

This generality, however, has its shortcomings. We must remember that our motivation for these studies is usually not directed to the energy of oscillation, as such. Systems do not make too much noise, or fatigue, or have component malfunction because they contain too much energy, but because they move, are strained or are stressed too much or too often. Consequently, we must interpret the energy of oscillation in terms of dynamical variables of engineering interest such as velocity, strain, pressure, etc. We have already done this to a degree in earlier chapters, particularly in paragraphs 3.4. In paragraph 4.1 we shall expand on this important aspect of response estimation.

The calculations and formulas of energy prediction in Chapter 3 are directed toward predicting average energies, the average taken over the set of systems forming the ensemble of "similar" systems constructed according to the SEA model described in paragraph 3.2. Since we have a population of systems, the energy actually realized by any one system (the one sitting in the laboratory, for example) will not be precisely equal to the average energy as calculated by the SEA formulas. We can get an idea of how much our system may deviate from the average system by calculating a standard deviation (s.d.) of the system energy. This has, in fact, been done for some cases, and we can give some general ideas for estimating the variance (square of the s.d.) for practical situations. The calculation of response variance is discussed in paragraph 4.2.

The calculation of standard deviation is reassuring to one using the mean energy as an estimate only if the s.d. is a small fraction of the mean. Then we know that most realizations of the system (again including the one sitting in the laboratory) will have a response that is close to the mean. But what if the standard deviation is equal to the mean, or greater? In this event, the probability that any one realization is close to the mean will be small and the mean value loses its worth as an estimate.

In such a situation, it is common to calculate the probability that a realization will occur within some stated interval of response amplitudes. The probability associated with this "confidence interval" is called the "confidence coefficient". In order to carry out these calculations, however, we need a probability distribution for the energy. This distribution is introduced and the estimation procedures are derived from it in paragraph 4.3.

The distributions used for generating confidence intervals tend to be fairly accurate of the "middle range" of response levels, but are generally fairly inaccurate in predicting response having a very small probability. If we have a special interest in either the greatest or smallest response levels that may occur, we usually must take a different approach. Paragraph 4.4 describes the estimation of the probability of very large values (having small probability) in a particular situation. It may be argued that the problem treated in paragraph 4.4 is not properly an SEA problem, because it deals with the coherent properties of modal response. Nevertheless, the entire approach of describing modal responses by their m.s. values and adopting a statistical view of the population of systems being considered is very much within the SEA framework.

4.1 Mean Value Estimates of Dynamical Response

We can make useful estimates of the mean square amplitude of response based on the energy calculations in Chapter 3. We first consider systems like those discussed in paragraphs 2.3 and 3.4, in which only one mode of one system is interacting with a group of modes of another. This one mode may actually be a single dof system like that shown in Fig. 2.1 or it may be a mode that we single out for particular examination within a multi-modal system. We may do this by

restricting the excitation bandwidth so that this is the only
system mode that is resonant.

Single Mode Response

Suppose that the multimodal system 1 is excited by noise
sources of bandwidth $\Delta\omega$. System 1 has total energy $E_{1,tot}$
and N_1 modes that resonate in this bandwidth. System 2 has
only one mode with mode shape $\psi_2(x_2)$. According to paragraph
3.2, the energy of this mode will be given by

$$E_2 = E_1 \; \frac{\eta_{21}}{\eta_2 + \eta_{21}} \qquad\qquad (4.1.1)$$

where η_{21} may be determined from averages over interaction
parameters as in Eq. (3.2.8) or from boundary impedances as
in paragraph 3.3. Let us assume that this has been done, and
we now want to find the actual response at location x_2.

From Eq. (2.2.3) we have

$$y_2(x,t) = Y_2(t) \; \psi_2(x), \qquad\qquad (4.1.2)$$

with the general result that

$$E_2 = \int_X dx \; \rho \langle \dot{y}^2 \rangle_t = M_2 \langle \dot{Y}_2^2(t) \rangle_t = M_2 \omega_2^2 \; \langle Y_2^2(t) \rangle_t$$

$$(4.1.3)$$

where M_2 is the mass of system 2 and ω_2 is its resonance frequency (which might just as well be taken as ω, since ω_2 is
within $\Delta\omega$ by hypothesis). Thus

$$\langle y_2^2(x,t)\rangle_t = \frac{E_2}{\omega^2 M_2}\, \psi_2^2\,(x) = \left[\frac{E_{1,tot}}{N_1}\, \frac{\eta_{21}}{\eta_2 + \eta_{21}}\right]\frac{1}{\omega^2 M_2}\, \psi_2^2(x)$$

$$(4.1.4)$$

Thus, our statistical model of the system does not restrict us in any way from determining the spatial distribution of response. Of course, the normalization of the ψ_2 function still ensures that

$$\langle y_2^2 \rangle_{\rho,t} = E_2/\omega^2 M_2.$$

In Fig. 4.1 we show the form of the distribution of y_2^2 for a typical mode shape of a beam. The distribution of $\langle y_2^{\prime\prime}\rangle^2$ which would be proportional to the surface strain in the beam, is also sketched.

Multi-Modal Response.   When there are several resonant modes (N_2) of (indirectly excited) system 2 in the band $\Delta\omega$, then Eq. (4.1.1) has the form

$$\frac{E_{2,tot}}{N_2} \equiv E_2 = \frac{E_{1,tot}}{N_1}\, \frac{\eta_{21}}{\eta_2 + \eta_{21}} \qquad (4.1.5)$$

and the m.s. response of the system in time and space (or mass) is

$$\langle y_2^2 \rangle_{\rho,t} = E_{2,tot}/\omega^2 M_2, \qquad (4.1.6)$$

which we may regard as our estimate of the system response.
Note, however, that if such an estimate were made in the
situation of Fig. 4.1, it would be rather poor, since the
actual value of $\langle y^2 \rangle_t$ only equals the estimate $\langle y^2 \rangle_{\rho, t}$ at a
finite number of points and oscillates around the estimate
without converging on it.

In the case of two or three dimensional systems with many
modes of vibration, the function $\langle y_2^2 \rangle_t$ tends to converge on
the estimate of Eq. (4.1.6) in many important instances. To
demonstrate this, we consider a rectangular supported plate
which has modes

$$\psi(x) = 2 \sin k_1 x_1 \sin k_2 x_2 \qquad (4.1.7)$$

where

$$k_1 = \sigma_1 \pi / \ell_1, k_2 = \sigma_2 \pi / \ell_2; \sigma_1, \sigma_2 = 1, 2 \ldots$$

The modes that resonate within the interval $\Delta k = \Delta \omega / c_g$ are
shown in the hatched region of Fig. 4.2.

As a result of our model of the system interaction,
spelled out in paragraph 3.2, we can make the following
observations about the response of system 2.

(1)   The response may be written in the form

$$y_2(x,t) = \sum_\sigma Y_\sigma(t) \, \psi_\sigma(x) \qquad (4.1.8)$$

(2)   The modal response amplitudes $Y_\sigma(t)$ are incoherent;

$$\langle Y_\sigma(t) Y_\tau(t) \rangle_t = \langle Y_\sigma^2(t) \rangle_t \delta_{\sigma,\tau} \qquad (4.1.9)$$

(3)   The modes of system 2 are equally energetic

$$\langle Y_\sigma^2(t)\rangle_t = \langle Y_2^2(t)\rangle_t = E_2/\omega^2 M_2 \qquad\qquad (4.1.10)$$

Thus, we can say that

$$\langle Y_2^2(x,t)\rangle_t = \sum_{\sigma\ \tau} \langle Y_\sigma(t) y_\tau(t)\rangle_t \psi_\sigma(x)\ \psi_\tau(x)$$

$$= \langle y_2^2(t)\rangle_t \sum_{\sigma}\ \psi_\sigma^2(x), \qquad\qquad (4.1.11)$$

where the allowed values of σ are determined by the modes of
system 2 that resonate in $\Delta\omega$.

The result of Eq. (4.1.11) is quite general. For the
specific two-dimensional system at hand, we can write,
therefore,

$$\langle y_2^2\rangle_t/\langle y_2^2\rangle_{t,\rho} = 4N_2^{-1}\ \sum_{\sigma}\sin^2 k_1 x_1\ \sin^2 k_2 x_2$$

$$= N_2^{-1}\ \sum_{\sigma}(1-\cos 2k_1 x_1)(1-\cos 2k_2 x_2)$$

$$(4.1.12)$$

We evaluate the sum by replacing it with an integral over ϕ
in the hatched region of Fig. 4.2. We can see that if dN_2
is the number of modes in the rectangle bounded by Δk and $d\phi$,
then $dN_2/N_2 = 2d\phi/\pi$, since the modes are uniformly distributed
in ϕ. We may then write

$$\langle y_2^2 \rangle_t / \langle y_2^2 \rangle_{t,\rho} = \frac{2}{\pi} \int_0^{\pi/2} \left[1 - \cos(2kx_1 \sin\phi)\cos(2kx_2 \cos\phi) \right.$$

$$\left. + \cos(2kx_1 \sin\phi)\cos(2kx_2 \cos\phi) \right] d\phi$$

$$= 1 - J_0(2kx_1) - J_0(2kx_2) + J_0(2kr)$$

(4.1.13)

where $r = \sqrt{x_1^2 + x_2^2}$. We have evaluated the integral for the region near x_1, $x_2 = 0$. Since the sum in Eq. (4.1.12) is unaffected by the substitution $x_1 \to \ell_1 - x_1$, $x_2 \to \ell_2 - x_2$, the total pattern is as shown in Fig. 4.3.

Thus, we can estimate a spatially varying temporal mean square response of a system even with rather extensive assumptions regarding the system interactions. Patterns of the kind shown in Fig. 4.3 have been developed for a variety of systems, and are available from the literature. They are usually of interest if one wishes to avoid locating an item or making measurements at a point where the response is excessively high or otherwise unrepresentative.

Wave estimates. The use of wave models for SEA response estimates extends the power of SEA greatly. We can develop response patterns like the one shown in Fig. 4.3 equally well using a diffuse wave field model like that discussed in paragraphs 2.4 and 3.3. The more important application of wave notions employs the use of average impedance functions at system boundaries for the evaluation of coupling loss factors as discussed in paragraph 3.3.

As an example, consider the system shown in Fig. 4.4a, a beam cantilevered to a flexible plate. Along the line of connection of these systems there will be very little displacement (we shall assume there is none) because of the very high in-plane impedances of these systems. Nevertheless, power will be transmitted by torques (moments) and rotational motion of the junction. In order to evaluate the coupling loss factor, we must find the power transferred between the systems.

The "decoupled" systems are produced by eliminating the rotation of the contact line as shown in Fig. 4.4b. The modes of the beam for this condition are readily found - the beam has clamped - free boundary conditions. The plate, however, has a very complicated boundary condition - the boundary conditions around its outer edges and clamped along a finite line in the interior! It would require a complicated computer routine to find its modes.

Nevertheless, we can invoke two principles from our work thus far that allow us to solve this problem. The first is that of reciprocity of the coupling loss factor

$$N_p \eta_{pb} = N_b \eta_{bp} \qquad (4.1.14)$$

which allows us to use η_{bp}, the beam to plate coupling loss factor (which is easier to calculate) to evaluate η_{pb}. Secondly, the moment impedance "looking into" the plate may be evaluated by considering the plate to be infinite. This has the effect of eliminating reflected energy in the plate from returning to the junction. In a manner completely analogous to the results in paragraphs 2.2 and 2.3 for the admittance for a transverse force, the moment impedance, averaged over the location of the junction of the plate and over resonance frequencies of the interacting systems in the band $\Delta\omega$, is given by the infinite system impedance, which is known.

In paragraph 3.3, we obtained a general result that, in terms of actual boundary impedances, the power flow between systems is given by

$$\Pi_{bp} = \omega a_{bp} E_{b,tot}^{(b)} \qquad (4.1.15)$$

where

$$\omega \, a_{bp} = \frac{2}{\pi n_b} \frac{R_b R_p}{|Z_b + Z_p|^2} \qquad (4.1.16)$$

On the other hand, we have the general result that (see
Eq. 3.2.10)

$$\Pi_{bp} = \omega\eta_{bp}(E_{b,tot} - \frac{N_b}{N_p} E_{p,tot})$$ (4.1.17)

It must be emphasized that Eqs. (4.1.15) and (4.1.17) are
fully equivalent and apply to all conditions of system size,
damping, etc. In general η_{bp} and a_{bp} are different functions.
However, if the receiving system (the plate in this instance)
is made very large, while its loss factor η_p is held fixed,
the term $N_b E_{p,tot}/N_p$ will vanish in Eq. (4.1.17). If we use
the infinite plate impedances in Eq. (4.1.16),

$$a_{bp} = \frac{2}{\pi\omega n_b} \frac{R_b R_p(\infty)}{|Z_b + Z_p(\infty)|^2}$$ (4.1.18)

The relation between η_{bp} and a_{bp} is then readily found.
Since the input power to the system is independent of the
junction,

$$\Pi_{1,in} = E_{1,tot}^{(b)} \Delta_1 = E_{1,tot}\Delta_1 + \omega a_{12} E_{1,tot}^{(b)}$$ (4.1.19)

we have

$$\omega\eta_{bp} E_{b,tot} = \omega\eta_{bp} \left[1 - \frac{\omega a_{bp}}{\Delta_b} \right] E_{b,tot}^{(b)}$$

$$= \omega a_{bp} E_{b,tot}^{(b)}$$ (4.1.20)

and, consequently,

$$\eta_{bp} = a_{bp} (1 - \omega a_{bp}/\Delta_b)^{-1}$$ (4.1.21)

When we deal with a single beam mode, then $\Delta_b = R_b/M_b$ and the average of a_{bp} over the band is

$$\langle a_{bp}\rangle_{\Delta\omega} = \frac{2}{\pi\omega} \frac{R_b R_p(\infty)}{(R_b + R_p)^2} \cdot \frac{\pi}{2} \frac{R_b + R_p}{M_b}. \qquad (4.1.22)$$

Substituting into Eq. (4.1.21) then gives

$$\omega\eta_{bp} = R_p(\infty)/M_b \qquad (4.1.23)$$

On the other hand, when the modes of the beam are dense so that $n_b\Delta_b \gg 1$, the beam "looks infinite", and one has

$$\frac{\omega a_{bp}}{\Delta_b} = \frac{2}{\pi} \frac{R_b(\infty) R_p(\infty)}{|Z_b(\infty)+Z_p(\infty)|^2} \frac{1}{n_b\Delta_b} \ll 1 \qquad (4.1.24)$$

and referring to Eq. (4.1.21), we get

$$\eta_{bp} \to a_{bp} = \frac{2}{\pi n_b} \frac{R_b(\infty) R_p(\infty)}{|Z_b(\infty)+Z_p(\infty)|^2}. \qquad (4.1.25)$$

Where we have used $Z_p(\infty)=R_p(\infty)-i\,X_p(\infty)$ to emphasize that these are input impedance functions for the infinitely extended system. Extending the receiving system affects the value of a_{bp}, but does not affect the value of η_{bp}, since η_{bp} depends on a mode-mode coupling coefficient $\langle B_{\alpha\sigma}\rangle$ in the form [see Eq. (3.2.10)]

$$\omega \eta_{bp} = \ <B_{bp}> \ n_p \ \ \Delta\omega. \tag{4.1.26}$$

As the plate gets larger, the coupling $<B_{bp}>$ diminishes inversely with plate area since the quantities $\mu^2, \gamma^2, \kappa^2$ are inversely proportional to the mass of the receiving system according to Eq. (3.1.4) (or plate area) and n_p is proportional to plate area. Thus, while Eq. (4.1.16) applies to any situation, Eq. (4.1.21) is only correct when infinite system parameters are used for the receiving system.

In the system of Fig. 4.4, the infinite system junction impedances are

$$z_b^M = \rho_b c_b^2 S_b \kappa_b^2 c_{fb}^{-1} (1+i) \tag{4.1.27}$$

and

$$z_p^M = 16 \rho_s \kappa_p^2 c_p^2 / \omega (1-i\Gamma) \tag{4.1.28}$$

where ρ_b, c_b, κ_b are the material density, longitudinal wave-speed, and radius of gyration for the beam with the same parameters for the plate with a "p" subscript. S_b is the cross-sectional area of the beam, $c_{fb} = \sqrt{\omega \kappa_b c_b}$ is the flexural phase speed on the beam and ρ_s the mass per unit area of the plate. The quantity Γ is a moment susceptance parameter that depends on the shape of the junction. If we assume that $z_p^M \gg z_b^M$, then $R_p |z_p|^{-2} \rightarrow G_p$, the plate moment conductance and the result is independent of Γ. With this assumption and assuming the beam and plate are cut from the same material, we get

$$\eta_{bp} \ \rightarrow \ w/4\ell, \tag{4.1.29}$$

a surprisingly simple result. We have also used the modal density of the beam

$$n_b(\omega) = \ell/2\pi c_{fb} \quad .$$

(4.1.30)

The coupling loss factor from plate to beam is, according to Eq. (3.2.12)

$$\eta_{pb} = n_b \eta_{bp}/n_p \quad ,$$

(4.1.31)

where $n_p = A_p/4\pi\kappa_p c_p$.

Using Eq. (4.1.29), we obtain an estimate for the average response of the beam for bands of noise,

$$\langle v_b^2 \rangle = \langle v_p^2 \rangle \frac{M_p}{M_b} \frac{n_b}{n_p} \frac{\eta_{bp}}{\eta_b + \eta_{bp}}$$

(4.1.32)

In Fig. 4.5 we show a comparison of this simple theoretical result for the two limiting experimental conditions, first for $\eta_{bp} \gg \eta_b$ and then for $\eta_{bp} \ll \eta_b$. Modal energy equipartition between the beam and plate would be expected to obtain in the first instance, but not in the second.

Strain Response. As an illustration of how energy response may be expressed in strain or stress of the system, we have already shown that the pressure within a sound field is given by

$$\frac{\langle p^2 \rangle V_R}{\rho_o c^2} = \langle v^2 \rangle \ M_R$$

(4.1.33)

where $\langle v^2 \rangle$ is the m.s. velocity of the fluid (air) particles. Using $\rho_0 c^2 = \gamma P_0$, this expression may also be written as

$$\frac{\langle p^2 \rangle}{(\gamma P_0)^2} = \langle (\frac{\delta \rho}{\rho_0})^2 \rangle = \frac{\langle v^2 \rangle}{c^2} \qquad (4.1.34)$$

where $\delta \rho / \rho_0$ is the volumetric strain (dilatation) of the fluid since γP_0 is the volumetric stiffness (bulk modulus) of an ideal gas. The equation states that the m.s. strain equals the m.s. mach number (ratio of particle velocity to sound speed) of the particles.

Quite a similar result also obtains for plates. In a plate of thickness h, the strain distribution across the plate thickness is given by (see Fig. 4.6)

$$\varepsilon(z) = \frac{2z}{h} \varepsilon_m \qquad (4.1.35)$$

where ε_m is the strain at the free surface of the plate, where it is a maximum. The strain energy density of a layer of plate material dz thickness is $1/2 \, dz \varepsilon^2(z) Y_0$, where Y_0 is Young's modulus, so that the total strain energy density is

$$PE \text{ density} = Y_0 \int_{-h/2}^{h/2} dz \, \frac{1}{2} z^2 \, \frac{4}{h^2} \varepsilon_m^2 = \frac{2\kappa^2}{h} Y_0 \langle \varepsilon_m^2 \rangle \qquad (4.1.36)$$

If this is equated to the kinetic energy density $\frac{1}{2}\rho_p h \langle v^2 \rangle$, we get

$$<\varepsilon_m^2> \frac{h^2}{4\kappa^2} \frac{<v^2>}{c_\ell^2} = 3 \frac{<v^2>}{c_\ell^2} \quad \text{(homogeneous)} \qquad (4.1.37a)$$

where we have used $\kappa^2 = h^2/12$ for a homogeneous plate. For a sandwich plate with all its stiffness at the surface, $\kappa = h/2$ and one has

$$<\varepsilon_m^2> = <v^2>/c_\ell^2 \quad \text{(sandwich)}. \qquad (4.1.37b)$$

The results of Eq. (4.1.37) are entirely analogous to that of Eq. (4.1.34) and allow us to develop mean square strain estimates from energy (or velocity) estimates.

4.2 Calculation of Variance in Temporal Mean Square Response

In paragraphs 3.3, 3.4 and 4.1 we have shown how average energy estimates of system response may be developed and how m.s. response estimates are derived from the energy. We have seen, however, in Figs. 4.1, 4.3 and 4.5 that the average energy may not be a good estimate of m.s. response at a particular location on the system or in a frequency band in which the number of modes is too low. In this section, we shall develop estimates for the deviation of realized response from the mean in terms of the standard deviation (square root of variance) from the mean square response. Note that the variance that we are discussing is from one "similar" system to another, or from one location to another, not temporal variance. If we are dealing with noise signals, we assume that the averaging time is sufficiently long to remove variations due to this cause.

The variance in response between the mean and any member of the population is produced by four major effects.

   (1)   The modal energies of the directly excited system
          may not be equal, because of the spatial
          dependence of the actual sources of excitation
          and the internal coupling of the blocked system
          may not be great enough to ensure equipartition.

(2) The actual number of resonant interacting modes
     in each realization of the coupled system will
     vary.

(3) The strength of the coupling parameters will
     fluctuate from one realization to another because
     of slight changes in modal shape at the junction
     of the systems.

(4) The response at the selected observation position
     will vary because the observation position is
     randomly located and because the mode shapes change
     from one realization to another.

The way in which the variance of response depends on
these four factors is not obvious, and indeed, has not been
worked out for many cases of interest. There is a considerable area of interesting research work that needs to
be done in analyzing variance of interacting systems. We
should also remark here that we are required to use modal
analyses to calculate variance. The wave-mode duality is
useful for mean value estimates, but a wave analysis of
variance by its nature disregards spatial coherence effects
that are essential to the calculation of variance.

Modal Power Flow and Response.   To calculate variance
in the modal energies, we return to the system of Fig. 3.4.
The power flow from all of the resonators of system 1 to
resonator σ is (system 2 has no external excitation)

$$\Pi_{1\sigma} = \sum_{\alpha} A_{\alpha\sigma} E_{\alpha}^{(b)} = \sum_{\alpha} B_{\alpha\sigma} (E_{\alpha}-E_{\sigma}) \qquad (4.2.1)$$

where $A_{\alpha\sigma}$ is given by Eq. (3.1.20) and $B_{\alpha\sigma}$ is given by
Eq. (3.1.26) with appropriate subscripts entered. If the
damping of the "σ" is Δ_{σ}, then the dissipation of energy in
this mode is $E_{\sigma}\Delta_{\sigma}$ and we have

$$E_{\sigma} = \sum_{\alpha} E_{\alpha}^{(b)} (A_{\alpha\sigma}/\Delta_{\sigma}) = (\sum_{\alpha} E_{\alpha} B_{\alpha\sigma}) (\Delta_{\sigma}+\sum_{\alpha} B_{\alpha\sigma})^{-1}. \qquad (4.2.2)$$

If we are interested in the velocity response, then we saw
from Eq. (4.1.10) that

$$<v_\sigma^2>_t = (E_\sigma/M_2)\psi_\sigma^2(x_2) \tag{4.2.3}$$

where x_2 is the observation position. Thus, the total m.s.
response of system 2 observed at position x_2 is just

$$<v_2^2(x_2,t)>_t = \frac{1}{M} \sum_{\sigma_2} \psi_\sigma^2(x_2) \sum_\alpha E_{(\alpha)}^{(b)} A_{\alpha\sigma}/\Delta\sigma \tag{4.2.3a}$$

$$= \frac{1}{M_2} \sum_\sigma \psi_\sigma^2(x_2) \left(\sum_\alpha E_\alpha B_{\alpha\sigma}\right) \Big/ \left(\Delta_\sigma + \sum_\alpha B_{\alpha\sigma}\right) \tag{4.2.3b}$$

In writing Eq. (4.2.3) we have now relaxed the assumption
of Chapter 3 that the modes of system 1 are equally energetic,
but we continue to assume that modal responses are temporally
incoherent.

The four factors listed above that contribute to variance
are specifically shown in Eq. (4.2.3). The variance in modal
energies (1) is in $E_\alpha^{(b)}$ or E_α. The coupling variance strength
(3) is represented by the quantities $A_{\alpha\sigma}$ or $B_{\alpha\sigma}$, and the
observation variance (4) is incorporated in ψ_σ^2. The variation
in number of interacting modes (2) is represented by the
summation in α. Our problem is to find the variance in this
sum on a knowledge (or reasonable assumptions) about the
statistics of the terms entering the equation.

To simplify matters a bit, let us first concentrate on
the single mode response $<v_\sigma^2>_t$ so that only a single sum over
α is involved. The appearance of $B_{\alpha\sigma}$ in the denominator of
Eq. (4.2.3b) is awkward from an analytical point of view, so
that we would prefer to use Eq. (4.2.3a). However, since the
coupling factors also enter the denominator of $A_{\alpha\sigma}$ (but not of
$B_{\alpha\sigma}$), we cannot avoid this problem. Another reason for pre-
ferring Eq. (4.2.3a) is that we might be able to estimate the
statistics of $E_\alpha^{(b)}$ better than E_α, since the former does not

contain contamination by coupling effects. Nevertheless,
we are essentially forced to use Eq. (4.2.3b) since it alone
allows us to use an important result in the statistics of
summed random processes that is discussed in the following
paragraphs.

There are two situations in which Eq. (4.2.3b) can
simplify sufficiently to allow one to make calculations.
The first is where $\sum_\alpha B_{\alpha\sigma} >> \Delta_\sigma$. In this event, we have

$$\langle v_\sigma^2 \rangle_t = (\psi_\sigma^2 (x_2)/M_2) \; \sum_\alpha E_\alpha B_{\alpha\sigma}/\sum_\alpha B_{\alpha\sigma} \; . \qquad (4.2.4)$$

In this case, the energy of the σ mode is a weighted average
of the energy of the modes of system 1 with weighting co-
efficients $b_{\alpha\sigma} = B_{\alpha\sigma}/\sum_\alpha B_{\alpha\sigma}$. Notice that this condition, which
is essentially an equipartition condition, does not require
that $B_{\alpha\sigma} >> \Delta_\sigma$, and is thus a much weaker requirement on the
coupling than when one has two single dof systems interacting.

The second situation for which one can calculate variance
is when $\Delta_\sigma >> \sum_\alpha B_{\alpha\sigma}$. In this case

$$\langle v_\sigma^2 \rangle_t = (\psi_\sigma^2/M_2 \Delta_\sigma) \; \sum_\alpha E_\alpha B_{\alpha\sigma} \qquad (4.2.5)$$

We note that $B_{\alpha\sigma}$ is given by Eq. (3.1.26) with appropriate
subscripts. We will write it here in the form

$$B_{\alpha\sigma} = \frac{\lambda_{\alpha\sigma}}{\xi_{\alpha\sigma}^2 + 1} \; \frac{\Delta_\alpha \Delta_\sigma}{\Delta_\alpha + \Delta_\sigma} \qquad (4.2.6)$$

where $\lambda_{\alpha\sigma}$ is given by Eq. (3.1.33) and $\xi_{\alpha\sigma} = 2 \; (\omega_\sigma - \omega_\alpha)/(\Delta_\alpha + \Delta_\sigma)$,
in a manner analogous to the development in paragraph 2.2.
If we now assume $\Delta_\alpha = \Delta_1 =$ const (all modes of system 1 have
equal damping), then according to Eq. (2.2.21),

$$\langle B_{\alpha\sigma}\rangle_{\omega_\alpha} = \langle \lambda_{\alpha\sigma}\rangle_{\omega_\alpha} \frac{\Delta_1 \Delta_\sigma}{\Delta_1 + \Delta_\sigma} \frac{\pi(\Delta_1 + \Delta_\sigma)}{2\Delta\omega} \tag{4.2.7}$$

which reproduces Eq. (3.2.8)

In the present instance, we are interested in the variance of the sum in Eq. (4.2.5). The assumption $\Delta_\sigma \approx \Delta_1$ has the effect of causing each of the functions $(\xi_{\alpha\sigma}^2 + 1)^{-1}$ to have the same shape as a function of frequency. We can plot Eq. (4.2.5) then as shown in Fig. 4.7. It appears as a sum of pulses, of strength

$$C_{\alpha\sigma} = \frac{(\psi_\sigma^2/M_2)\Delta_1\Delta_\sigma}{\Delta_1 + \Delta_\sigma} E_\alpha \lambda_{\alpha\sigma}. \tag{4.2.8}$$

If we assume that the resonance frequencies are uniformly probable over the interval $\Delta\omega$ and are independent of each other, then the standard deviation of Eq. (4.2.5) may be shown to be

$$\sigma^2_{v_\sigma^2} = \frac{\pi}{2} n_1(\omega) (\Delta_1 + \Delta_\sigma) \left[\frac{(\psi_\sigma^2/M_2)\Delta_1\Delta_\sigma}{\Delta_1 + \Delta_\sigma}\right]^2 \langle E_\alpha^2\rangle_\alpha \langle \lambda_{\alpha\sigma}^2\rangle_\alpha. \tag{4.2.9}$$

This is the "important result" referred to above. Consequently, using Eq. (4.2.7) in the average of Eq. (4.2.5), we get

$$\frac{\sigma^2_{v_\sigma^2}}{m^2_{v_\sigma^2}} = \left\{n_1 \frac{\pi}{2} (\Delta_1 + \Delta_2)\right\}^{-1} \frac{\langle E_\alpha^2\rangle_\alpha}{\langle E_\alpha\rangle^2_\alpha} \frac{\langle \lambda_{\alpha\sigma}^2\rangle_\alpha}{\langle \lambda_{\alpha\sigma}\rangle^2_\alpha}, \tag{4.2.10}$$

showing the explicit dependence of variance on uncertainty in the modal energy of the directly excited system and in the coupling parameters. This relation also shows the effect of the number of overlapping modes in reducing the ratio of variance to the square of the mean.

Finally, if we refer to Eq. (3.2.10) we see that the quantity $N_1 <B_{\alpha\sigma}>_\alpha \equiv \omega\eta_{21}$. The standard deviation in the sum $\sum_\alpha B_{\alpha0}$ will tell us something about the uncertainty in the coupling loss factor η_{21}. If for the present discussion only, we define $\omega\eta_{21} = \sum_\alpha B_{\alpha0}$ then $m_\eta = <\eta_{21}>_\alpha$ is the coupling loss factor we have been dealing with. Its variance is then given by

$$\sigma_\eta^2 = \frac{1}{\omega^2} <\lambda_{\alpha\sigma}^2> \left(\frac{\Delta_1\Delta_\sigma}{\Delta_1+\Delta_\sigma}\right)^2 \frac{\pi}{2} n_1 \ (\Delta_1+\Delta_\sigma)$$

and

$$\frac{\sigma_n^2}{m_\eta^2} = \frac{1}{\frac{\pi}{2} n_1 (\Delta_1+\Delta_2)} \frac{<\lambda_{\alpha\beta}^2>}{<\lambda_{\alpha\beta}^2>^2} \tag{4.2.11}$$

Returning to Eq. (4.2.5), if the observation position x_2 is located randomly in space, then we have additional factor of $<\psi_\sigma^4> / <\psi_\sigma^2>^2$ in Eq. (4.2.10) since ψ_σ^2 must be regarded as a random amplitude factor. If we assume that system 1 is excited by a randomly located point force at x_1, then according to Eq. (2.2.19), the energy of the αth mode will vary as $\psi_\alpha^2(x_1)$. If we assume that the systems are joined at a point x_1' in system 1 coordinates and x_2' in system 2 coordinates, then according to Eqs. (3.2.6) and (3.2.9), we may expect $\lambda_{\alpha\sigma}$ to vary as $\psi_\alpha^2 (x_1') \ \psi_\sigma^2(x_2')$. Thus, the greatest variation that we can expect in the response of a single mode of system 2 is

$$\frac{\sigma^2_{v_\sigma}}{m^2_{v_\sigma}} = \left\{ n_1 \frac{\pi}{2} (\Delta_1+\Delta_2) \right\}^{-1} \left[\frac{<\psi_1^4>}{<\psi_1^2>^2} \right]^2 \left[\frac{<\psi_2^4>}{<\psi_2^2>^2} \right]^2 \qquad (4.2.12)$$

where ψ_1 and ψ_2 are typical mode shapes of system 1 and system 2 respectively. We may expect that in any particular system, the variance will be less than that implied by Eq. (4.2.12).

Now suppose that we have a group of modes of system 2 represented by a modal density $n_2(\omega)$. We have a combination $n_2\Delta\omega$ independent response functions, each of which has a ratio of variance to square of mean given by Eq. (4.2.12). The total ratio of variance to square of mean for multi-modal response of system 2 is, therefore,

$$\frac{\sigma^2_{v_2}}{m^2_{v_2}} = \left\{ n_1 n_2 \frac{\pi}{2} (\Delta_1+\Delta_2) \Delta\omega \right\}^{-1} \left[\frac{<\psi_1^4>}{<\psi_1^2>^2} \right]^2 \left[\frac{<\psi_2^4>}{<\psi_2^2>^2} \right]^2$$

$$(4.2.13)$$

Notice that this result is completely symmetric in the properties of system 1 and 2. We should expect, therefore, that for systems excited and observed at point locations, the variance in response will be independent of which system is used as a "source" and which is used as a "receiver".

It will be apparent to the reader that a substantial number of assumptions have been made in order to get the result expressed by Eq. (4.2.13). Most of these assumptions, with notable exception of the assumptions of uniform damping, $\Delta_\alpha=\Delta_1$ and $\Delta_\alpha=\Delta_2$, should have the effect of increasing response variance. We, therefore, may reasonably regard Eq. (4.2.13) as a conservative estimate of the variance in the work that follows in that it estimates more variance than may, in fact, occur.

4.3   Calculation of Confidence Coefficients

We see from Eq. (4.2.13) that increasing the number of modal interactions by increasing the combined modal bandwidth $\pi/2$ $(\Delta_1+\Delta_2)$, and increasing the number of modes observed by increasing the excitation bandwidth $\Delta\omega$ will both cause a reduction in variance of response. In this section, we show how the variance estimates may be used to predict the fraction of measurements that will fall within a certain interval of values, generally based on the predicted mean.

The Probability Density. If the total probability density $\overline{\phi(\theta)}$ for the observed response $<v^2>_t=\theta$ were known, then the probability that θ would fall within the "confidence interval" $\theta_1 < \theta < \theta_2$ is simply

$$CC = \int_{\theta_1}^{\theta_2} \phi(\theta) \; d\theta \qquad\qquad (4.3.1)$$

where the (fiducial) probability CC is called the confidence coefficient. Since we do not know what the function $\phi(\theta)$ is, this may seem rather useless. We do know, however, that θ has positive values only, and we can estimate its mean m_θ and its standard deviation σ_θ from paragraph 4.2.

A probability density that meets our requirements and has tabulated integrals is the "gamma density"

$$\phi(\theta) = \theta^{\mu-1}\exp(-\theta/\lambda)/\lambda^\mu\Gamma(\mu) \qquad\qquad (4.3.2)$$

where $\mu =m_\theta^2/\sigma_\theta^2$ and $\lambda=\sigma_\theta^2/m_\theta$ and $\Gamma(\mu)$ is the gamma function. If we change variables to $y = \theta m_\theta/\sigma_\theta^2$, then Eq. (4.3.1) becomes

$$CC = \frac{1}{\Gamma(\mu)} \int_{\theta_1 m_\theta / \sigma_\theta^2}^{\theta_2 m_\theta / \sigma_\phi^2} dy \; y^{\mu-1} e^{-y} = \Gamma^{-1}(\mu) \left[\gamma\left(\mu, \frac{\theta_2 m_\theta}{\sigma_\theta^2}\right) - \gamma\left(\mu, \frac{\theta_1 m_\theta}{\sigma_\theta^2}\right) \right]$$

(4.3.3)

where $\gamma(\mu, B)$ is the incomplete gamma function defined by

$$\gamma(\mu, B) \equiv \int_0^B y^{\mu-1} e^{-y} \, dy$$

(4.3.4)

and is tabulated in standard mathematical handbooks.

Confidence Intervals. First, let us consider a simple "exceedance" type of confidence interval. That is, search for a value of $\theta_{max} \equiv r m_\theta$ such that the probability (CC) that any realized value of $\theta < \theta_{max}$ is known. The equation to be solved then is

$$CC = \Gamma^{-1}(\mu) \; \gamma(\mu, r\mu)$$

(4.3.5)

and the result is a line of constant CC as a function of r and μ. The solution to Eq. (4.3.5) as found numerically is graphed in Fig. 4.8.

From Fig. 4.8 we see that if the ratio of the standard deviation to the mean (as computed from Eq. (4.2.13) for example) is equal to unity, then we may expect a measured response, selected at random, to have a value less than 2.5dB more than the mean in 80% of the cases. In 95% of the cases the measurement should be less than 5 dB more than the mean and less than 7 dB more than the mean in 99% of the cases. In many situations, this is precisely the way that we would want the estimate stated, particularly if we were concerned about the response exceeding some prescribed value that might lead

to component failure, or excessive speech interference, or
acoustical detectability.

Also note, that since the calculation of $\sigma_\theta^2/m_\theta^2$ is to be
carried out in frequency bands $\Delta\omega$, or for individual modes
that resonate at different frequencies, the form of the confidence interval is that of a spectrum. One would usually
plot an "average estimate", based upon the methods of
section 4.1 as in Fig. 4.5, then plot a second spectrum
determined by 10 log r for each frequency band. One would
than look upon the second spectrum as an "upper bound" for
the data, with a degree of confidence determined by the
selected confidence coefficient.

As a second example of a confidence interval, consider
one that brackets the calculated mean. Such interval might
be stated as "in the 500 Hz octave band, I estimate that in
95% of the cases, the measured acceleration level is within
±5 dB of the average estimate of -10 dB re 1g." The "cases"
referred to here, of course, are repeated measurements with
the loading and observation points varied consistent with
the calculations of mean and variance, and for a population
of systems. Obviously, one rarely has this population to
deal with; there is only one system in the laboratory, or at
most 2 or 3. The system that one has might be very good or
very poor from a viewpoint of variance, but we must hope
that the estimate for the population will be useful in dealing with it. This situation is directly analogous to gambl-
ing. One only has a particular realization of all possible
hands of cards before him, but the pattern of betting, calculation of odds and strategy is made on the basis of all
possible hands and draws.

The evaluation of this "bracketing" interval is made by
setting $\theta_2=rm_\theta$ and $\theta_1=m_\theta/r$ in Eq. (4.3.3). The equation to
be solved in this case is

$$CC = \Gamma^{-1}(\mu) \{\gamma(\mu,\mu\ r)-\gamma(\mu,\mu/r)\}. \qquad (4.3.6)$$

The solution to Eq. (4.3.6) is shown in Fig. 4.9. For
example, the value of normalized variance that would produce
the hypothetical estimation interval of the preceding

paragraph is $\sigma^2/m^2 = 0.2$. An interval of this type is very useful in determining the value of the mean alone as an estimate of response. Since one can usually only make measurements of structural vibration and acoustical noise to a repeatability of 1 or 2 dB, then if an estimation interval of ± 1 dB has a high confidence coefficient (80%, say) then the mean value is a very good estimate. We see from Fig. 4.9 that this requires that $\sigma^2/m^2 \leq 0.1$.

Example. As an illustration of the discussion in paragraph 4.2 and 4.3, we consider the system shown in Fig. 4.4 excited at a point and observed at a point. We derived an average energy estimated for this system in paragraph 4.1, which we may use to obtain m_θ. We now apply Eq. (4.2.13) to obtain an estimate of variance and then use Figs. 4.8 and 4.9 to determine confidence intervals.

First, the beam (System 2) is a one-dimensional system with modes of the form $\sin kx$, and the plate (System 1) is two dimensional with modes of the form $\sin k_1 x_1 \sin k_2 x_2$. If we regard the locations x_1, x_2 as uniformly located over the structural surfaces, we find that

$$\frac{\langle \psi_2^4 \rangle}{\langle \psi_2^2 \rangle^2} = \frac{\langle (\sin kx)^4 \rangle_x}{\langle (\sin kx)^2 \rangle_x^2} = \frac{3}{2} \quad \text{(one-dimensional)} \tag{4.3.7a}$$

$$\frac{\langle \psi_1^4 \rangle}{\langle \psi_1^2 \rangle^2} = \frac{\langle (\sin kx_1)^4 \rangle \langle (\sin kx_2)^4 \rangle}{\langle (\sin kx_1)^2 \rangle \langle (\sin kx_2)^2 \rangle^2} = \frac{9}{4} \quad \text{(two-dimensional)} \tag{4.3.7b}$$

Obviously, the corresponding quantity for three-dimensional systems is $(3/2)^3 = 27/8$.

Eq. (4.3.7) gives us two of the factors in Eq. (4.2.13). To obtain the third, we must say more about the system. However, note that the spatial variance alone contributes a factor $(3/2)^2 (9/4)^2 = 11$, so that we must have of the order

of 10 modes interacting or sampled to bring σ^2/m^2 down to
the order of 1 or 100 modes to get down to $\sigma^2/m^2 = 0.1$. In
general, the higher the dimensionality of the system, the
greater is the contribution to variance from the spatial
sampling factors. To counteract this, however, the higher
dimensional systems normally have greater modal density, and
this reduces variance.

Suppose that the beam in Fig. 4.4 is made of aluminum
and is 1.3 m long, .05 m wide, and .003 m thick. The average
frequency separation between modes for this beam is according
to (4.1.23)

$$\delta f_b = c_{fb}/\ell = 130 \sqrt{f/1000} \ . \qquad (4.3.8)$$

Thus, at 1000 Hz, the average separation is 130 Hz and at
250 Hz it is 65 Hz. We assume that the plate is also of
aluminum, is .003 m thick, and has an area of 1.7 m^2. The
average frequency separation for plate modes is then

$$\delta f_p = hc_\ell/\sqrt{3}A = 5H_z \ . \qquad (4.3.9)$$

Also, a reasonable value for loss factor for both systems is
$\eta_1 = \eta_2 = .01$. Thus,

$$\frac{\pi}{2}(\Delta_1 + \Delta_2) = \frac{\pi}{2} f (\eta_b + \eta_p) = 10^{-2} \pi f, \ Hz \qquad (4.3.10)$$

Let us assume that our data is taken in octave bands, so that
$\Delta f = f/\sqrt{2}$. Thus, combining Eqs. (4.3, 8, 9, 10), we get

$$\left\{ n_1 n_2 \frac{\pi}{2}(\Delta_1 + \Delta_2) \Delta \omega \right\}^{-1} = \frac{130 \sqrt{f/1000} \ 5}{10^{-2} \pi f. \ f/\sqrt{2}} = 1000 f^{-3/2} \qquad (4.3.11)$$

Thus, at f=100 Hz, this factor is 1 and at 10,000 Hz it is 10^{-3}. The product of this modal count term and the spatial factors is presented in Fig. 4.9 for the example. We have also sketched the variance expected when there is no uncertainty in modal amplitude on the beam, reducing the variance by a factor of $(2/3)^2$, from Eq. (4.3.7a). This is appropriate if for example we make response measurements at the free end of the beam and connect the plate to the base of the beam.

Now we can use this estimate of variance to establish a mean bracketing confidence interval in each frequency band. If this is done for a confidence coefficient of 0.8, we obtain the result shown in Fig. 4.11. We have used the estimate for the mean from Fig. 4.5 in the case where $n_{bp} < n_b$, which is not quite the case for our example since $n_{bp} = w/4\ell \simeq .05/5 \simeq .01 \simeq n_b$. The confidence interval in no way depends on the estimate of the mean, however.

Altogether then, our analysis of variance, in combination with a mean estimate and the calculation of confidence intervals based on normalized variance, provides us with a way of making estimates of response for real systems that includes variations in a realistic manner. The weakest link of this chain at present is the estimation of variance. There is good reason to believe, for example, that the location of resonance frequencies among the population of "similar" systems is not well described by a distribution that assumes that modes occur with equal probability along the frequency axis. Such a distribution predicts a high probability for very closely separated modes, whereas the normal coupling among modes will tend to "split" them, as demonstrated by Fig. 3.2. Thus, interacting groups of modes should have a lower probability of very small differences in resonance frequencies, and this should result in smoother distributions of response and lower variance.

4.4   Coherence Effects - Pure Tone and Narrow Band Response

In this final section, we consider the effects on our analysis of the bandwidth of excitation $\Delta\omega$. When $\Delta\omega$ is very broad, then we have shown that certain frequency integrations such as those of paragraphs 2.1 and 3.1 become quite simple, and that the expected variance in system response calculated

in paragraph 4.2 is reduced as the noise bandwidth increases.
When $\Delta\omega$ gets smaller than the average spacing between modes,
we may expect that variance will increase because of the
smaller number of modes included in the averages. When $\Delta\omega$
becomes of the order of a modal bandwidth, other complications
(or possibly simplifications) may develop.

    Narrow band excitation of resonator. Let us suppose
that the excitation $\ell(t)$ of the single dof system of Fig. 2.1
is the pure tone signal $L\,e^{-i\omega t}$. The response, according to
Eq. (2.1.20) is

$$
v(t) = LY\,e^{-i\omega t} = Le^{-i\omega t}(-i\omega_o M)^{-1}\left[i\eta + (\frac{\omega}{\omega_o} - \frac{\omega_o}{\omega})\right]^{-1}
$$

$$(4.4.1)$$

In Chapter 2, we averaged $|Y|^2$ over an interval $\Delta\omega$ (large
compared to $\omega_o\eta$) to determine the noise response of the
resonator. Let us now suppose that ω is fixed and we are
selecting resonators out of a population having resonance
frequencies ω_o uniformly distributed over the interval $\Delta\omega$
(this is one feature of the SEA model introduced in
paragraph 3.2). Then the average m.s. response will be

$$
\langle v^2\rangle_{\omega_o} = \frac{1}{2}|L|^2\,\langle|Y|^2\rangle_{\omega_b} = \frac{1}{2}|L|^2\,\frac{\Delta_e}{\Delta\omega}\,\frac{1}{(\omega\eta\,M)^2} \tag{4.4.2}
$$

where $\Delta_e = (\pi/2)\omega\eta$, the effective bandwidth. This is quite a
slowly varying function of ω, so that if we now say that $\ell(t)$
is a narrow band signal of bandwidth $\delta\omega$ so that $1/2\,L^2 \to 2S_\ell(\omega)\delta\omega$,
we have (remember that spectral densities in ω are defined from
$-\infty < \omega < +\infty$).

$$
\langle v^2\rangle_\omega = \frac{2S_\ell(\omega)\,(\delta\omega/\Delta\omega)\,\Delta_e}{(\omega\eta\,M)^2} \tag{4.4.3}
$$

which should be compared to Eq. (2.1.39). We may say that
the effect of restricting the bandwidth to $\Delta\omega$, is to reduce
the effective spectral density of the excitation in the ratio
$\delta\omega/\Delta\omega$. When $\delta\omega \to \Delta\omega$, we obtain the original response of the
resonator to the noise of bandwidth $\Delta\omega$. The response of a
resonator to a signal of any bandwidth is of the same form as
a noise response, if we average over the resonance frequency
of the resonator.

Interacting Modes. The result of Eq. (4.4.3) allows us
to interpret the result of Eq. (3.1.32) in an interesting
way. The interaction of mode "2" with a single mode "1" was
found to be equivalent to excitation of mode by a white noise
source. Let us re-examine this result in the light of Eq.
(4.4.3). To do this, imagine for simplicity that the
resonators in Fig. 3.1 have gyroscopic coupling only. Then,
noise excitation with spectral density S_{ℓ_1} of resonator 1 with
$y_2 \equiv 0$, results in a m.s. velocity of this resonator given by

$$< v_1^2 > = \frac{\pi S_{\ell_1}(\omega)}{\Delta_1 M_1^2} = \frac{S_{\ell_1}(f)}{4\Delta_1 M_1^2} \qquad (4.4.4)$$

This m.s. velocity has an equivalent bandwidth $\Delta_{e,1} = \pi\Delta_1/2$,
since its spectral form is that of the admittance of resonator 1.

From Eq. (3.1.3b) a m.s. velocity $<v_1^2>$ will produce a m.s.
force on resonator 2 given by $<v_1^2> G^2 = <v_1^2> \gamma^2 M_1 M_2$. This m.s.
force is "spread" over a bandwidth $\Delta_{e,1}$, so that its spectral
density is

$$2 S_{\ell_2}(\omega) = <v_1^2> \gamma^2 M_1 M_2/\Delta_{1e} = \frac{\pi S_{\ell_1}(\omega)}{\Delta_1 M_1^2} \cdot \frac{\gamma^2 M_1 M_2}{\Delta_{e,1}} \qquad (4.4.5)$$

Thus

$$S_{\ell_2} = S_{\ell_1} \frac{M_2}{M_1} \frac{\gamma^2}{\Delta_1 \Delta_2} \left(\frac{\pi}{2} \Delta_2 \right) \frac{1}{\Delta_{e,1}} \qquad (4.4.6)$$

is the spectral density. Since, according to Eq. (4.4.3), such
a signal is equivalent to a broad band noise over a band $\Delta\omega$,
with spectrum $S_{\ell_2} \Delta_{e1} / \Delta\omega$, the equivalent broad band excitation
produced by the interaction is

$$S_{\ell_2} \frac{\Delta_{e,1}}{\Delta\omega} = S'_{\ell_2} = S_{\ell_1} \frac{M_2}{M_1} \frac{\pi\Delta_2/2}{\Delta\omega} \lambda \qquad (4.4.7)$$

where $\lambda = \gamma^2/\Delta_1\Delta_2$ for gyroscopic coupling only, which is consistent with Eq. (3.1.32) for weak coupling.

We have shown two important features of the interaction
with this calculation. First, we can properly account for
the interaction when the coupling is weak ($\lambda \ll 1$) as excitation
of the indirectly excited system by noise, through a filter
of bandwidth Δ_{1e} and transfer magnitude determined by the
coupling strength, if the resonance frequency of the receiving
system is allowed to take random values over the interval $\Delta\omega$.
Secondly, if the effective bandwidth of the excitation of
system 2 is Δ_{e1}, then if more than one mode may be excited
simultaneously, we may expect important coherence effects to
arise from such narrow band excitation. We now discuss just
what those effects may be.

Coherence Effects in Multi-Modal Response to Band-Limited
Noise. We suppose that the system described by Eq. (2.2.1) is
excited by noise source of bandwidth $\Delta\omega$. This source may be a
vibration exciter or a damped resonator driven by noise as discussed in the preceding paragraphs. The displacement response
is of the form given in Eq. (2.2.3). The density average of m.s.
displacement is, of course

$$\langle y^2 \rangle_{\rho,t} = \sum_m \langle Y_m^2(t) \rangle_t \qquad (4.4.8)$$

We now assume that all modes in the interval $\delta\omega$ have the same
damping, with effective bandwidth Δ_e. If we excite this system
at a location where all modes have nearly the same amplitude
(on a free plate, this would correspond to a corner), then we
can assume that the m.s. response of each mode is identical.

Accordingly,

$$<y^2>_{\rho,t} = N <y^2>_t \tag{4.4.9}$$

where N is the number of resonantly excited modes and $<y^2>_t$ is the m.s. response of each.

Eq. (4.4.9) gives the average response of the system in the spirit of paragraph 4.1, assuming completely incoherent response. Let us suppose, however, that the response of every mode were perfectly coherent, as it would be under sinusoidal excitation ($\Delta\omega << \Delta_e$). In this event, the m.s. response is at any point is

$$<y^2>_t = \sum_{m,n} < Y_m(t) \ Y_n(t) >_t \ \psi_m(x)\psi_n(x). \tag{4.4.10}$$

If there is a location where every mode has an antinode ($\psi=\psi_{max}$) and every modal vibration is in phase, then the m.s. response at that location is

$$\max <y^2>_t = N^2 \ <Y^2>_t \ \psi^2_{max} \tag{4.4.11}$$

The ratio of the rms response at this location to the incoherent rms response is

$$\frac{y_{max}}{y_{rms}} = \psi_{max} \ \sqrt{N} \tag{4.4.12}$$

Thus, if N is large, this coherent "hot spot" of response may be substantially greater than the incoherent estimate.

Let us now assume that the noise bandwidth $\Delta\omega$ is large compared to Δ_e. The modal density of the system is $n(\omega)$. Thus, the size of each group of coherent excited modes is $n\Delta_e$, which by presumption must be large for coherent effects to be of any importance. The coherent response maximum, therefore, according to Eq. (4.4.11), is $(n\Delta_e)^2 <Y^2>_t \psi_{max}^2$, and a m.s. response of $n\Delta_e<Y^2>_t$, according to Eq. (4.4.9). Each possible coherent peak, therefore, exists in an incoherent "background" given by $n(\Delta\omega-\Delta_e)$, so that the total response at a hot spot is

$$\max <y^2>_t = (n\Delta_e)^2 \psi_{max}^2 <y^2>_t + n(\Delta\omega-\Delta_e) <y^2>_t$$

$$(4.4.13)$$

The ratio of maximum to average response then is

$$\frac{\max<y^2>_t}{<y^2>_{x,t}} = n \frac{\Delta_e^2}{\Delta\omega} \psi_{max}^2 + (1- \frac{\Delta_e}{\Delta\omega}) \quad ; \quad \Delta\omega>\Delta_e \qquad (4.4.14a)$$

$$= n\Delta_e\psi_{max}^2 \qquad\qquad ; \quad \Delta\omega<\Delta_e \qquad (4.4.14b)$$

As an example, consider the plate of the example in section 4.3. In that example $n\Delta_e=(\pi/2) fn_p/\delta f_p=10^{-2}\pi f/10$. At 1000 Hz, the number of coherent modes would be ~ 3. At 10,000 Hz, we would have 31 modes in each coherent group. The "statistical response concentrations" for this plate under 1/3 OB and pure tone response ($\psi_{max}=2$ for 2-dimensional mode) at 1000 Hz are

$$(1/3 \text{ OB})\frac{\max <y^2>_t}{<y^2>_{x,t}} = \pi \cdot \frac{5\pi}{250}\cdot 4 +(1-\frac{10^{-2}\pi\cdot 10^3}{250}) \approx 1.8$$

$$\text{(pure tone)} \quad \frac{\max<y^2>_t}{<y^2>_{x,t}} = \pi \cdot 4 = 13$$

At 1000 Hz, there is a small response concentration due to coherence for 1/3 OB excitation, but a concentration factor of about 3.5 will occur for pure tone excitation.

At 10,000 Hz, we have

$$\text{(1/3 OB)} \quad \frac{\max<y^2>_t}{<y^2>_{x,t}} = \frac{5\pi^2 \cdot 10^2 \cdot 4}{2500} + (1 - \frac{\pi \cdot 10^{-2} \cdot 10^3}{2 \cdot 2500})$$

$$\simeq \quad 8 + 1 = 9$$

$$\text{(pure tone)} \quad \frac{\max<y^2>_t}{<y^2>_{x,t}} = \pi \cdot 10 \cdot 4 = 126$$

The concentration for the band of noise is still moderate but for pure tone excitation, the response concentration is greater than 11.

It is clear that statistical response concentrations are sizeable. We have not discussed the probability that such a coherent peak, as represented by Eq. (4.4.12), will occur. Although the point is not completely resolved, it appears that such concentrations have a good likelihood of occurring. As a practical matter, such coherence effects will be important when large systems with high modal density and moderate to high damping are excited by pure tones. When these "statistical" concentrations occur, they may be considerably larger than the "stress concentrations" that occur near boundaries as exemplified by Fig. 4.3.

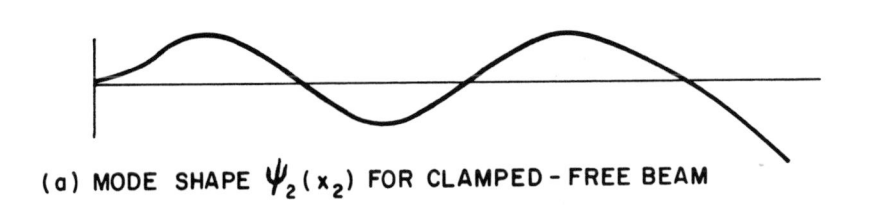

(a) MODE SHAPE $\psi_2(x_2)$ FOR CLAMPED - FREE BEAM

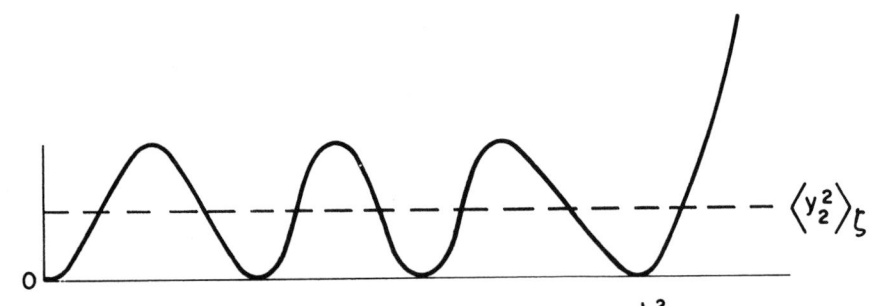

(b) M.S. DISPLACEMENT OF CLAMPED - FREE BEAM; $\psi_2^2(x_2)$

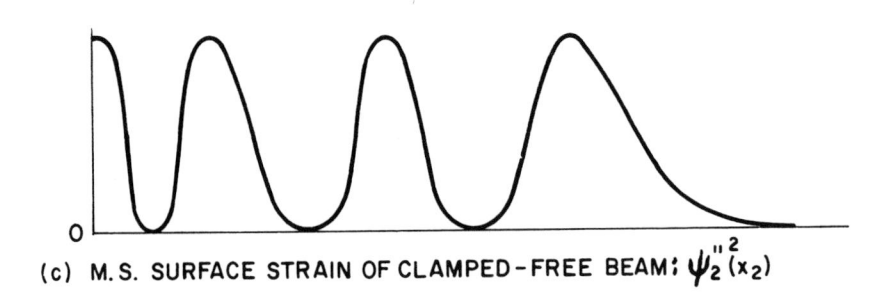

(c) M.S. SURFACE STRAIN OF CLAMPED - FREE BEAM; $\psi_2''^2(x_2)$

FIG 4.1

SPATIAL DISTRIBUTION OF RESPONSE FOR CLAMPED-FREE BEAM

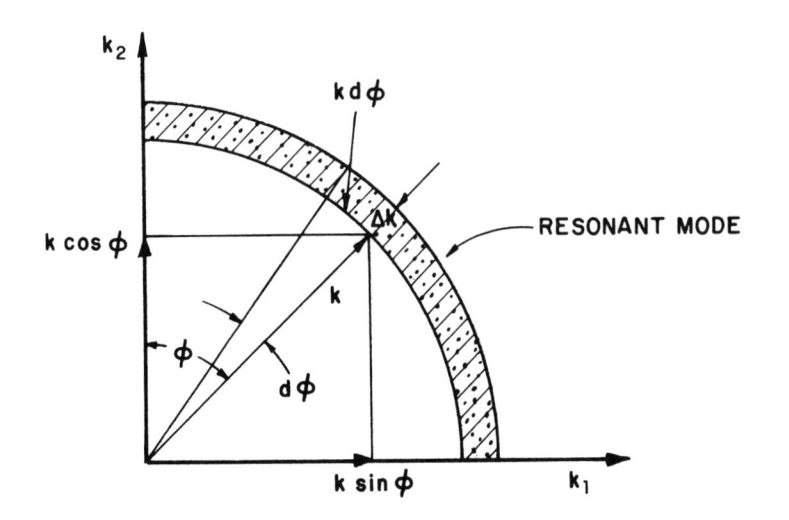

FIG. 4.2

DISTRIBUTION OF 2-DIMENSIONAL MODES EXCITED BY BAND OF NOISE $\Delta\omega$

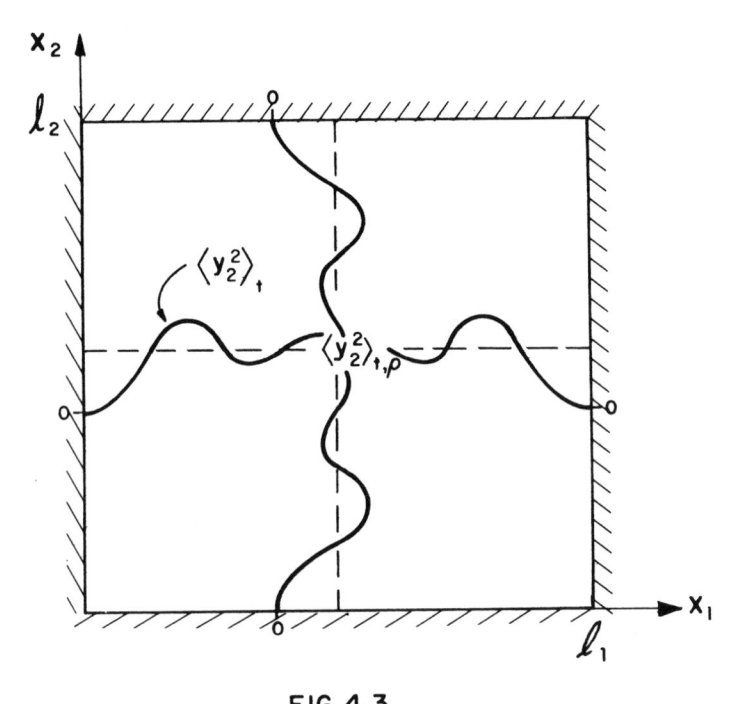

FIG. 4.3

MULTI-MODAL TEMPORAL MEAN SQUARE RESPONSE OF RECTANGULAR PLATE

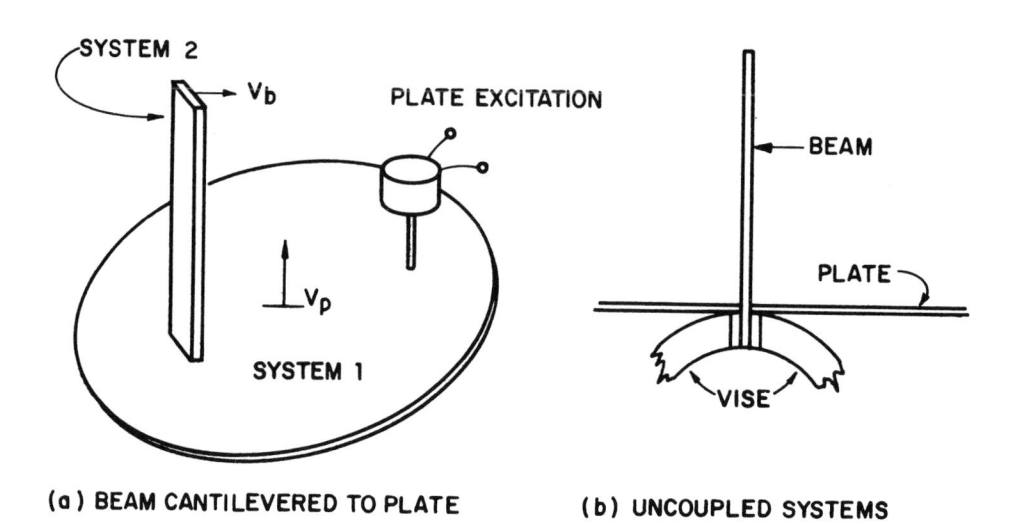

(a) BEAM CANTILEVERED TO PLATE   (b) UNCOUPLED SYSTEMS

FIG 4.4

BEAM−PLATE SYSTEM, BEAM INDIRECTLY EXCITED

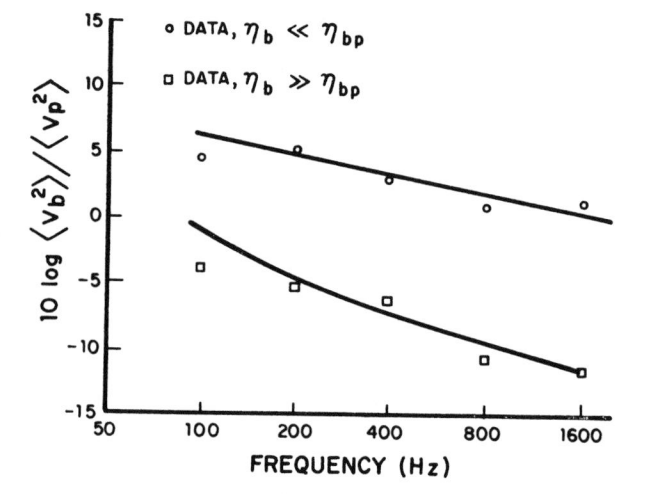

FIG 4.5

COMPARISON OF MEASURED RESPONSE RATIOS WITH THOSE
PREDICTED BY EQ. (4.1.25)

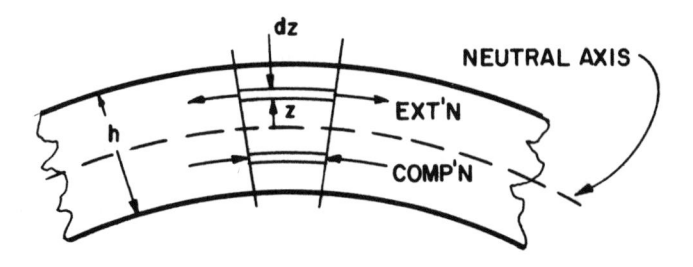

FIG 4.6

BENDING STRAIN IN A BEAM OF THICKNESS h

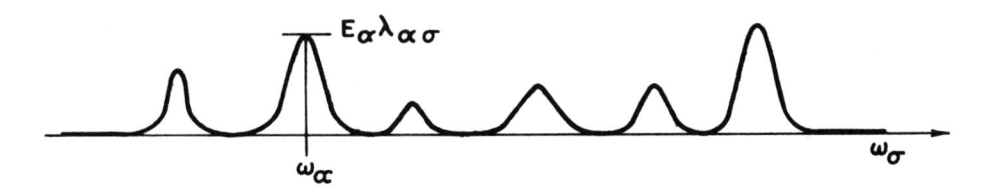

FIG 4.7

THE SUM IN EQ. (4.2.5) AS A FUNCTION OF ω_σ

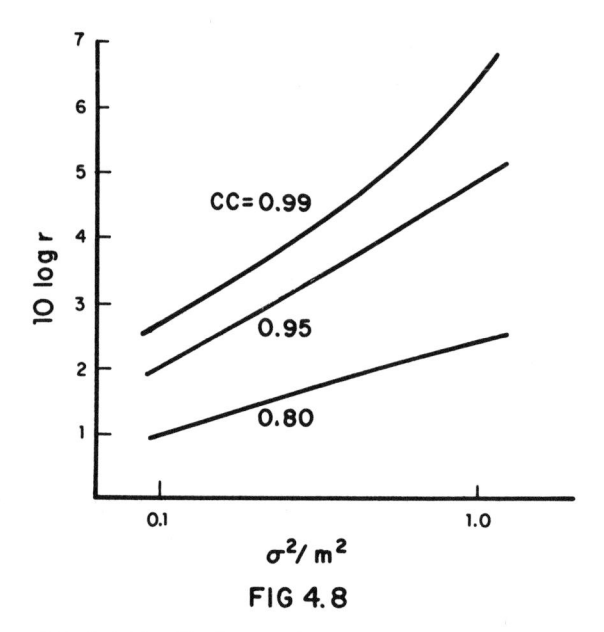

FIG 4.8

**UPPER BOUND OF SIMPLE EXCEEDANCE ESTIMATION INTERVALS
AS A FUNCTION OF NORMALIZED VARIANCE**

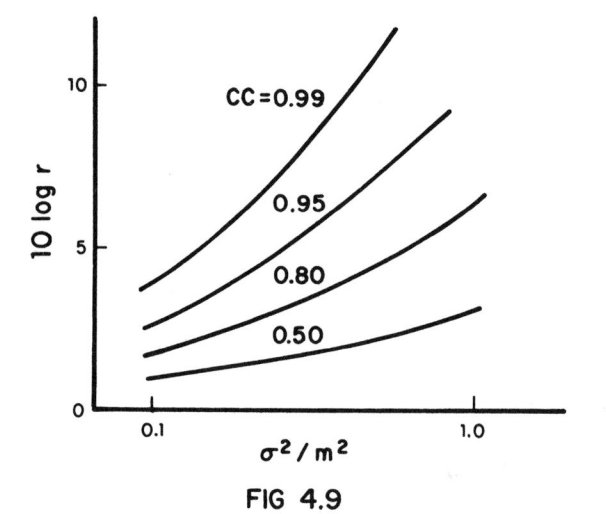

FIG 4.9

**HALF-WIDTH OF MEAN-BRACKETING ESTIMATION INTERVALS
AS A FUNCTION OF NORMALIZED VARIANCE**

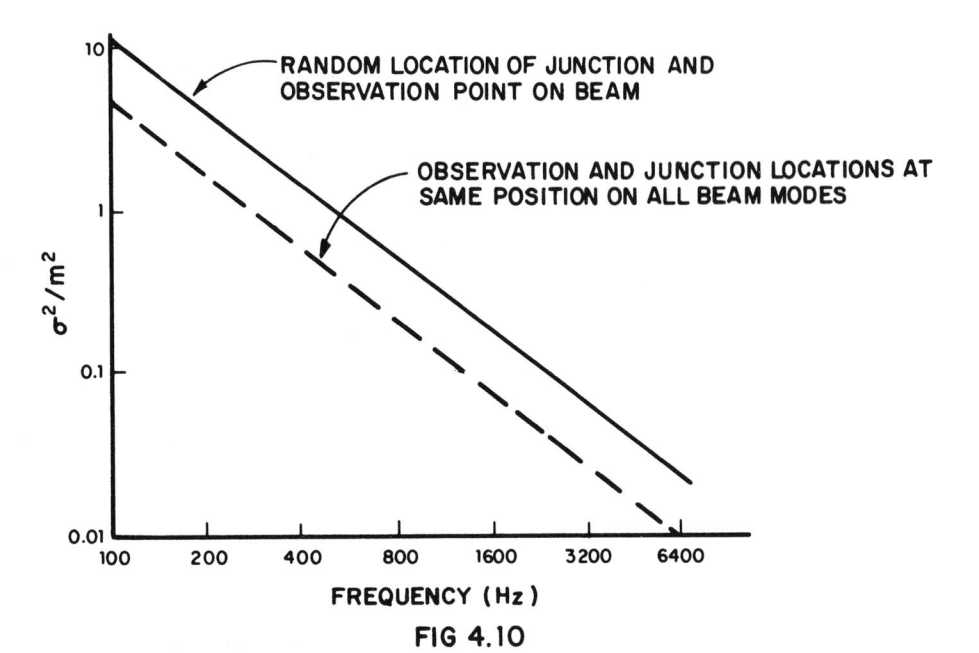

FIG 4.10

NORMALIZED VARIANCE AS A FUNCTION OF FREQUENCY FOR
BEAM – PLATE EXAMPLE IN SECTION 4.3

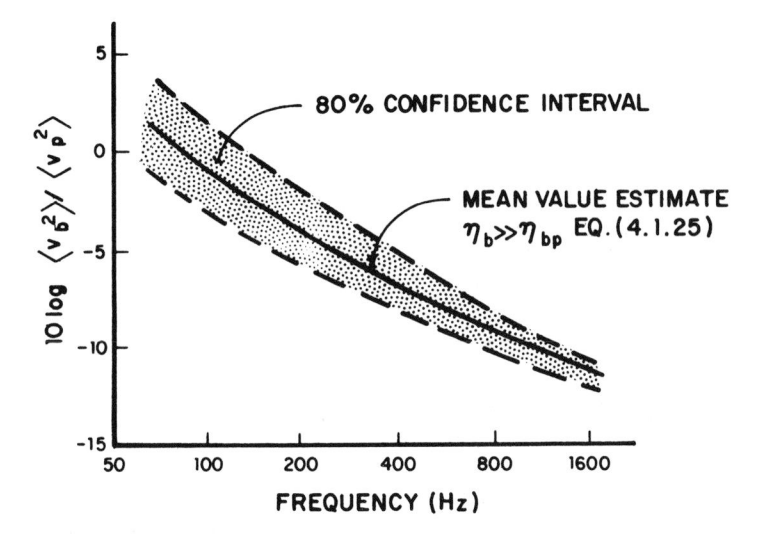

FIG 4.11

MEAN BRACKETING 80% CONFIDENCE INTERVAL BASE ON FIG 4.9
FOR SYSTEM OF FIG 4.4, MEAN VALUE ESTIMATE OF FIG 4.5
AND VARIANCE OF FIG 4.10 (DASHED LINE)

ANNOTATED BIBLIOGRAPHY FOR PART I

Chapter 1.   The Development of SEA

1.0   Introduction

A general review of the procedures of SEA is given in

(aa)   R. H. Lyon, "What Good is Statistical Energy Analysis,
Anyway?" Shock and Vibration Digest, Vol. 3, No. 6,
pp 1-9 (June 1970).

The use of statistics in random vibration is presented in

(ab)   S. H. Crandall and W. D. Mark, Random Vibration in
Mechanical Systems (Academic Press, New York, 1963).

(ac)   Y. K. Lin, Probabilistic Theory of Structural Dynamics
(McGraw-Hill Book Co., Inc., New York 1967).

A well known "classical" reference in vibration
analysis is

(ad)   J. P. Den Hartog, Mechanical Vibrations - Fourth Edition
(McGraw-Hill Book, Co., Inc., 1958).

Problems involved in calculating resonance frequencies
and mode shapes of higher order modes are reviewed in

(ae)   R. Bamford, et al., "Dynamic Analysis of Large Structural
Systems," contribution to Synthesis of Vibrating Systems,
Ed. by V. H. Neubert and J. P. Raney (A.S.M.E.,
New York 1971).

(af)   E. E. Ungar, et al., "A Guide for Predicting the
Vibrations of Fighter Aircraft in the Preliminary Design
Stages" AFFDL-TR-71-63, April 1973.

The statistical theory of room acoustics from a modal
viewpoint is presented in

(ag)   P. M. Morse and R. H. Bolt, "Sound Waves in Rooms"
Rev. Mod. Phys. 16, No. 2, pp. 69-150 (April 1944).

and is presented from a wave viewpoint in

(ah)   L. L. Beranek Acoustics (McGraw-Hill Book Co., Inc.
       New York, 1954).

The statistical mechanics of groups of resonators is
important in the theory of specific heats of crystals.
An introductory reference is

(ai)   F. K. Richtmyer and E. H. Kennard, Introduction to
       Modern Physics (McGraw-Hill Book Co., Inc. 1947).

The representation of a thermal reservoir as a noise
generator is discussed in

(aj)   A. van der Ziel, Noise (Prentice Hall, Inc. Englewood
       Cliffs, N. J. 1955).

Measures of acoustical characteristics of rooms that
are thought to affect the listening quality of a room
are presented in

(ak)   L. L. Beranek, Music, Acoustics and Architecture (John
       Wiley & Sons, Inc., New York, 1962).

The quotation by Mehta is to be found in the introduction to

(al)   M. L. Mehta, Random Matrices and the Statistical Theory
       of Energy Levels, (Academic Press, New York 1967).

1.1   Beginnings

The earliest work in SEA is to be found in

(am)   R. H. Lyon and G. Maidanik, "Power Flow Between Linearly
       Coupled Oscillators" J. Acoust. Soc. Am., Vol. 34, No. 5,
       pp. 623-639 (May 1962).

(an)   P. W. Smith, Jr., "Response and Radiation of Structural
       Modes Excited by Sound" J. Acoust. Soc. Am., Vol 34,
       No. 5 pp. 640-647 (May 1962).

The papers concerned with removal of the "light
coupling" restriction are

(ao)   E. E. Ungar, "Statistical Energy Analysis of Vibrating
       Systems" Trans. ASME, J. Eng. Ind. pp 626-632 (Nov. 1967).

(ap)   T.D. Scharton and R. H. Lyon, "Power Flow and Energy
       Sharing in Random Vibration" J. Acoust. Soc. Am.
       Vol. 43, No. 6 pp 1332-1343 (June 1968).

       Reports that contain the earliest results on variance
       analysis are

(aq)   R. H. Lyon, et al., "Random Vibration Studies of Coupled
       Structures in Electronic Equipments" Report No. ASD-TDR-
       63-205. Wright Patterson Air Force Base, Dayton, Ohio.

(ar)   R. H. Lyon, "A Review of the Statistical Analysis of
       Structural Input Admittance Functions. Report No.
       AD-466-937. Wright Patterson Air Force Base, Dayton,
       Ohio.

       These variance calculations were used to develop
       estimation intervals in

(as)   R. H. Lyon and E. Eichler, "Random Vibration of Con-
       nected Structures" J. Acoust. Soc. Am., Vol..36, No. 7,
       pp 1344-1354 (July 1964).

       The extension of the SEA formulation to three systems
       in tandem is in

(at)   E. Eichler, "Thermal Circuit Approach to Vibrations in
       Coupled Systems and the Noise Reduction of a Rectangular
       Box", J. Acoust. Soc. Am., Vol. 37, No. 6, pp 995-1007
       (June 1963).

       Three element transmission for the plate-beam-plate
       system is given in

(au)   R. H. Lyon and T. D. Scharton, "Vibrational Energy
       Transmission in a Three Element Structure", J. Acoust.
       Soc. Am., Vol. 38, No. 2, pp 253-261 (August 1965).

       The calculation of plate-edge admittances is given in

(av)   E. Eichler, "Plate-Edge Admittances", J. Acoust. Soc.
       Am., Vol. 36, No. 2, pp 344-348 (February 1964).

       The effect of reinforcing beams and constrained edges on
       the radiation resistance of flat plates is given in

(aw)   G. Maidanik, "Response of Ribbed Panels to Reverberant
       Acoustic Fields", J. Acoust. Soc. Am., Vol. 34, No. 6,
       pp 809-826 (June 1962).

A similar calculation for cylinders is given in

(ax)   J. E. Manning and G. Maidanik, "Radiation Properties of
       Cylindrical Shells", J. Acoust. Soc. Am., Vol 36, No. 9,
       pp 1691-1698 (September 1964).

(ay)   L. Cremer and M. Heckl, Structure-Borne Sound (Springer
       Verlag, Berlin 1973). Translated by E. E. Ungar.

The input impedances in matrix form for foundations
resting on a visco-elastic half space may be found in

(az)   L. Kurzweil, "Seismic Excitation of Footings and Footing-
       Supported Structures", Ph.D. Thesis, MIT Dept. of Mech.
       Eng., September 1971.

Modal densities for acoustical spaces are derived in
(ag). Modal densities for flat and curved structures
may be found in (ay) and in

(ba)   V. V. Bolotin, "On the Density of the Distribution of
       Natural Frequencies of Thin Elastic Shells", J. Appl.
       Math Mech., Vol. 27, No. 2, pp 538-543 (Trans. from
       Soviet J.: Prikl. Mat. Mekh., Vol. 27, No. 2,
       pp 362-364 (1963).

Other modal density calculations may be found in

(bb)   M. Heckl, "Vibrations of Point Driven Cylindrical Shells",
       J. Acoust. Soc. Am., Vol. 34, No. 10, pp 1553-1557 (1962).

(bc)   J. E. Manning, et al., "Transmission of Sound and
       Vibration to a Shroud-Enclosed Spacecraft". NASA Report
       CR-81688, October 1966.

(bd)   K. L. Chandiramani, et al., "Structural Response to
       Inflight Acoustic and Aerodynamic Environments". BBN
       Report 1417, July 1966.

(be)   D. K. Miller and F. D. Hart, "Modal Density of Thin
       Circular Cylinders", NASA Contractors Report CR-897 (1967).

(bf)   E. Szechenyi, "Modal Densities and Radiation Efficiencies
       of Unstiffened Cylinders Using Statistical Methods", J.
       Sound Vib., Vol. 19, No. 1, pp. 65-81 (1971).

The modal density of cones is presented in

(bg)   J. E. Manning, et al., "Vibration Transmission in the
       RVTO Phase 1A Reentry Vehicle", BBN Report 1533,
       March 1968.

       and for dished shells in

(bh)   J. P. D. Wilkinson, "Modal Densities of Certain Shallow
       Structural Elements," J. Acoust. Soc. Am., Vol 43, No. 2,
       pp 245-251 (1968).

       Structural damping is expressed as an edge absorption
       coefficient in

(bi)   M. Heckl, et al., "New Methods for Understanding and
       Controlling Vibrations of Complex Structures", Wright
       Patterson Air Force Base Technical Note ASD-TN-61-122
       (1962).

       and the air-pumping mechanism for damping of structures
       is presented in

(bj)   E. E. Ungar, "Energy Dissipation at Structural Joints,
       Mechanisms and Magnitudes," U.S. Air Force FDL-TDR-64-98
       (1964).

       The application of SEA to sound transmission through
       double walls is reported in the following articles:

(bk)   M. J.Crocker, et al., "Sound and Vibration Transmission
       Through Panels and Tie Beams Using Statistical Energy
       Analysis", Trans. ASME: J. Engineering Ind. (Aug 1971)
       pp 775-782.

(bl)   P. H. White and A. Powell, "Transmission of Random
       Sound and Vibration Through A Rectangular Double Wall",
       J. Acoust. Soc. Am., Vol. 40, No. 4, pp 821-832 (1965).

(bm)   A. Rinsky, "The Effects of Studs and Cavity Absorption
       on the Sound Transmission Loss of Plasterboard Walls",
       Sc.D. Thesis, MIT Dept. of Mech. Eng., February 1972.

       An attempt to elucidate implications of the SEA model in
       terms of classical vibration analysis is presented in

(bn)   J. L. Zeman and J. L. Bogdanoff, "A Comment on Complex
       Structural Response to Random Vibrations", AIAA Journal,
       Vol. 7, No. 7, pp 1225-1231 (July 1969).

       and also in

(bo)   W. Gersch, "Mean Square Responses in Structural Systems",
       J. Acoust. Soc. Am., Vol 48, No. 1 (Pt. 2) pp 403-413
       (1970).

(bp)   W. Gersch, "Average Power and Power Exchange in
       Oscillators", J. Acoust. Soc. Am., Vol. 46, No. 5
       (Pt. 2) pp 1180-1185 (1969).

The work of Lotz referred to is

(bq)   R. Lotz, "Random Vibration of Complex Structures", Ph.D.
       Thesis, MIT Dept. of Mechanical Engineering, June 1971.

A comparison of the effect of power flow of various
assumptions regarding modal frequency statistics (or
occurrence in frequency) is given in (ap). A report
in which calculations for a deterministic system may
be compared with the SEA calculations is

(br)   J. E. Manning and P. J. Remington, "Statistical Methods
       in Acoustics", J. Acoust. Soc. Am., Vol. 57, No. 2,
       (Feb., 1975).

1.2   The General Procedures of SEA

Discussions of "similar" mode groups may be found in
(aa, bc) and in

(bs)   J. E. Manning, "A Theoretical and Experimental Model-
       Study of the Sound-Induced Vibration Transmitted to a
       Shroud-Enclosed Spacecraft". BBN Report 1891, submitted
       May 1, 1970 to NASA.

The equivalent RC circuit for energy flow is discussed
in (am, at). The power injecting properties of various
random loading environments are discussed in (bd) and
also in

(bt)   R. H. Lyon "Boundary Layer Noise Response Simulation
       with a Sound Field", Chapter 10 of Acoustical Fatigue
       in Aerospace Structures, Ed. by W. J. Trapp and
       D. M. Forney (Syracuse University Press, Syracuse,
       N. Y., 1965).

(bu)   R. H. Lyon, Random Noise and Vibration in Space Vehicles
       Shock and Vibration Information Center, U. S. Dept. of
       Defense, 1967).

(bv) I. Dyer, "Response of Plates to a Decaying and Convecting
      Random Pressure Field". J. Acoust. Soc. Am., Vol. 31, No. 7
      pp 922-928 (1965).

      The relative magnitudes of time averaged and oscillating
      power flow between coupled resonators is discussed in

(bw) D. E. Newland, "Calculation of Power Flow Between Coupled
      Oscillators", J. Sound Vib., Vol. 3, No. 3, pp 262-276
      (1966).

      A brief summary of the problems of measuring various SEA
      parameters is given in

(bx) R. H. Lyon, "Analysis of Sound-Structural Interaction by
      Theory and Experiment", Contribution to Proceedings of
      Purdue Noise Control Conference, July 14-16, 1971,
      Purdue University.

      Modal density measurements of cylinders are presented
      in (bb, bs). Measurements for flat plates are given in
      (ar, at), and for mass loaded plates in

(by) R. W. Sevy and D. A. Earls, "The Prediction of Internal
      Vibration Levels of Flight Vehicle Equipments Using
      Statistical Energy Methods," U.S. Air Force Technical
      Report AFFDL-TR-68, June 1968.

      Calculations of modal densities are given in (ba through
      bh) for a variety of structural elements or subsystems.
      Input impedances to beams and plates are found in (ao,
      ay, av). Input impedances to sound fields may be found
      in (am, aw, ax, bd, bg), and in

(bz) R. H. Lyon, "Statistical Analysis of Power Injection and
      Response in Structures and Rooms" J. Acoust. Soc. Am.,
      Vol. 45, No. 3, pp 545-565 (1969).

      The derivation of transmission loss and its relation to
      coupling loss factor may be found in

(ca) I. L. Ver and C. I. Holmer, "Interaction of Sound Waves
      with Solid Structures", Chapter 11 of Noise and
      Vibration Control, Ed. by L. L. Beranek (McGraw-Hill
      Book Co., Inc. New York 1971).

(cb) C. I. Holmer, "Sound Transmission Through Structures:
      A Review" M.Sc. Thesis, John Carroll University,
      Cleveland, Ohio 1969.

A useful "dictionary" of transmission loss for various practical wall structures may be found in

(cc)   R. D. Berendt, G. E. Winzer and C. B. Burrough, "A Guide to Airborne, Impact and Structure Borne Noise Control in Multifamily Dwellings." U. S. Dept. of Housing and Urban Development, Washington, D. C., September 1967.

The relation of vibrational velocity to stress in a variety of structural elements is presented in

(cd)   F. V. Hunt, "Stress and Strain Limits on the Attainable Velocity in Mechanical Vibration", J. Acoust. Soc. Am., Vol. 32, No. 9, pp 1123-1128 (1960).

(ce)   E. E. Ungar, "Maximum Stresses in Beams and Plates Vibrating at Resonance", J. Eng. Ind. Vol. 84, No. 1, pp 149-155 (February 1962).

Analysis of response variance is developed in (aq, ar, as, bz). A basic discussion of confidence intervals and coefficients may be found in

(cf)   A. M. Mood, Introduction to the Theory of Statistics (McGraw-Hill Book Co., Inc., New York, 1950). Chapter 11.

1.3   Future Developments of SEA

The range of professional workers making use of SEA is discussed in (aa). That reference prompted additional discussion.

(cg)   J. L. Zeman and J. L. Bogdanoff, Letter to the Editor. The Shock and Vibration Digest, Vol. 3, No. 1, p 2 (1971).

(ch)   C. T. Morrow, "Can Statistical Energy Analysis be Applied to Design?" The Shock and Vibration Digest, Vol. 3, No. 5, Pt. 1 (1971).

and a reply

(ci)   R. H. Lyon, Letter to the Editor. The Shock and Vibration Digest, Vol. 3, No. 10, pp 3-4 (1971).

A computer solution of simultaneous equations in SEA
has been carried out and is reported in

(cj)   D. M. Wong, "Transmission of Noise and Vibration Between
       Coupled Cylinders," Lockheed Sunnyvale Report, 1969.

The development of a range of models more complex than
the SEA models presented here that have greater
"information capacity" has been discussed in

(ck)   R. H. Lyon, "Application of a Disorder Measure to Acoustical
       and Structural Models" J. Eng. Ind., Trans. ASME, Vol. 93,
       Ser. B, No. 3, pp 814-818 (August 1971).

The allowance for modal interference effects the modal
resonance frequency spacing has been discussed in (al,
bz).

1.4 Organization of Part I

Various surveys of SEA have been published for various
audiences. They include (aa, ao, at, bq, br, bx) and
also

(cl)   R. H. Lyon and G. Maidanik, "Statistical Methods in
       Vibration Analysis", AIAA Journal, Vol. 2, No. 6,
       pp 1015-1024 (1964).

(cm)   P. A. Franken and R. H. Lyon, "Estimation of Sound-
       Induced Vibrations by Energy Methods, with Applications
       to the Titan Missile", Shock and Vibration Bulletin,
       No. 31, Part III, pp 12-16 (1963).

(cn)   R. H. Lyon, "Basic Notions in Statistical Energy Analysis,"
       Contribution to Synthesis of Vibrating Systems, Ed. by
       V. H. Neubert and J. P. Raney, (ASME, New York, 1971).

(co)   P. W. Smith, Jr. and R. H. Lyon, "Sound and Structural
       Vibration", NASA Contractor Report CR-160, March 1965.

Chapter 2. Energy Description of Vibrating Systems

2.0 Introduction

Additional discussion of energy variables in vibration

may be found in (co). The relation of impedance functions to energy variables is discussed in

(cp) E. A. Guillemin, Introductory Circuit Theory, (Chapman and Hall, Ltd, New York 1953) Chapter 7.

Discussion of the relation between modal and wave descriptions may be found also in (ag, am, co).

2.1 Modal Resonators

Solutions for the single dof vibrator may be found in (ab, ad, co, cp). Relations among various measures of damping may be found in (ay). Reverberation time as a measure of the damping of sound waves is discussed in (ah, co). The idea of viewing various frequency regions of response is "mass-controlled", etc. is discussed in (ca, co). Random excitation of simple resonators is extensively discussed in (ab, ac). The notion of a white noise generator as a constant power source is in (am). The removal of the divergent mass law response of a mode to white noise may be found in

(cq) R. H. Lyon, "Shock Spectra for Statistically Modeled Structures", Shock and Vibration Bulletin, No. 40, Part 4, pp 17-23, December 1969.

2.2 Modal Analysis of Distributed Systems

The theory of the vibration of distributed systems by using modal expansions may be found in (an) and also in

(cr) P. M. Morse and H. Feshbach, Methods of Theoretical Physics (McGraw-Hill Book Col, Inc., New York, 1953).

When damping is not proportional to the mass density an irreducible modal coupling results which is extensively discussed in (bn). Modal lattices are used extensively in acoustics (ag, ah) and in x-ray diffraction theory and solid state physics:

(cs) L. Brillouin, Wave Propagation in Periodic Structures (McGraw-Hill Book Co., Inc., New York, 1946).

An elementary discussion of energy and group velocity
may be found in (ai). The expansion of the structure
admittance function as a sum of modal admittances is
given in (bz). The point input resistance of an infinite
plate is derived in (ay, co). The use of conductance
measurement as a means of evaluating modal density was
first proposed in (bx).

2.3 Dynamics of Infinite Systems

The development of dispersion relations and their interpretation in terms of phase and group velocities may
be found in

(ct) L. M. Brekhovskikh, Waves in Layered Media (Academic
Press, New York 1960). Derivations for the "stiffness
operator" for strings, beams, and sound fields may be
found in

(cu) P. M. Morse, Vibration and Sound, Second Edition,
(McGraw-Hill Book Co., Inc. 1948). The theory of
Fourier integrals and the use of contour integrals
to evaluate the inverse transform may be found in
(cr). The derivation of mean free path for a two
dimensional system is to be found in (co). The discussion of resonator interaction with a finite plate
is adapted from (bo).

2.4 Modal-Wave Duality

The decomposition of a wave field into "direct" and
"reverberant" components is discussed in (ag, ah, ay,
bz, co). The formation of the direct field as a
result of modal coherence may be found in (ag).
Various integral expressions for and relations between
bessel functions may be found in (cr).

Chapter 3. Energy Sharing by Coupled Systems

A discussion of the environmental excitation of
structures may be found in (bd, bu) and in

(cv)   R. H. Lyon, "An Energy Method for Prediction of Noise
       and Vibration Transmission", Shock and Vibration
       Bulletin, No. 33, Part II pp 13-25 (February 1964).

       The definition of uncoupled systems has been an important
       element in several studies (bc, bk, br, bs). An
       interesting application of reciprocity to develop
       engineering formulas for noise transmission is

(cw)   M. A. Heckl and E. J. Rathe, "Relationship Between
       the Transmission Loss and the Impact-Noise Isolation
       of Floor Structures", J. Acoust. Soc. Am., Vol 35, No. 11,
       pp 1825-1830 (1963).

3.1   Energy Sharing Among Resonators

       The discussion in this section is taken from (ap),
       which in turn is based on

(cx)   T. D. Scharton, "Random Vibration of Coupled Oscillators
       and Coupled Structures," Sc.D. Thesis, MIT Dept. of
       Mechanical Engineering, 1965.

       Additional discussion of the effects of blocking
       vibrating systems may be found in (bq, bn) and in

(cy)   D. E. Newland, "Power Flow Between A Class of Coupled
       Oscillators," J. Acoust. Soc. Am., Vol. 43, No. 3,
       pp 553-559 (1968).

       The example of two resonators coupled by stiffness only
       is taken from (cn). The behavior of coupled resonators
       in free vibration may be found in (ad). The analysis
       of power flow in this section and in (ao) is carried
       out in the "frequency domain". For a time domain
       analysis, see (am). Integrations over the frequency
       response functions in Eq. (3.1.14) may be found in
       (ab). The result of Eq. (3.1.19) was found in (am)
       for weak coupling, although the result (for weak
       coupling) was also applied to relations like Eq. (3.1.25).
       The essential correctness of Eq. (3.1.25) must be
       found from (ao, ap, cx). The equivalence of average
       modal interaction and white noise excitation was first
       put forward in (am).

3.2 Energy Exchange in Multi-Degree-of-Freedom Systems

Additional discussion of the assumptions involved in
forming the "uncoupled" modes for the blocked condition
may be found in (bn). A discussion of the treatment
of damping as a random parameter is to be found in
(cg). Quite a thorough analysis of sound-structural
coupling, carried out on a completely modal basis is
to be found in

(cz) F. Fahy, "Vibration of Containing Structures by Sound
in the Contained Fluid", J. Sound Vib., Vol. 10, No. 3
pp 490-512 (1969).

(da) F. Fahy, "Response of a Cylinder to Random Sound in the
Contained Fluid", J. Sound Vib., Vol. 13, No. 2,
pp 171-194 (1970).

Discussions of the generation of multi-modal interactions from the two-mode interaction are to be found
in (am, ao, at). The first published use of a diagram
like that of Fig. 3.5 appears to be (cv).

3.3 Reciprocity and Energy Exchange in the Wave Description

For a discussion of diffusion, reverberation and
energy density in sound fields, see (ah, cu). The
general conditions for reciprocity in continuous
systems are given in (cr, ct) and for "lumped"
systems in (cp). The use of reciprocity to develop
response estimates of the SEA form may be found in
(cv). The use of impedance relations to determine
coupling loss factors may be found in (am, as, au,
bc). A study of two plates connected at a point is
presented in (bq). The relation between coupling
loss factor and transmission coefficients (or
transmission loss) is developed in (bk, cn) and in

(db) R. H. Lyon, Aerospace Noise Vibration. Notes for a
lecture series offered by Bolt Beranek and Newman,
1966.

Transmission loss data are to be found in (cc).

3.4 Some Sample Applications of SEA

This discussion of the excitation of a resonator by a

sound field may be compared to that in (an). The
analysis of coupled beams may be compared with
similar discussions in (ap, br) and

(dc)   H. G. Davies, "Exact Solutions for the Response of Some
       Coupled Multimodal Systems", J. Acoust. Soc. Am., Vol. 51,
       No. 1 (Pt. 2), pp 387-392 (1971).

The relation between "mass law" transmission loss and
the coupling loss factor is discussed in (bk, db).

Chapter 4.   The Estimation of Response in Statistical Energy
Analysis.

4.0   Introduction

Additional discussion of the requirements for response
estimation may be found in (aa, ao, bu, cg, ch, ci)
and in

(dd)   J. L. Bogdanoff, "Meansquare Approximate Systems and
       Their Application in Estimating Response in Complex
       Disordered Linear Systems", J. Acoust. Soc. Am.,
       Vol. 38, No. 2, pp 244-252 (1965).

General source books for statistical estimation procedures are (bf) and

(de)   H. Cramer, Mathematical Methods of Statistics (Princeton
       University Press, Princeton, 1946).

4.1   Mean Value Estimates of Dynamical Response

The mean square pressure in reverberant sound fields
has been studied extensively in

(df)   R. V. Waterhouse, "Statistical Properties of Reverberant
       Sound Fields", J. Acoust. Soc. Am. Vol. 43, No. 6,
       pp 1436-1444 (1968).

Spatial coherence effects in multi-modal response in
the neighborhood of the drive point are discussed in
(bz) and in

(dg)   L. Wittig, "Random Vibration of Point-Driven Strings
       and Plates", Ph.D. Thesis, MIT Dept. of Mech. Eng.,
       May 1971.

       The problem of adequate sampling of a reverberant
       field for reducing uncertainty in the energy content
       of the field has been discussed by

(dh)   R. V. Waterhouse and D. Lubman, "Discrete Versus
       Continuous Averaging in a Reverberant Sound Field",
       J. Acoust. Soc. Am. Vol. 48, No. 1 (Pt.1) pp 1-5
       (1970).

       The beam plate system studied in this section is similar
       to that in (as). There has been much confusion regarding
       the use of infinite system impedances to evaluate coupling
       loss factors in finite systems, as was done in (am, as,
       an, aw). Hopefully the discussion in this section will
       clarify the procedure. The expression of system strain
       in terms of a particle "Mach number" is found in (cd,
       ce).

4.2   Calculation of Variance in the Temporal Mean Square
      Response.

       The analysis of variance is an extensive and important
       topic in statistical theory (cf, de). The procedures
       used in this section were first developed in (ag, ar,
       as) and are based on the theory of variance of Poisson
       pulse processes as presented in

(di)   S. O. Rice, "Mathematic Analysis of Random Noise".
       Contribution to Noise and Stochastic Processes, Ed.
       by N. Wax (Dover Publications, Inc. New York, 1954).

       The effects of temporal variation in multi dof system
       variance are discussed by

(dj)   H. Andres, "A Law Describing the Random Spatial
       Variation of Sound Fields in Rooms and Its Application
       to Sound Power Measurements." Acustica, Vol. 16,
       p 278-294 (1965). Translation by I. L. Ver, BBN-TIR-70,
       (1968).

       Eq. (4.2.9) is found from a relation in (di). Additional
       discussion of the role of mode shapes and sampling

strategies for locations of excitation and observation points in affecting response variance may be found in (as, bz). The use of multiple observation points or line averages to reduce variance are discussed in (dh). Modifications to the variance calculation for non-Poisson intervals between resonance frequencies are discussed in (bz). Note that the effective number of modal interactions given in Eq. (4.2.13) is not the total number of possible interactions N_1N_2 as suggested in (cz, da).

4.3 Calculation of Confidence Coefficients

The approach and terminology in this section are taken from (cf). The gamma and related distributions are discussed in (cf, de) and in

(dk)   E. Parzen, Stochastic Processes (Holden-Day, Inc., San Francisco, 1962).

The two particular confidence intervals derived in this section were first presented in (as, cv). The incomplete gamma function is discussed in

(dl)   A. Erdelyi, et al., Higher Transcendental Functions, Vol. 2 (McGraw-Hill Book Co., Inc. 1953). Chapter IX.

Confidence coefficients for the determination of environmental levels based on loads and response data are used in

(dm)   P. T. Mahaffey and K. W. Smith, "A Method for Predicting Environmental Vibration Levels in Jet-Powered Vehicles", S. & V. Bulletin, No. 28, Pt. 4, pp 1-14 (1960).

4.4 Coherence Effects - Pure Tone and Narrow Band Response

The discussion in this section is based on

(dn)   R. H. Lyon, "Spatial Response Concentrations in Extended Structures". Trans. ASME, J. Eng. Ind., November 1967.

Additional discussion of extreme value statistics
for multi-dimensional sinusoids may be found in

(do)   R. H. Lyon, "Statistics of Combined Sine Waves"
       J. Acoust. Soc. Am., Vol. 48, No. 1 (Pt. 2)
       pp 145-149 (1970).

PART II   ENGINEERING APPLICATIONS

INTRODUCTION TO PART II

Part II focuses on engineering applications of statistical energy analysis in the prediction of vibration. The procedures are based upon the theoretical discussions of Part I, but they are intended to be independent of that theoretical background. Thus, if one is willing to accept the basic formulas as presented, the estimates can be performed without reference to Part I.

Part II is subdivided into three sections. Section I discusses the use of SEA parameters in preliminary design. Many of these parameters represent information about the system that is not commonly evaluated at this stage. The emphasis in the prediction methods introduced is on the average value of response, but some information concerning the variability or uncertainty in the response estimate is included because an understanding of this potential uncertainty can be quite important in interpreting departures from the predicted values of response. For example, a consistent departure from the prediction of several frequency bands is a signal that some aspect of the SEA model is inadequate and should be modified, whereas a scatter in the data around the predicted average that is within the calculated uncertainty is an indication that the model cannot be improved upon easily.

Section II is concerned with evaluating the SEA parameters-- loss factor, mode count, and coupling loss factor. Empirical, experimental, and theoretical methods of evaluation are described separately although in practical situations it may be necessary to use a mix of procedures to get all the desired values. Mode counts, for example, tend to be easy to calculate and difficult to measure, whereas damping is simpler to measure and more difficult to calculate.

We have tried to preserve an engineering-applications-oriented approach in the formulas and techniques discussed, and also in the notation in which the results are presented. In many cases, formulas are given in a form slightly modified from that of the original source if the change simplifies the computation required without sacrificing a great deal of accuracy. Also, for example, cyclic frequency (in hertz) is used rather than radian frequency since this is the quantity one tends to use in experimental evaluation. This preference is not faithfully followed in all instances, but it has been generally adhered to.

We cannot, of course, cover all possible cases of interest in design work in this book, partly because of space limitations, but also partly because all the important cases of interest have not yet been worked out. We have tried to indicate sufficient sources of additional information so that the reader will be able to extend the discussions here to problems of interest.

Section III presents an example of the use of SEA in the prediction of the vibratory response of a hypothetical high speed flight vehicle and the transmission of that vibration to an internally mounted equipment shelf. As the reader will note, a variety of models, techniques, and analyses are required to cover the range of frequencies and subsystem behaviors. In the final analysis, there is still room for improvement of the model, but the procedures indicate the direction such improvements will have to take and leave it to the designer to decide whether the next step is worthwhile.

This example should be looked upon as an indication that hard estimation problems can be solved with reasonable effort, rather than as a formula for success in other estimation efforts. The various approaches to evaluating mode counts, injected powers, and coupling loss factors may be of little direct value in the next estimation problem, but a willingness to look at several different models and to use experiment when calculation cannot resolve an issue is a lesson of great value.

This part of the book does not close the door on SEA developments but, rather, opens it up. As SEA becomes more widely used by designers, the data on parameters, the models that have been analyzed, and the correlations of predictions with data will grow. Thus, in using SEA, designers are not only availing themselves of a new tool, but are contributing directly to the growth in usefulness of that tool.

CHAPTER 5. RESPONSE ESTIMATION DURING PRELIMINARY DESIGN

In the early stages of design of a high speed flight vehicle, it is often necessary to develop estimates for vibration amplitudes and spectra of major sections of the vehicle. This may be to anticipate possible fatigue problems as a result of the mission profiles and their mix for the aircraft, or to provide data to the environmental specialists in the development of proper test specifications. [1]

This requirement for response estimation will usually first arise during a fairly early stage of the vehicle development. At this point, only some mission parameters and major structural geometry may be known, but details of construction, fastening, etc. may well not be known. Nevertheless, it is necessary that one make the vibration estimates in the face of this uncertainty and do it as well as possible in order to avoid overdesign and its attendant extra expense, or conversely, underdesign and the potential for malfunction or failure.

It is perhaps worth commenting on three procedures for estimating response that have been used. They are the Mahaffey-Smith procedure [2], the Franken method [3], and the Eldred procedure. [4] Of these, the Mahaffey-Smith procedure has been the most widely used because it is quite simple to apply and provides results in a form that is directly usable by the engineer. It suffers from a complete lack of inclusion of acoustical or structural parameters, meaning that the estimate cannot be improved as more becomes known about the vehicle.

The Eldred procedure improves on the Mahaffey-Smith method by incorporating structural damping as a parameter. This is done by making an analogy with the response of a simple 1 dof system to noise excitation as described in Chapter 2 of Part I. We should note, however, that the damping of different built-up aerospace structures is not very different so that the inclusion of this parameter does not greatly strengthen our hand in developing response estimates. The Franken method does not include damping, but

does include surface mass density of the structure and its
radius of curvature. We can look therefore for certain changes
in response as we vary the structure. Thus, although the
basis of the Franken method is empirical, the data
normalization used allows its application to quite a variety
of structures.

The major advantage of SEA is that the degree of
parameterization of the prediction problem can be varied as
more structural and loads information becomes available.
Thus, it is possible to continually refine our response
estimates as the design becomes more fixed. An important
consequence of this is that we can see the effect of various
design alternatives in changing noise and vibration levels.
When SEA is used, there is no "discontinuity" between the
empirical procedures used in the preliminary phase and the
much more detailed analyses that may be used at later stages
of design.

In Part I of this book , we followed a pedagogical
approach in which the theory and applications of SEA were
developed along a "conventional" path. We began with the
model and then worked out its consequences. In this
Section I of Part II, we take a rather different approach.
The SEA procedure is described as a complete process in
Chapter 6. The steps involved are itemized and explained
as an engineering procedure rather than as a conceptual
development as was done in Part I. We first focus on the
output of the procedure - the response estimate. Generally,
the SEA calculation gives results in terms of energy of
vibration, which must then be converted to stress, or some
other dynamical variable. We then proceed to determine how
that energy estimate was arrived at. The use of energy
estimates to develop estimates for other response variables
is discussed in Chapter 7.

The calculation of the energy estimate itself requires
use of the SEA model for energy flow. As discussed in
Part I and reviewed in Chapter 8, this model contains energy
storage, loss and transfer parameters - which are the modal
density, loss factor and coupling loss factor, respectively.
In addition, we must know the input power injected into the
structure by the environment. Chapter 8 shows how the energy
estimate is developed from knowledge of these parameters when
they are expressed in either analytical or graphical form.

The SEA parameters that are important in determining
the energy for the actual system under design must then
be evaluated. Chapter 9 shows how this is done by using
experimental methods, theoretical predictions and empirical
values. Two kinds of coupling factors are discussed: those
appropriate for blocked energies, and those for actual system
energies. Modal densities are discussed in both theoretical
and experimental terms. Input and junction impedances are
introduced for the purposes of evaluating coupling loss
factors and input power.

The final chapter of Section I deals with the construction of system models. This is one of the most important
steps in SEA. The general model configuration is likely to
be fairly similar for various detailed designs of a given
system. Thus, a particular class of high speed vehicles
are all likely to have similar system models, so that once
defined, only relatively minor changes to the model will be
necessary.

CHAPTER 6. PROCEDURES OF STATISTICAL ENERGY ANALYSIS

6.0 Introduction

In this Chapter, we survey the entire SEA estimation
procedure as an engineering process, as contrasted with the
theoretical approach taken in Part I. The process consists
of (1) model definition, (2) parameter evaluation, (3) calculating energy distributions and (4) response estimation.
In general, it will not be necessary to repeat all parts
of this process for every new situation, but only the latter
parts may have to be revised. For that reason, in
Chapters 3 through 6, we have reversed the order of
presentation, discussing the latter stages of the process
first since they are more directly connected with what the
engineer wants to know and are also the parts of the procedure that are most likely to change with each new
situation.

In this Chapter, however, we follow the more customary
sequence in which we start with the modeling discussion and
end up with response estimation. We try to indicate as well
as we can just what the real engineering considerations are
at each step. In the later chapters, more detailed discussion of the procedures will be presented.

6.1 Modeling the System

As an example of a system that we might wish to model,
consider an aircraft and its attached equipment pod shown
in Figure 6.1. Such a system will have its dominant
response determined by "resonant", that is, damping controlled vibration and, hence is a candidate for SEA
prediction. We also assume that whatever nonlinear effects
there are may be neglected for the purposes of estimating
response at this stage.

Modeling this physical system requires that we identify
the following features of energy flow and storage in the
system:

   (1)   Sources of input power and mode groups (subsystems)
        on which they directly act. In the case of
        Fig. 6.1, these would be the acoustical noise
        field and turbulent boundary layer acting on both
        the aircraft and the pod.

(2) Groups of "similar" and "significant" modes that
     store energy and result in response that affects
     the estimates we wish to make. In the case of
     Fig. 6.1, we might make a first try by grouping
     all flexural modes of the aircraft panels to-
     gether, and flexural modes of the pod together.
     If it turned out later that some group of modes
     originally ignored was important, then one would
     have to add it to the diagram.

(3) It is automatically assumed that every group of
     energy-storing modes will have a finite amount of
     damping. Thus, the identification of the dis-
     sipation of energy represents no additional task.

(4) The junctions through which appreciable energy
     may flow from one mode group to another. In the
     case of Fig. 6.1, this is the pair of connecting
     spars between the aircraft and the pod. The
     energy flow links are usually identifiable by a
     direct interface between the energy storage boxes
     or mode groups.

When all of the energy storage and power input, dissipation and transfer processes are defined, the SEA modeling
is essentially complete and results in a diagram like that
shown in Fig. 6.1 (but is generally more complicated!). At
this point, the SEA parameters that we need to solve for the
energy values will be evident. In the next section, we
review briefly the procedures by which the parameter
evaluation is made.

6.2 Evaluating the Parameters

The parameters that define the four processes of
paragraph 6.1 are:

(1) the input conductance (or input power),

(2) modal densities (or mode count) to determine the
     number of modes in a "box",

(3) the dissipation loss factor that relates energy
     stored to power dissipated,

(4) the coupling loss factor that relates differences
     of modal energy of subsystems to the power flow.

In this section, we give a brief overview of the parameter
evaluation process which we shall cover more fully in
Chapter 9.

If the environmental excitation produces a force on a
system, then the parameter governing the input power resulting
from that force is in the form of an input "conductance".
If the load is a prescribed motion (such as the injection of
volume velocity into an acoustical space), then the proper
parameter is an input resistance of the system. These
parameters which collectively are known as "immittance"
functions [5] are known for structural excitation by some
environmental loads, such as shakers, turbulent boundary
layers, sound fields, and separated flows on vehicles. In
any particular situation, however, a measurement or new calculation of the appropriate immittance may be necessary.

Modal densities are known for many of the elements that
make up the systems that interest us. Modal density is
usually written n(f), meaning the average number of modes
that resonate in a 1 Hz band around the frequency f.
Elements for which modal densities are known include flat
plates (isotropic and anisotropic), curved plates,
acoustical spaces, strings and beams. [6] Since modal
densities of combined systems are additive, any sytem that
is composed of these elements also has a known modal
density. Modal densities can be measured by counting
resonances as the excitation frequency is varied, but considerable care is necessary to avoid missing a sizable
number of them in the count.

Care must also be taken to use the proper modal density
for the kinds of modes that are important for the purpose
at hand. For example, in Fig. 6.1 the modes of the aircraft that contain the most energy and are best coupled to
the turbulence and noise are those of panel flexure. Thus,
even though torsional or in-plane motion modes of the
fuselage may resonate in the frequency range of interest, we
would very likely ignore these non-flexural modes. It turns
out that a 10% error in the mode count is only about 1 dB
uncertainty in the response estimate, so that it is not
necessary to have great precision in the modal density.

The damping of the mode group is expressed in SEA work
by the loss factor. This quantity is the reciprocal of the
electrical engineer's Q or quality factor and is twice the
mechanical engineer's critical damping ratio. Calculation of
the damping from first principles is generally quite un-
reliable so that one usually relies on measurements to predict

loss factor as a function of frequency. Typical values for
the loss factor are in the range 0.1 to 0.001.

A damping measurement is usually accurate enough to
determine loss factor to two significant figures, but not
much more than that. Two commonly employed ways of measuring
damping are a free decay of the system, or a measurement of
modal response bandwidth when the excitation is a pure tone.
The bandwidth technique is a bit more accurate when it works,
but it must be done on a mode-by-mode basis, and it fails when
the modal density is high and the modes get too close to each
other along the frequency axis. Occasionally, one can also
measure damping by injecting a known amount of power into the
system and observing the steady-state response.

The transfer of power between mode groups is determined
by the coupling loss factor. This parameter, in turn, is
related to or determined by other parameters that are
probably more familiar to engineers who work with the systems
involved. For example, the coupling loss factor for power
flow between two rooms can be readily related to the transmission loss of the wall, which is a familiar parameter to
workers in building acoustics. Similarly, the coupling loss
factor governing the interaction of a panel and a sound field
may be related to the radiation damping (or mechanical
radiation resistance) of the panel, a quantity that is known
to structural vibration engineers. Thus, although the
coupling loss factor itself is not a well known parameter,
in particular situations it is often relatable to other
parameters that are more familiar. Since these other parameters have been calculated or tabulated in many instances,
the coupling loss factor can also be evaluated.

6.3 Solving for the Energy Distribution

The SEA model allows one to calculate the equilibrium
energy of each mode group from a knowledge of the parameters
involved. The simultaneous equations for the energy are
linear, algebraic and the solutions give each energy in
terms of all the input powers to the system and the various
loss and coupling loss factors. Normally, the input power
to each mode group and the parameters such as modal density
and coupling and dissipative loss factors are assumed to be
known. One then solves for the energy values.

It is also possible in principle to use the energy
equilibrium equations to try to evaluate all of the SEA
parameters, and the coupling loss factors in particular.

In this instance one tries to control the input power and to measure the energy equilibrium values as well as possible. The equations are then solved for the coupling loss factors. This is not a totally satisfactory way of determining these parameters, however, since the solutions seem to be quite sensitive to rather small errors in the measured energy values. Nevertheless, this approach has been used with limited success in a few situations. [7,8]

The result of the energy calculation is the total energy of resonant vibration (or oscillation) in a frequency band Δf wide for each "box" of the system. This information is then used to predict the actual dynamical response that is of direct interest.

6.4 Evaluating Response from Energy Estimate

In Chapter 7, we will develop the procedures used in converting the energy found from the calculations just discussed into useful estimates of vibration, pressure, strain or some other appropriate response variable. In this section, we merely summarize the procedure, which has two major estimation outputs: the mean energy or rms response and, where appropriate, an estimation interval.

The mean value estimate of energy gives the mean square acceleration in some frequency band (an octave band, for example). This mean square response is also a spatial average, so that it applies to a region of the structure. For example, an energy estimate might be found for the aircraft fuselage to be 10 joules (watt-seconds) in the 1 kHz octave band. If the total mass of the fuselage were 1000 Kg, then one estimate of the m.s. velocity in the band would be 10^{-2} (m/sec)^2 and the rms acceleration would be approximately 60 g's. This average result is both a spatial average and an ensemble average, i.e., it applies to a hypothetical group of structures, similarly constructed, of which our structure is a representative sample.

An important modification of this average estimate is to take account of "response concentration", that is, to recognize that particular regions of the structure will have a time mean square response that is greater than the spatial average. For example, the m.s. pressure near the wall of a room is twice the space average value. As another example, the m.s. velocity at the free tip of a beam is just four times the spatial m.s. velocity of the beam as a whole. For other boundary conditions on panel type structures, Ungar and Lee have evaluated several of these "response concentration" values [9].

Since we admit to variation, we must have a procedure
for making estimates that incorporate this uncertainty in
a useful and realistic way. This is done by forming an
estimation interval. Such an interval is a familiar idea.
It is frequently seen in cases in which vertical bars are
placed on a mean value measurement to either indicate the
range of observed data or the mean, plus or minus the
standard deviation. We can be more definitive than this
in our use of estimation intervals in Chapter 7. We calculate the range of values of response that are expected
to "capture" the observations, say 95% of the time. This
form of estimate is of value when applying SEA to
reliability analyses since the probability of levels
occurring that exceed a particular value may be quite
important.

The estimation intervals are derived from a calculation
of the standard deviation of the m.s. response. We have
explained in Part I that the theory in support of this
calculation is not so well established as it is for the
mean value theory. Nevertheless, the importance of this
aspect of estimation is great enough that we must use the
limited results that are available to calculate the
standard deviation for the purposes of interval estimation.

The result of the estimation then is to say that, for
example, the rms value of fuselage acceleration in the
1 kHz band is 60 g's. In addition, if we bracket this
mean value with a range from 20 to 200 g's ($\pm$ 10 dB), we
would expect to have 95% of all observations of response
fall within this interval. Such an estimate can then be
used to compute fatigue accumulation or some other response
related failure rate.

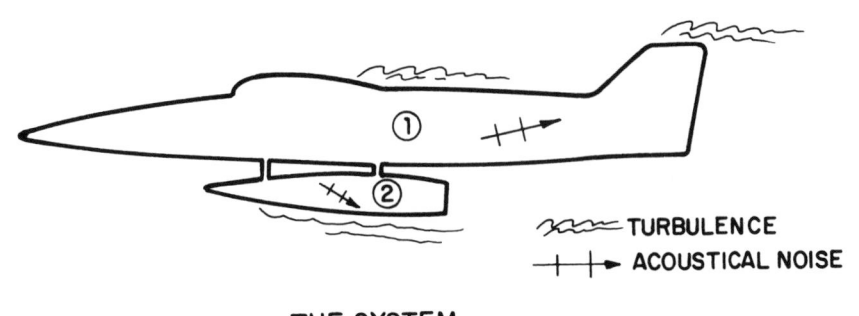

TURBULENCE
ACOUSTICAL NOISE

THE SYSTEM

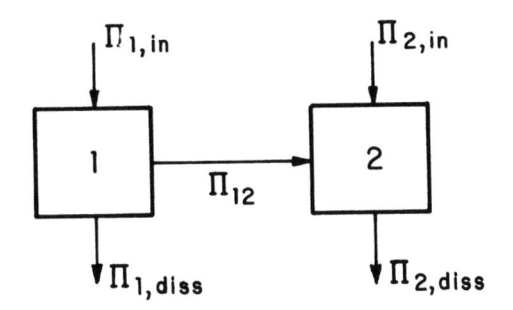

FIG. 6.1

AN SEA MODEL OF THE SYSTEM

CHAPTER 7. ESTIMATION OF DYNAMICAL RESPONSE

7.0 Introduction

This chapter is concerned with the estimation of the dynamical response of a system in terms of such variables as stress, acceleration, pressure, etc. The starting point is the knowledge of system vibrational energy, typically in the standard frequency bands. The goal is a prediction of mean square response for the same frequency bands. We include uncertainty in the estimate by using estimation intervals, and response concentration effects due to boundaries or impedance variations in the system.

The energy is assumed to be known, it will have been found previously according to the methods of Chapter 8. However, the discussion of this chapter is sufficient in itself to allow one to make many useful conversions from one response variable to another. For example, the average energy of vibration of a system may be determined by measurements of acceleration. The methods of this chapter allow one to convert this acceleration measurement to an estimate of stress. Further, one could estimate the increase in variance due to restricting the bandwidth of measurement. Much of the data manipulation that is useful in particular instances has its basis in the discussions of this chapter, quite apart from their broader usefulness in SEA.

7.1 Representations of the Energy Estimate

In engineering usage, the "power spectral density" (psd) of the energy of vibration may be denoted by $\mathcal{E}(f)$, and the energy of vibration associated with a frequency band Δf in the range $f_1 < f < f_2$ would simply be

$$E_{\Delta f} = \int_{f_1}^{f_2} \mathcal{E}(f) \ df. \tag{7.1.1}$$

This integration is shown graphically as the hatched area in Fig. 7.1. If $f_2 - f_1 = 1$ Hz, then the integral is plotted as the energy psd, which is $\mathcal{E}(f)$ in Fig. 7.1. Other

"constant frequency bandwidth" presentations of data in aerospace applications are typically 10 Hz or 50 Hz.

It is also common practice to present data or the
results of calculations in terms of "constant percentage
bandwidths". In this instance, the nominal "center
frequency" of the band is the geometric mean of the limits:

$$f_2 = \sqrt{f_1 f_2}$$

and the bandwidth $\Delta f = f_2 - f_1$ is always a constant fraction r
of the center frequency: $\Delta f = r f_c$. If $r = 1/\sqrt{2}$, then $f_2 = 2f_1$ and
we have "octave bands". If $r = 0.26$, then the bands are
called "1/3 octave bands". A table of the center frequencies
for the standardized octave and 1/3 octave bands is shown
in Fig. 7.2.

The mks unit for energy is the joule (or watt-second)
which is the amount of work involved when a force of 1
newton acts along a distance of 1 meter. The unit for the
psd of energy $\mathcal{E}(f)$ is joule-sec, since the unit of frequency
is cycles per second or hertz (Hz).* Since the energy in a
band $E_{\Delta f}$ is the integration of $\mathcal{E}(f)$ over frequency, the band
energy unit is also the joule.

We may also express the energy E as a "level" defined
by

$$L_E \equiv 10 \log E/E_{ref}, \text{ dB} \qquad (7.1.2)$$

The standard reference for energy is 10^{-12} joules. A band
energy of 0.1 joule, therefore, would correspond to a band
energy level of 110 dB, normally written

---

*Clarity of presentation would insist that there is no
 difference between the unit "Hz" and $(second)^{-1}$, but
 engineering usage of Hz is widespread.

$$L_E = 110 \text{ dB re } 10^{-12} \text{ joules}$$

In Fig. 7.3, we show how the octave and 1/3 octave band levels of energy might be graphed in a typical situation. The 1/3 octave band levels will generally be of the order of 5 dB lower than the octave band levels since the response of three adjacent 1/3 octave bands must be combined to account for the response in a single octave band. Also, note that only the band values have significance, the connecting lines between the points are only an aid to seeing the overall trend of the data as a function of frequency.

7.2   Conversion from Energy to Other Variables

The relation between vibratory energy and the space-time mean square velocity of the structure or sound field is

$$E = M \langle v^2 \rangle = M v_{rms}^2 \qquad (7.2.1)$$

where M is the structural mass in kilograms.

By taking 10 log of this expression, we can express this as a relation between energy levels and velocity levels,

$$L_E = 10 \log E/E_{ref} = 10 \log M + L_v + 120 \qquad (7.2.2)$$

where $L_v \equiv 20 \log v_{rms}/v_{ref}$, and $v_{ref} = 1$ m/sec. Thus, for example if a structure has a mass of 10 kg, and the band level of energy is 97 dB, (the 125 Hz octave band level of Fig. 7.3) then the velocity level is

$$L_v = 97 - 10 \log 10 - 120 = -3 \text{ dB re 1 m/sec}$$
$$(7.2.3)$$

We can develop a velocity spectrum for this structure
using Eq. (7.2.2). This spectrum is presented in Fig. 7.4
and corresponds to the octave band spectrum of energy shown
in Fig. 7.3. Note that a velocity level of -40 dB corresponds
to an rms velocity of 1 cm/sec.

The displacement or acceleration spectra are readily
derived from the velocity spectrum, simply by noting that
for each band having a center frequency f,

$$<a^2> = 4\pi f^2 \quad <v^2>; \quad <d^2> = <v^2>/4\pi^2 f^2 \qquad (7.2.4)$$

If we define

$$\left. \begin{array}{l} L_a = 20 \log a_{rms}/a_{ref} \\[2em] L_d = 20 \log d_{rms}/d_{ref} \end{array} \right\} \qquad (7.2.5)$$

in which $a_{ref} = 10$ m/sec^2 ($\simeq$1g) and $d_{ref} = 1$ m, then

$$L_a = 10 \log 4\pi^2 + 20 \log f + 20 \log v - 20 \log 10$$

$$= L_v + 20 \log f - 4, \text{ dB re 1 g}$$

and

$$L_d = L_v - 20 \log f - 16, \text{ dB re 1 m.}$$

As an example, these are also graphed in Fig. 7.4 for the
velocity spectrum shown there. Of course, the relative nu-
merical values of the band levels depend on the reference
value chosen, but the comparative shapes of the spectra are
not. Note that the displacement spectra tend to be large
at low frequencies and the acceleration spectra are higher at
high frequencies. A conversion nomograph between levels is
shown in Fig. 7.5.

The mean square velocity we have been dealing with has
been called the "kinetic velocity", [10] a velocity defined
in terms of the kinetic energy of the system. If the
structure is fairly homogeneous (beams, cylinders, plates
and sound fields meet this requirement), the kinetic velocity
is also equal to the space-time mean square velocity of the
system.

In addition, these systems also allow us to use a simple
relationship between the mean square velocity and the strain
in the structure. One can readily show that for a variety
of mechanical motions (flexure, torsion, compression, etc.)
the mean square strain $<\varepsilon^2>$ is simply related to the kinetic
velocity. [11, 12]

$$<\varepsilon^2> = K <v^2>/c_\ell^2 \qquad\qquad (7.2.6)$$

in which the constant K depends on the type of motion and
system geometry but varies over a small range near unity.
For estimation purposes we can set K = 1.

A useful reference strain is $\varepsilon_{ref}=10^{-6}$, or "1 microstrain". Defining the strain level as

$$L_\varepsilon = 10 \log <\varepsilon^2>/\varepsilon_{ref}^2$$

one has

$$L_\varepsilon = L_v - 20 \log c_\ell + 120$$

$$\text{[dB re } 10^{-6}, \text{ or} \\ \text{1 microstrain]}$$

$$= L_E - 10 \log M - 20 \log c_\ell \qquad (7.2.7)$$

Thus, if $L_V = -30$ dB re 1 m/sec and if $c_\ell = 5000$ m/sec, $L_\epsilon = -30 - 74 + 120$ or 16 dB re 1 microstrain, corresponding to an rms strain of 6.5×10^{-6}

We can also relate the velocity to stress τ by noting that the strain times stiffness modulus is stress. For most structures the appropriate stiffness is the Young's modulus Y_o,

$$\langle \tau^2 \rangle = Y_0^2 \langle \epsilon^2 \rangle \tag{7.2.8}$$

The stress level is defined as

$$L_\tau \equiv 10 \log \langle \tau^2 \rangle / \tau^2_{ref.} \tag{7.2.9}$$

Any convenient reference stress τ_{ref} can be used. If we use $\tau_{ref} = 10^6$ newtons/meter^2 then

$$L_\tau = 20 \log Y_o - 120 + L_\epsilon$$

$$= L_v + 20 \log Y_o - 20 \log c_\ell$$

$$= L_v + 20 \log \rho c_\ell \tag{7.2.10}$$

using $Y_o = \rho c_\ell^2$ where ρ is the density and c_ℓ is the longitudinal wave speed. Thus, the form of the velocity spectrum also determines the form of the strain and stress spectra. In Table 7.1, we summarize the relations between the variables discussed and the level defined in the preceding paragraphs. These relations as a group allow us to

Variable	Relation	Level	Formula	Reference
energy E	1	$L_E = 10 \log E/E_{ref}$	$L_E = 10 \log E + 120$	$E_{ref}=10^{-12}$
velocity v	$\langle v^2 \rangle = E/M$	$L_v = 10 \log \langle v^2 \rangle / v^2_{ref}$	$L_v = L_E - \log M - 120$	$v_{ref}=1$ m/sec
displacement d	$\langle d^2 \rangle = \langle v^2 \rangle / \omega^2$	$L_d = 10 \log \langle d^2 \rangle / d^2_{ref}$	$L_d = L_v - 20 \log f-16$	$d_{ref} = 1$ mm
acceleration a	$\langle a^2 \rangle = \omega^2 \langle v^2 \rangle$	$L_a = 10 \log \langle a^2 \rangle / a^2_{ref}$	$L_a = L_v + 20 \log f-4$	$a_{ref} = 10 \frac{m}{sec^2}$
strain ε	$\langle \varepsilon^2 \rangle = k\langle v^2 \rangle / c_\ell^2$	$L_\varepsilon = 10 \log \langle \varepsilon^2 \rangle / \varepsilon^2_{ref}$	$L_\varepsilon = L_v - 20\log c_\ell + 120$	$\varepsilon_{ref}=10^{-6}$
stress τ	$\langle \tau^2 \rangle = Y_0^2 \langle \varepsilon^2 \rangle$	$L_\tau = 10 \log \langle \tau^2 \rangle / \tau^2_{ref}$	$L_\tau = L_v + 20 \log \rho c_\ell$	$\tau_{ref}=10^6 \frac{newtons}{m^2}$

$[Y_0$ = Young's modulus in newt/m², c_ℓ = longitudinal wave speed in m/sec M = mass in kg, ω ≅ radian frequency, f = cyclic frequency, k = constant of order unity, ρ = mass density in kg/m³]

TABLE 7.1

RELATIONS BETWEEN DYNAMICAL VARIABLES AND LEVELS

generate average (meaning space-time square) response of
the system if the <u>average</u> energy of vibration or any of
the other variables are known.

7.3 Response Concentration Factors

The average response values are useful as a first step
in estimation of vehicle vibration, but there are at least
two reasons that we have to modify the average estimates
for engineering purposes. The first reason is that of
"response concentration", which is well known in the case
of the stress concentration factors commonly used in
structural design. Such response concentrations are tabulated
and available for various common configurations. [13]

The second kind of concentration factor or effect is
the local maximum of response due to coherence effects
between modes which many modes are excited simultaneously.
This coherence may be spatial or temporal. The variation of
and peaks in local rms response due to coherence between
modes was discussed in Chapter 4 . In this chapter, we merely
use some of those formulas to estimate extreme values of
response that result.

Modal Stress Concentration

We can estimate the spatial mean square of stress for
example, according to Eq. (7.2.8) and its antecedents. We
can also estimate the effect of edge fixation on the stress
of the boundary of a plate by methods reported by Ungar and
Lee [9]. These authors use the "dynamic edge effect" method
of Bolotin to calculate stresses in situations for which
there are no exact solutions to the thin-plate equations.

We imagine a modal vibration pattern like that sketched
in Fig. 7.6. The displacement w near the boundary defined
by x_2 = constant is written

$$w(x_1,x_2)=\sin k_1x_1[\sin k_2x_2+B_1 \cos k_2x_2+B_2 \, e^{-k_2'x_2}]$$

$$(7.3.1)$$

where

$$k_2' = \sqrt{2k_1^2 + k_2^2}$$

and

$$k_1^2 + k_2^2 = \omega^2/c_b^2 = k_b^2$$

where

$$c_b = \sqrt{\omega \kappa c_\ell}$$

is the free bending wave speed.

The maximum response of a single 2-dimensional mode in the "interior" is

$$\langle v_{max}^2 \rangle = 4 \langle v^2 \rangle \qquad\qquad (7.3.2)$$

since the mode shape function is

$$\psi_m = 2 \sin k_{m_1} x_1 \sin k_{m_2} x_2 \qquad\qquad (7.3.3)$$

Thus, we have a concentration factor of 2 built into a single mode response. The major interest in the edge effect is whether the constraint due to the edge causes stress concentration there that are greater than the interior values.

The analysis by Ungar and Lee [9] is rather detailed and lengthy, but they find that the ratio R of edge to interior stress can be simply stated in two important cases

(A) "clamped-like" edges

$$R = \frac{4\alpha_r G \, [2\alpha_t \, G^3 - H \, J]}{M\{\,[4\alpha_r G(\alpha_t G - H) + J - 2G^2]^2 + \Delta^2\,\}^{\frac{1}{2}}}; \quad (\alpha_r > \alpha_t)$$

$$(7.3.4)$$

(B) "free-like" edges

$$R = \frac{2G}{M} \frac{(\nu + \beta^2)(2\alpha_r H + G) - \nu G(4\alpha_r \alpha_t G^2 + J)}{\{\Delta^2 + [4\alpha_r G(\alpha_t G - H) + J - G^2]^2\}^{\frac{1}{2}}}; \quad (\alpha_r < \alpha_t)$$

$$(7.3.5)$$

In these expressions

$$\alpha_r = K_r / 2D \, k_p \qquad\qquad G = \sqrt{1 + \beta^2}$$

$$\alpha_t = K_t / 2D \, k_p^3 \qquad\qquad H = \sqrt{1 + 2\beta^2}$$

$$\beta = k_1 / k_2 \qquad\qquad J = 1 + \nu\beta^2$$

$$\Delta = J \, H - 4\alpha_t G^2 (G + \alpha_r H)$$

$$k_p^2 = k_1^2 + k_2^2 = G^2 k_y^2 \qquad M = \max \, [J, \nu + \beta^2]$$

The elastic supports are defined by a rotational stiffness K_r and a translational stiffness K_t per unit length. The plate has a Poisson's ratio ν and a bending rigidity $D = \rho_s \kappa^2 c_\ell^2$.

The surfaces $R = 1$ for these two conditions are shown in Figs. 7.7 and 7.8 which was adapted from the report by Ungar and Lee [9]. The region within the surface represents conditions in the parameter space for which the stresses along

the edge are greater than at interior locations. The actual
stress ratio for any situation may, of course, be found from
Eqs. (7.3, 7.4, and 7.5). For example, if we find the stress
ratio due to a "normally incident" (β=0) mode on the x_2=0 edge
(shown in Fig. 7.6) when the edge is clamped ($\alpha_r,\alpha_t \to \infty$), we
find R=2. Thus, the simple clamped edge produces a stress concentration factor of 2 above the peak stresses that occur in
the interior of the plate due to the normal mode shape. This
same stress concentration also occurs for "grazing modes"
($\beta \to \infty$) in which the waves are travelling parallel to the edge of
the panel.

Multi-Modal Stress Concentration.   In Chapter 4 we
discussed the effect of coherence between modes in multi-modal
response. We found that it was possible to obtain that we
called "statistical response concentrations" of significant
value when many modes are coherently excited. Such a situation
can arise when structures are excited by a pure tone at a
frequency well above the lowest panel resonances. The expression for the response concentration for this situation is
[see Eq. (4.4.12)]

$$\frac{w_{max}}{w_{rms}} = \psi_{max}\sqrt{N}$$

which is $2\sqrt{N}$ for 2-dimensional modes.

As an example of the application of Eq. (7.3.6), consider the aircraft shown in Fig. 6.1 excited by the pure tone
from a bypass engine fan at 2.5 kHz. Assume the structural
loss factor is 10^{-2}, so that an equivalent modal bandwidth
is approximately 40 Hz. If the fuselage has an area of 50 m^2
and an average thickness of 2 mm, the average spacing between
modes is δf = 0.12 Hz. Thus, 500 modes will be excited
by this tone and the statistical concentration factor is

$$R_{stat} = 2\sqrt{500} \simeq 45, \qquad \text{(pure tone)} \qquad (7.3.7)$$

quite a large concentration factor. Detailed analysis shows
that this concentration has a nearly unity probability of
occurring, but that its position will shift depending on the
frequency of the excitation. Thus, if the tone frequency
varies at all, this concentration point will wander over the
structure.

When the system excitation is broad band, the statistical response concentration is almost never important. The response concentration in the case of excitation by a band Δf is given by

$$R_{stat} = \left[\frac{(\frac{\pi}{2} \eta f)^2}{\delta f \, \Delta f} \psi^2_{max} + (1- \frac{\pi}{2} \frac{\eta f}{\Delta f}) \right]^{\frac{1}{2}} \quad (\Delta f > \eta f)$$

(7.3.8)

Applying this to our example, we get $R_{stat} = 10$ when $\Delta f = 600$ Hz (the bandwidth of the 2.5 kHz third octave band). Thus, the "statistical response concentration" is a function of the bandwidth of the vibration, decreasing as the noise bandwidth increases.

We have shown how we can estimate extreme values of response for single and multi-modal response situations. Such extreme values are of particular interest when "exceedance" type failures, such as fracture, plastic deformation, or collisions are considered. We now turn to procedures for forming estimation intervals which are useful in amplitude related damage estimates.

7.4 Variance and Estimation Intervals

Variance is a measure of the likely departure of any single measurement of response from the average value. Since the shapes and the resonance frequencies of the modes will vary from one structure to another and the details of location of excitation and response measurement will also vary, we may think of the output of a particular experiment as a statistical sampling of a distribution of possible response values. We present here some formulas for the standard deviation to be expected in cases in which one system is excited directly by a point source and for the case in which a second system is excited by its attachment to the directly excited system.

Single System Response.   As explained earlier, the
energy estimate provides us with a prediction of m.s.
response (velocity, say) which we write as $<v^2>$.   When a
single system is excited by a point source, the theoretical
relation for the variance (square of the standard deviation
σ) is as follows [10]

$$\frac{\sigma^2_{v^2}}{<v^2>^2} = \frac{2F(Q)}{Q} \left\{ 1 + \frac{1}{2M} \frac{<\psi^4>^2}{<\psi^2>^4} \right\} \qquad (7.4.1)$$

where $Q = 2\Delta f/\pi f\eta$, $M = \pi f\eta/2\delta f$, and ψ is the mode shape.
Also Δf is the bandwidth of the noise excitation, η is the
loss factor, and δf is the average frequency spacing
between modes.   The parameters η and δf are discussed
further to Chapter 9.   The function $F(Q)$ is graphed in
Fig. 7.9.   The modal shape factor has the following values:

$$<\psi^4>/<\psi^2>^2 \;=\; 3/2 \text{ (one-dimensional systems)}$$

$$= 9/4 \quad \text{(two-dimensional systems)} \quad (7.4.2)$$

$$= 27/8 \text{ (three-dimensional systems)}$$

The parameter M is the modal overlap, the ratio of
effective modal bandwidth to average frequency spacing between
the modes.   The parameter Q is a ratio of excitation bandwidth to modal bandwidth.   If we divide our interest between
cases of narrow and broad excitation bandwidths and small
and large degrees of modal overlap, we get the following
simplified formulas for the "normalized variance" [10]

$$\frac{\sigma^2_{v^2}}{<v^2>^2} \;=\; \frac{\delta f}{\pi\eta f} \frac{<\psi^4>^2}{<\psi^2>^4} \qquad \begin{array}{l} \text{narrow band } (Q<<1) \\ \text{no overlap } (M<<1) \end{array}$$

$$= 1 \qquad \begin{array}{l} \text{narrow band } (Q<<1) \\ \text{high overlap} (M>>1) \end{array}$$

$$= \frac{\delta f}{\Delta f} \; \frac{<\psi^4>^2}{<\psi^2>^4} \qquad\qquad \text{band of noise (Q>>1)}$$

band of noise (Q>>1)

little overlap (M<<1)

$$= 2 \; \frac{\pi f \eta / 2}{\Delta f} \qquad\qquad \text{band of noise (Q>>1)}$$

band of noise (Q>>1)

high overlap (M>>1)   (7.4.3)

The quantities evaluated by Eqs. (7.4.3) will turn out to be quite important for determining estimation intervals or "safety factors" for response prediction.

Coupled System Response. When two systems are joined together and only one system is directly excited by an external source, then the mean value of energy is estimated by methods to be discussed in Chapter 8. The variance in response has also been derived for this situation for the simple case of point excitation, point connection between systems, and point observation. Such a calculation is thought to represent an upper bound on the variance to be encountered in other situations. Thus, the estimates of confidence that we discuss later are likely to be on the conservative side.

From Chapter 4 the normalized variance for an indirectly excited system is [see Eq. (4.2.13)];

$$\frac{\sigma^2_{v^2}}{<v^2>^2} \qquad \frac{<\psi_1^4>^a}{<\psi_1^2>^{2a}} \quad \frac{<\psi_2^4>^b}{<\psi_2^2>^{2b}} \quad \frac{\delta f_1 \; \delta f_2}{\Delta f \; \frac{1}{2} \; \pi f (\eta_1 + \eta_2)} \qquad (7.4.4)$$

In this expression, ψ_1 is a mode shape for the directly excited system, ψ_2 a mode shape for the indirectly excited system. The average frequency separations for modes in the two systems are δf_1 and δf_2. The excitation and analysis bandwidths are Δf, and the loss factor for systems 1 and 2 are η_1 and η_2 respectively. If one of the systems has a single mode in the band, then set $\delta f = \Delta f$ for that system.

The exponents   a   and   b   in Eq. (7.4.4) are concerned
with the nature of the excitation and observation locations.
If these are "average" interior locations on the structures
(or sound fields) then a = b = 2. In some instances we can
sense the response so that all resonant modes contribute
equally to the response. Examples include measurement of
motion at the free end of a beam or sound pressure at the
corner of a rectangular room. If such a location is picked
for the excitation, then a = 1, and if such a location is
picked for the observation of response, then b = 1.

We see from Eqs. (7.4.1) and (7.4.4) that the variance
is generally reduced as the modes get more dense (and,
therefore, δf is reduced) and as the excitation bandwidth
Δf is increased. From Eq. (7.4.2), systems of higher
dimensionality have a greater contribution to the variance
by this spatial factor, whereas the usually higher modal
density of such a system is usually sufficient to offset
this effect. In balance, one-dimensional systems such as
ring frames and beams have higher variance than do plates,
cylinders or acoustical spaces.

7.5 Using Variance to Calculate Safety Factors

The variance of response has two applications that are
of interest to the structural designer. The first is as a
simple metric to judge the scatter of data in response
simulation experiments. The second is the setting of
estimation levels that represent a reasonable bound to the
expected response in a variety of situations. Usually such
a bound will be several times the estimate of average
response so that the observed response may be reasonably
expected to fall within the bound. We may interpret such a
bound, therefore, as a "safety factor" on the mean value
estimate.

In order to calculate estimation intervals, we need the
probability of the response variables, $<v^2>_t$, $<\tau^2>_t$, $<p^2>_t$,
etc. We let any of these positive variables to represented
by the variable θ. The probability density of response is
denoted $\phi(\theta)$, which will have the general form sketched in
Fig. 7.10. The probability that the measured response θ
will lie in the interval $\theta_1 < \theta < \theta_2$ is called the confidence
coefficient, CC, and is graphically represented as the
hatched area shown in Fig. 7.11.

A very convenient form for the probabilty density is the so-called gamma density. If we assume that θ is distributed according to this density, then we can relate the confidence coefficient directly to the estimation interval for a given normalized variance as determined either by Eq. (7.4.1) or (7.4.4). For example, if we define an upper limit $\theta_2 = r<\theta>$ where r is a positive number greater than unity and a lower limit $\theta_1 = 0$ for the estimation interval, we obtain the relationship shown in Fig. 7.11. Thus, for example, assume that we want to be 99% sure of not exceeding the response estimate θ_2, where $\theta_2 = r<\theta>$. If, for example, $\sigma_\theta^2/<\theta^2> = 0.3$ and if $r = 2.5$, then we can be 99% confident of not exceeding a response of $\theta_2 = 2.5<\theta>$. We can think of the factor r, therefore, as a kind of safety factor; the greater r is, the less chance we have of exceeding the estimate.

A second kind of estimation interval, which is of greater importance in experimental studies, may be termed a "bracketing estimate". In this case, we set $\theta_2 = r<\theta>$ as before, and also set $\theta_1 = <\theta>/r$. Since the ratio of the limit to the mean is fixed, we can refer to the estimate in this case as the mean plus or minus 10 log r (dB). This is a convenient and natural form for the estimation interval, particularly if the response has been expressed in logarithmic terms (as a level.) Again, using the gamma density for $\phi(\theta)$, the relation between CC, r, and $\sigma^2/<\theta^2>$ for the "bracketing interval" is shown in Fig. 7.12. The use of this graph to define an estimation interval is identical to the procedure described in the preceding paragraph.

Summary. Starting with the estimate of expected vibrational energy in a band Δf, we have shown how to reinterpret this estimate in terms of variables of more direct interest. We have also discussed the effects that may cause extreme values of response at particular locations. In addition, we have shown how to develop estimation intervals for use in those cases in which the expected variation leads to too much uncertainty for the use of the mean value as an estimate. In the next chapter, we discuss the way in which we obtain the energy estimate from the SEA model.

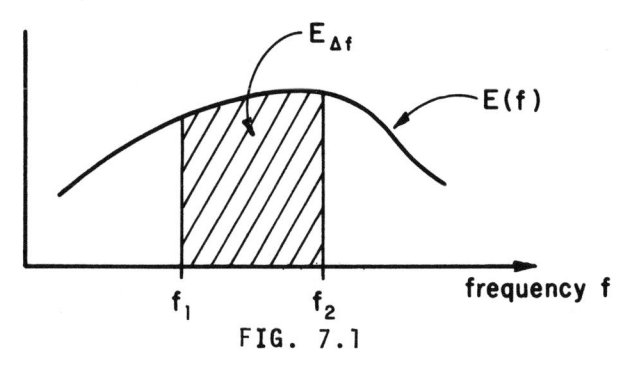

FIG. 7.1

INTERPRETATION OF ENERGY IN FREQUENCY BAND AS AREA UNDER SPECTRAL DENSITY CURVE

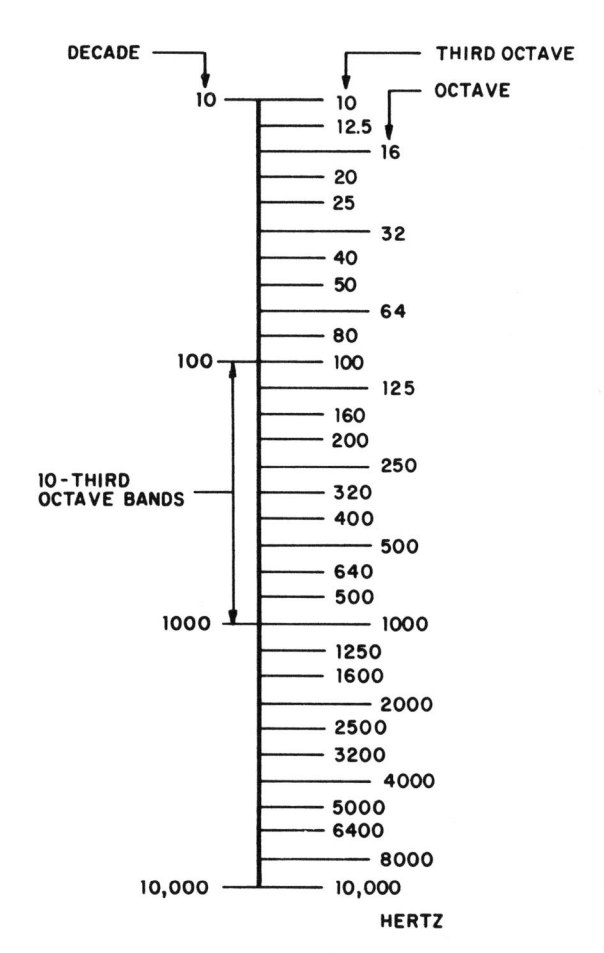

FIG. 7.2

STANDARD CENTER FREQUENCIES FOR ONE-THIRD AND FULL OCTAVE BANDS

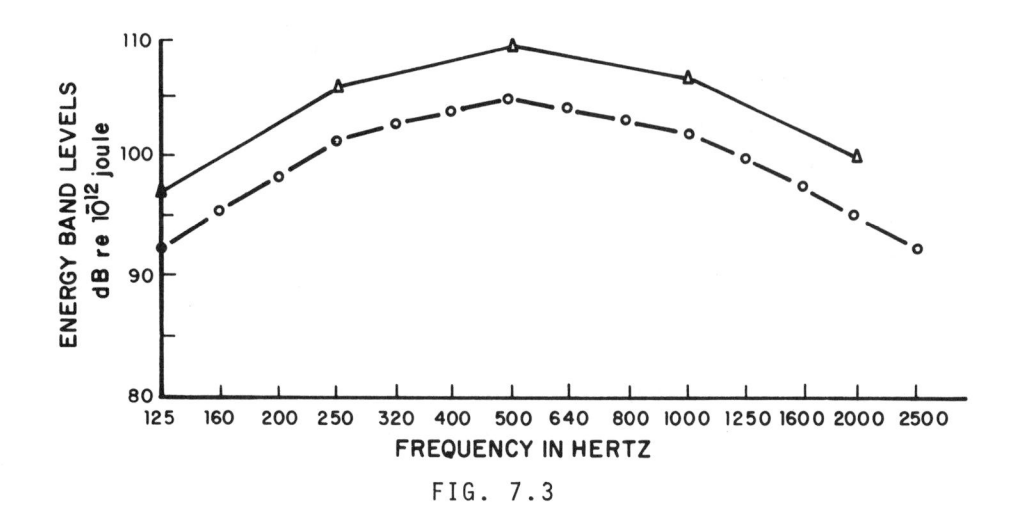

FIG. 7.3

TYPICAL BAND LEVELS SPECTRUM OF VIBRATIONAL ENERGY

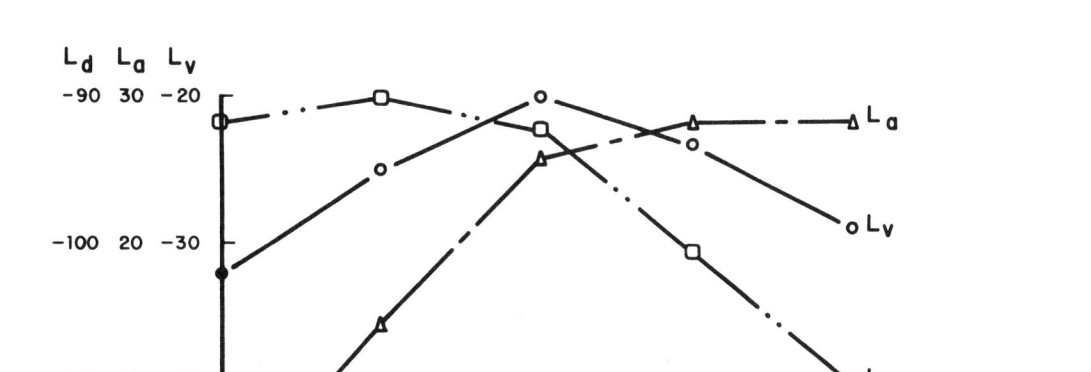

FIG. 7.4

COMPARISONS OF BAND LEVEL RESPONSE
SPECTRA FOR DIFFERENT VARIABLES

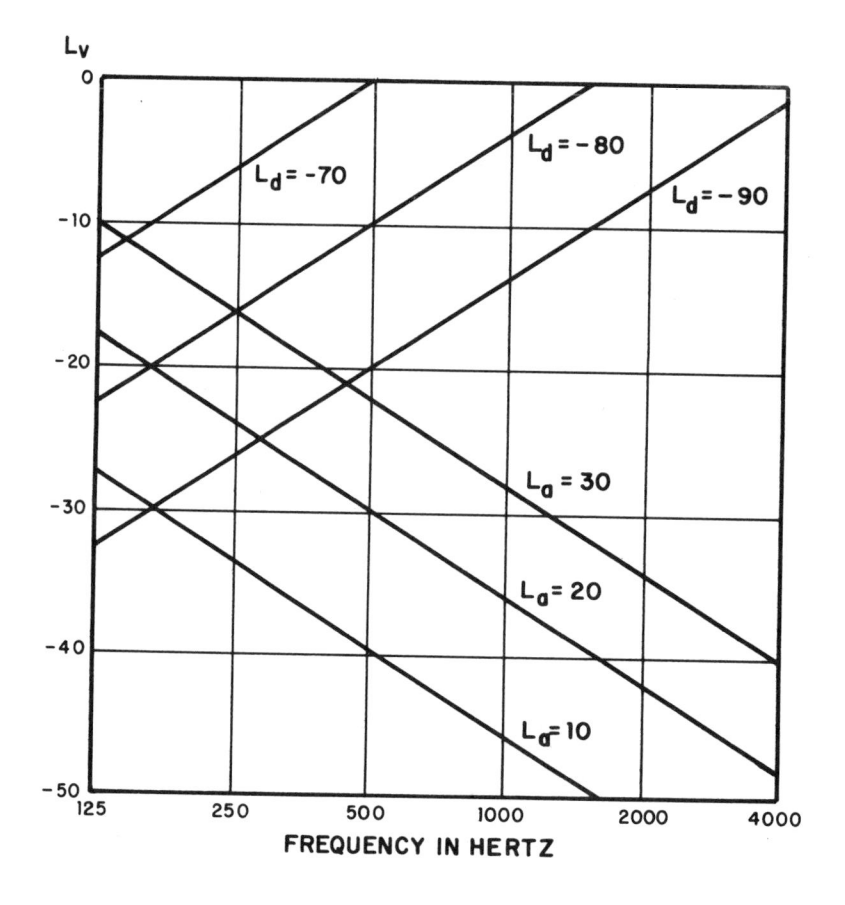

FIG. 7.5

**NOMOGRAPH FOR CONVERSION OF DISPLACEMENT
RELATED VARIABLES**

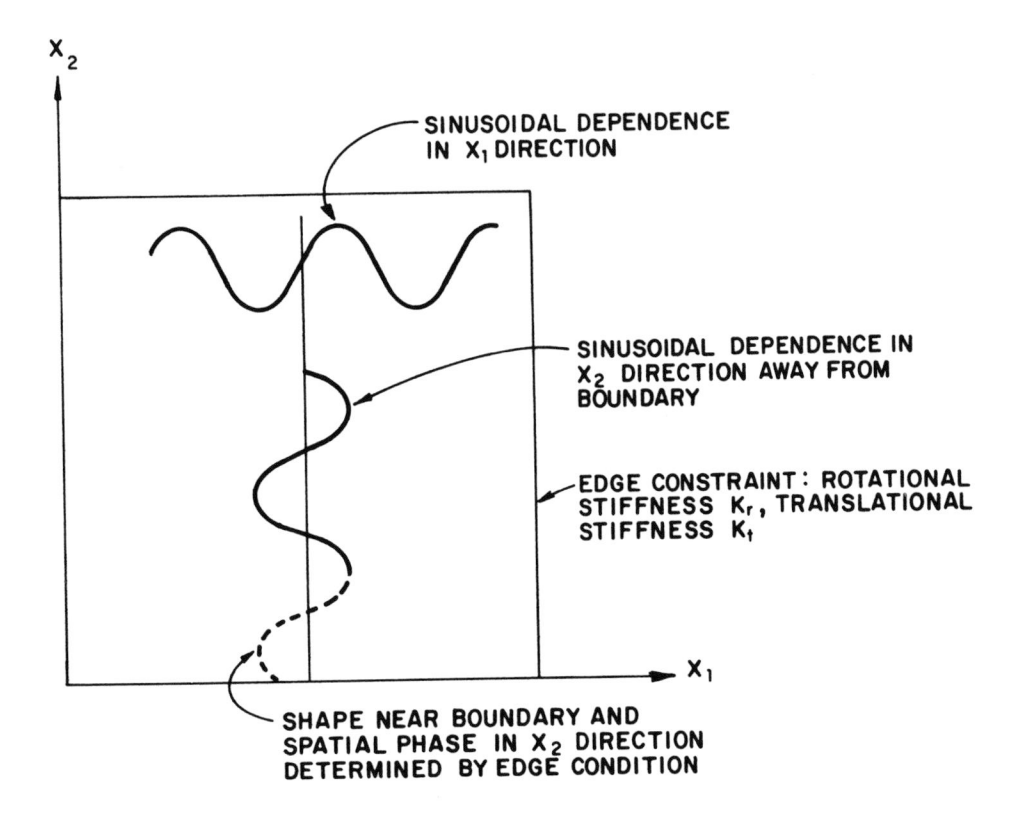

FIG. 7.6

SPATIAL DISTRIBUTION OF DISPLACEMENT ASSUMED
IN BOLOTIN "DYNAMIC EDGE EFFECT" MODEL

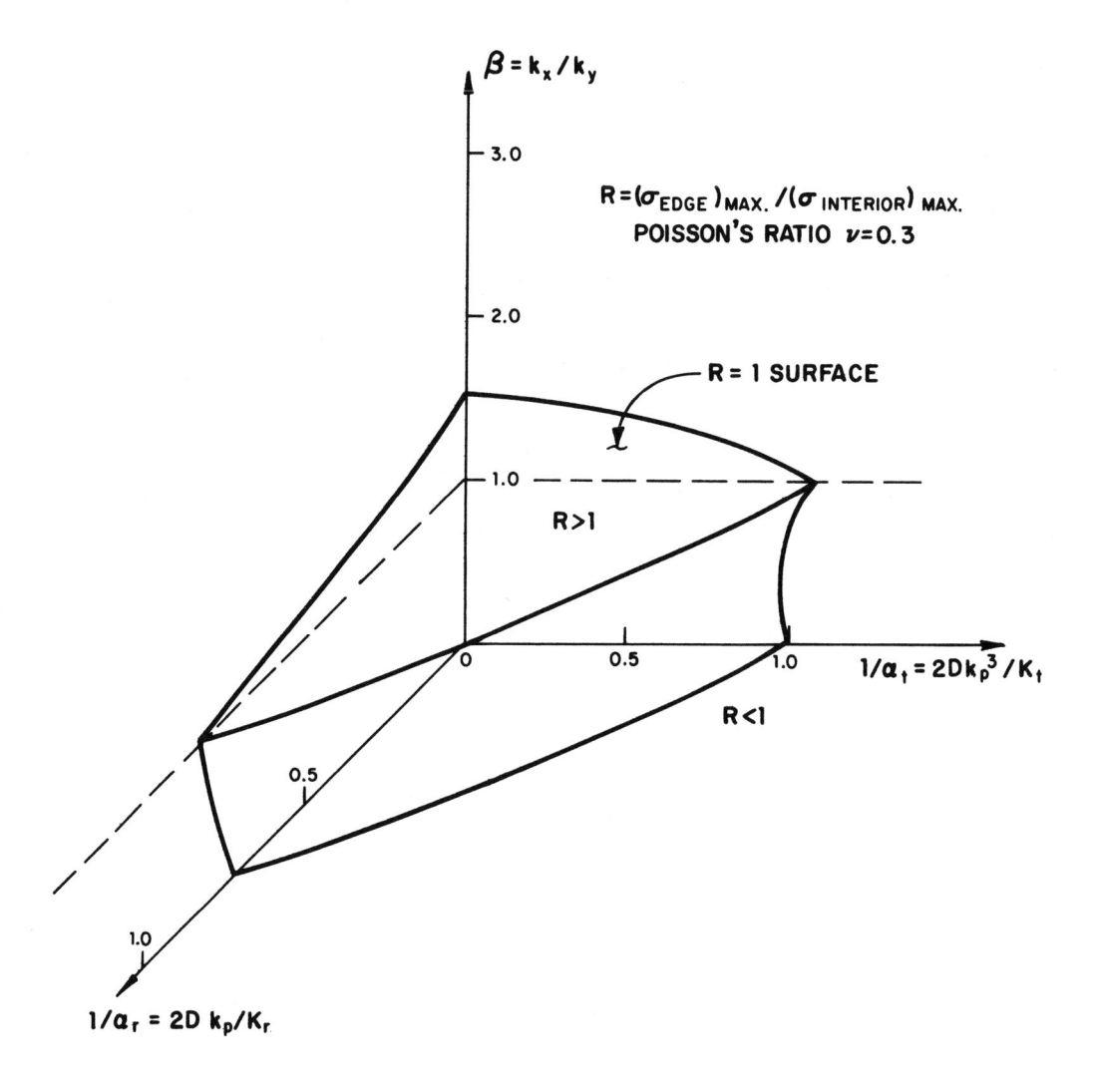

FIG. 7.7

REGIONS OF EDGE AND INTERIOR STRESS DOMINANCE IN
PLATES WITH ROTATIONALLY "CLAMPED-LIKE" EDGES $(a_r > a_t)$

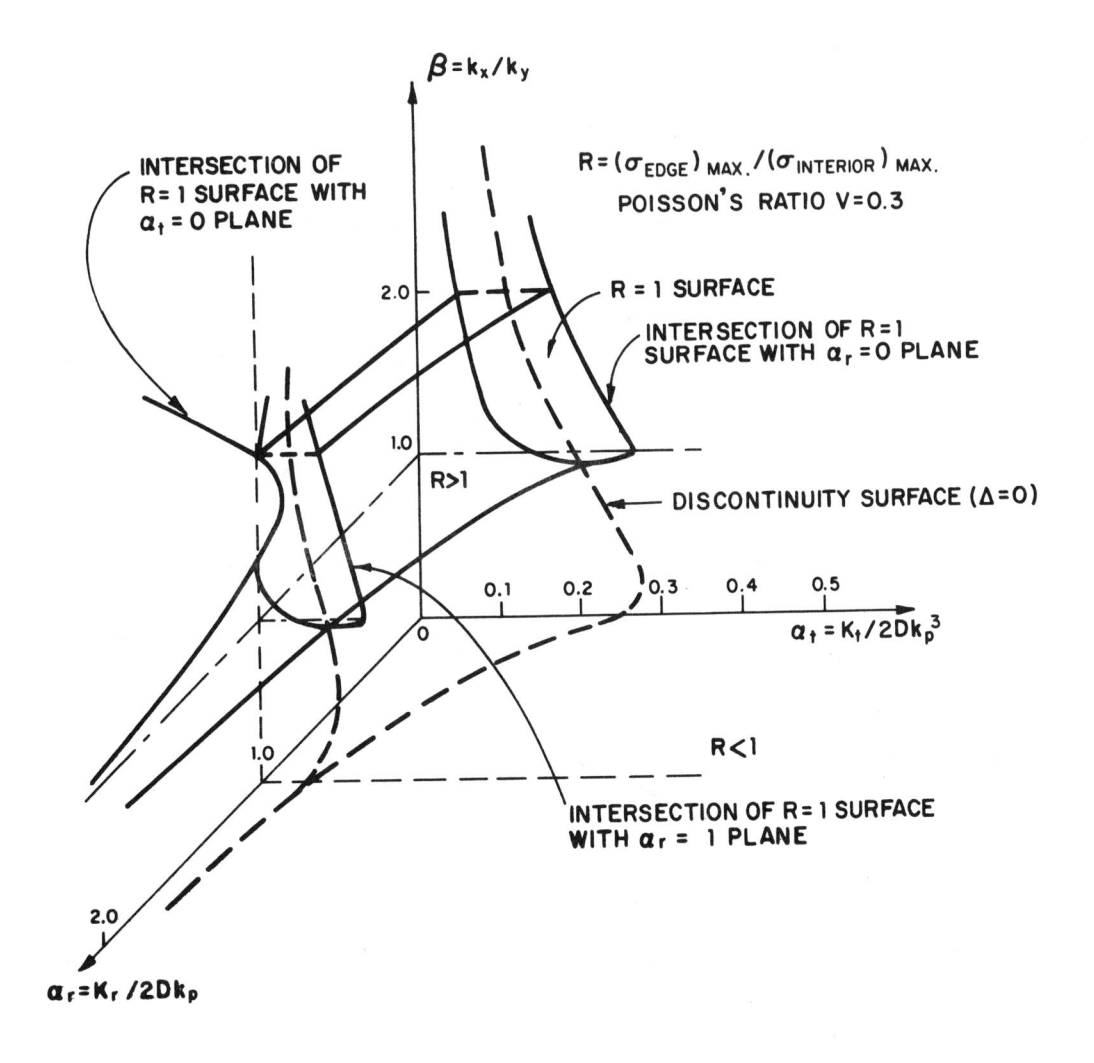

FIG. 7.8

REGIONS OF EDGE AND INTERIOR STRESS DOMINANCE IN
PLATES WITH ROTATIONALLY "FREE – LIKE" EDGES ($\alpha_r < \alpha_t$)

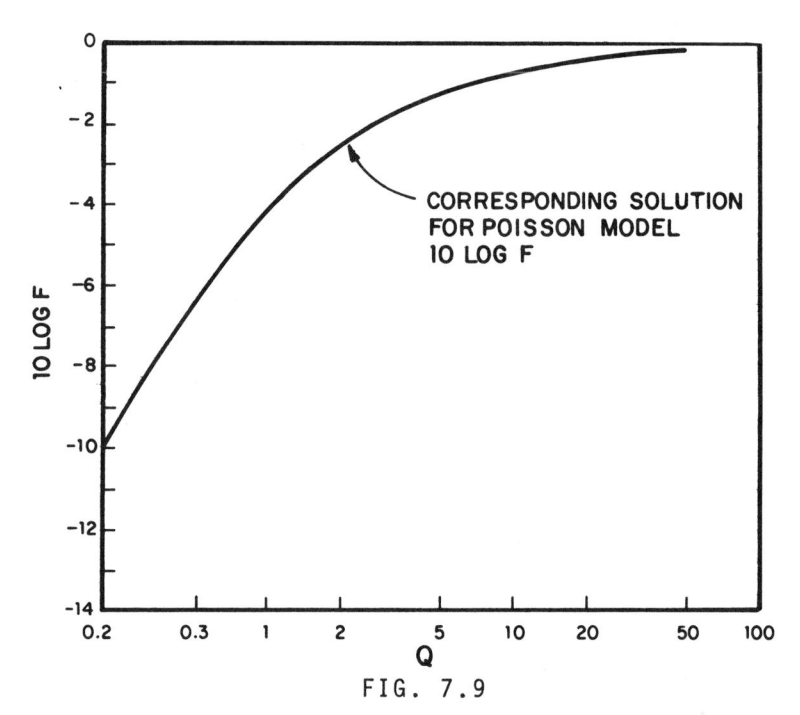

FIG. 7.9

EFFECT OF EXCITATION BANDWIDTH OF RESPONSE VARIANCE

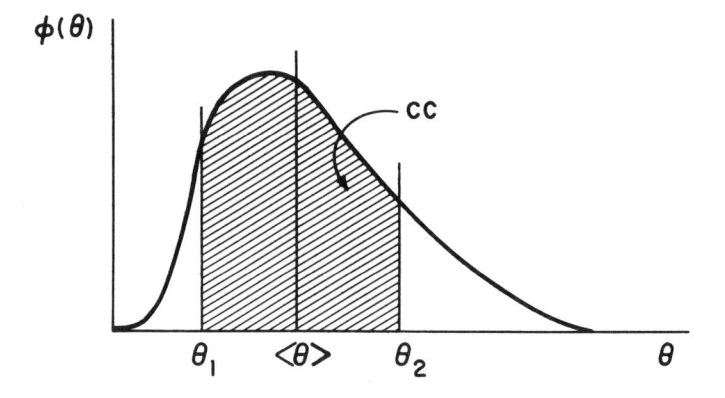

FIG. 7.10

FORM OF RESPONSE PROBABILITY DENSITY

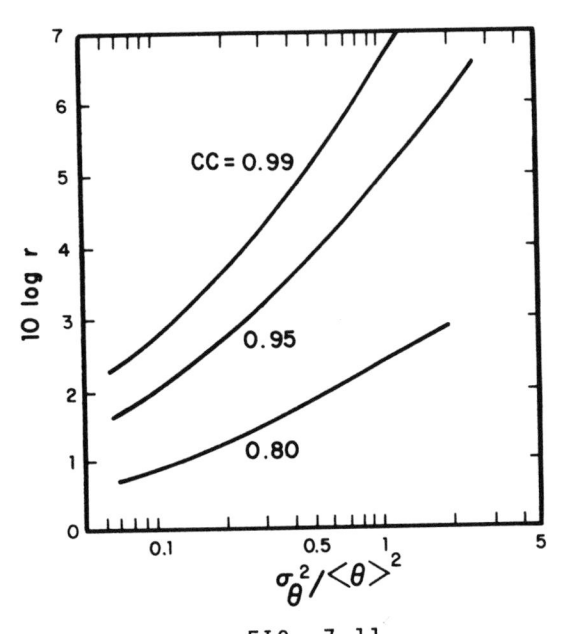

FIG. 7.11
**GRAPH OF "SAFETY FACTOR" r AGAINST
NORMALIZED VARIANCE**

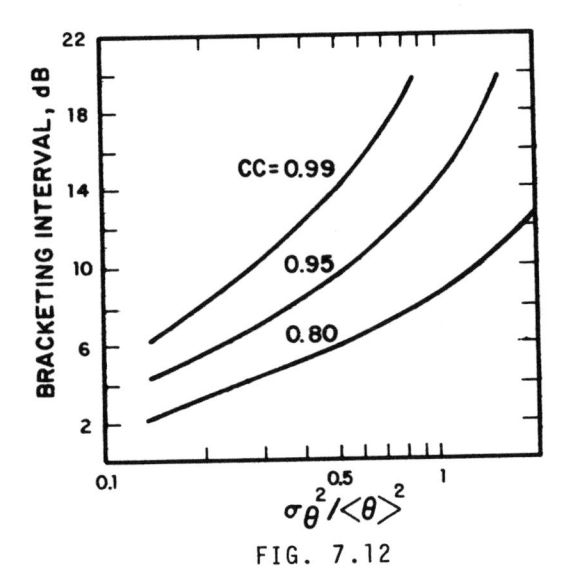

FIG. 7.12

GRAPH OF BRACKETING INTERVAL

CHAPTER 8.   ESTIMATING THE ENERGY OF VIBRATION

8.0 Introduction

This chapter is concerned with describing how one obtains an estimate of average system energy from the SEA model and a knowledge of its parameters. In this chapter, the model and its parameter values are assumed to be given. In Chapter 10 we will describe how the model is defined. Parameter values are discussed in Chapter 9. In practical situations, however, in which one deals with effects of minor changes in structural configuration or connections in an overall system that is the same from one situation to the next, then with each modified calculation of system energy, one is basically starting at the point where this chapter commences.

The energy that we solve for is the vibrational energy in the frequency band Δf of each "subsystem". This energy is found as a result of a set of linear simultaneous algebraic equations in which the energy of certain systems and input power to other systems are the known quantities. The unknown energies of the remaining systems are solved for in terms of these known quantities and the system parameters. The system parameters include the number N_i of resonant modes in the frequency band, Δf, of each system, the system damping as measured by the loss factor η, and the coupling loss factor, which we have discussed in Chapter 3 (of Part I) and will discuss further in Chapter 9.

Since Part II is concerned with engineering procedures, we do not derive the relationships that we use. The derivations of the basic relations were made in Part I. Here, the emphasis is on explaining what must be known to carry out the estimates, and how one applies that information to interpret the estimates.

8.1 How the Overall System is Described

A fairly general SEA model is depicted in Fig. 8.1. It consists of N subsystems, each of which may receive power Π_{in} from an external source (unspecified) and dissipate power Π_{diss} due to the damping of the subsystem.

In addition, power Π_{ij} is transferred between subsystems by the action of coupling forces at the junctions between these subsystems. Finally, the energy of each subsystem is denoted by E_{tot}, and it is this set of energy values that we solve for. Since the energy E_{tot} is the vibrational energy of the subsystem in a frequency band Δf, it is the energy that we began with to make our subsystem response estimates in Chapter 3.

The fundamental relation that we use is that of the conservation of energy for each subsystem, or a balance between power in and out of the subsystem. For the ith subsystem in Fig. 8.1, therefore, we have:

$$\Pi_{i,in} = \Pi_{i,diss} + \sum_{j=1}^{N_i} {}' \; \Pi_{ij} \qquad (8.1.1)$$

where the prime on the sum means that $j=i$ is excluded. We are able to solve for the system energies because both the dissipated power and transfer power can be related to subsystem energies. The dissipation relation is

$$\Pi_{i,diss} = \omega \eta_i \, E_{i,tot} \qquad (8.1.2)$$

where η_i is the loss factor previously introduced and $\omega = 2\pi f$, where f is the center frequency of the band of interest.

The power transferred from subsystem i to subsystem j was found in Part I to be [see Eq. (3.2.11)]:

$$\Pi_{ij} = \omega \eta_{ij} E_{i,tot} - \omega \eta_{ji} E_{j,tot} \qquad (8.1.3)$$

where the quantities η_{ij} are called coupling loss factors. They are not all independent because they must satisfy the consistency relation

$$N_i \eta_{ij} = N_j \eta_{ji} \tag{8.1.4}$$

where N_i, N_j are the numbers of resonant modes of subsystems i,j in the band Δf.

Placing Eqs. (8.1.2) and (8.1.3) into Eq. (8.1.1) results in the following set of equations:

$$\Pi_{1,in}/\omega = (\eta_1 + \Sigma_j' \, \eta_{1j}) E_{1,tot} - \Sigma_j' \, \eta_{j1} E_{j,tot}$$

$$\Pi_{2,in}/\omega = (\eta_2 + \Sigma_j' \, \eta_{2j}) E_{2,tot} - \Sigma_j' \, \eta_{j2} E_{j,tot}$$

$$\vdots$$

$$\Pi_{N,in}/\omega = (\eta_N + \Sigma_j' \, \eta_{Nj}) E_{N,tot} - \Sigma_j' \, \eta_{jN} E_{j,tot}$$

$$\tag{8.1.5}$$

The solutions for these simultaneous equations is found in the conventional way. We can, if we wish, express these equations in matrix form:

$$
\begin{bmatrix}
\eta_{1,tot} & -\eta_{21} & -\eta_{31} & \cdots & -\eta_{N1} \\
-\eta_{12} & \eta_{2,tot} & -\eta_{3,2} & & -\eta_{N,2} \\
-\eta_{13} & -\eta_{2,3} & \eta_{3,tot} & & -\eta_{N,3} \\
\vdots & \vdots & \vdots & \vdots & \\
-\eta_{1N} & -\eta_{2N} & -\eta_{3N} & & \eta_{N,tot}
\end{bmatrix}
\begin{Bmatrix}
E_{1,tot} \\
E_{2,tot} \\
E_{3,tot} \\
\vdots \\
E_{N,tot}
\end{Bmatrix}
=
\begin{Bmatrix}
\Pi_{1,in}/\omega \\
\Pi_{2,in}/\omega \\
\Pi_{3,in}/\omega \\
\vdots \\
\Pi_{N,in}/\omega
\end{Bmatrix}
$$

$$\tag{8.1.6}$$

This set of N simultaneous equations is appropriate if only values of input power are known. If, however, the energy $E_{i,tot}$ of any subsystem is known by measurement or some other means, then the energy balance equation (8.1.1) for that subsystem is removed, reducing the order of the set of equations and the known energy becomes a "source" term in each equation and is moved to the right hand side of the new set of simultaneous equations.

The diagonal terms of the coefficient matrix $\eta_{i,tot} = \eta_i + \Sigma_j \eta_{ij}$ may be thought of as total loss factors for each subsystem. This total loss factor includes not only dissipative losses, but also the effects of transfer losses to other subsystems. Of course, whether or not the loss to other subsystems actually occurs will depend on the energy that they contain.

If we call the coefficient matrix in Eq. (8.1.6) N, then it may be written

$$\mathbf{N} \cdot \vec{E}_t = \vec{\Pi}_{in}/\omega \qquad (8.1.7)$$

where $\vec{E}_t$ and $\vec{\Pi}_{in}$ are N-dimensional vectors having components $E_{i,tot}$ and $\Pi_{i,in}$ respectively, found by operating on Eq. (8.1.7) from the left with the inverse of N, defined as N^{-1}:

$$N^{-1} \cdot N \cdot \vec{E}_t = I \cdot \vec{E}_t = \vec{E}_t = N^{-1} \cdot \vec{\Pi}_{in}/\omega \qquad (8.1.8)$$

The elements of the inverse of N are $(-)^{i+j} M_{ij}/\Delta$, where M_{ij} is the minor determinant formed by eliminating the ith row and jth column of the transpose of the matrix N, and Δ is the determinant of N. The identity matrix I operating on the vector $\vec{E}_t$ leaves it unchanged.

The calculations involved in finding N^{-1} and, consequently, $\vec{E}_t$ must be carried out for each frequency band Δf of interest. Since the parameters such as Π_{in}, η_i, and η_{ij} will generally vary with frequency, even the set of algebraic calculations involved in solving Eq. (8.1.6) can get quite cumbersome. There are computer routines available, however, that can reduce this effort considerably if the number of subsystems N is large. When N is less than 4, however, the saving of effort is usually not worthwhile and formal algebraic solution is adequate.

Example: 2 Subsystems

The case in which there are only two subsystems applies to very many situations of practical interest; the situation depicted in Fig. 6.1, for example. Equations (8.1.6) in this case are

$$
\left\{
\begin{array}{cc}
\eta_1 + \eta_{12} & -\eta_{21} \\
 & \\
 & \\
-\eta_{12} & \eta_2 + \eta_{21}
\end{array}
\right\}
\left\{
\begin{array}{c}
E_{1,tot} \\
 \\
E_{2,tot}
\end{array}
\right\}
=
\left\{
\begin{array}{c}
\Pi_{1,in}/\omega \\
 \\
\Pi_{2,in}/\omega
\end{array}
\right\}
\qquad (8.1.9)
$$

The determinant of the coefficient matrix is

$$
\Delta = (\eta_1 + \eta_{12})(\eta_2 + \eta_{21}) - \eta_{12}\eta_{21}
$$

$$
= \eta_1\eta_2 + \eta_2\eta_{12} + \eta_1\eta_{21}
$$

(8.1.10)

Thus, the expression for $E_{1,tot}$ is

$$
E_{1,tot} = \left[\frac{\Pi_{1,in}}{\omega}(\eta_2 + \eta_{21}) + \frac{\Pi_{2,in}}{\omega}\eta_{21} \right]\Delta^{-1} \quad (8.1.11a)
$$

$$E_{2,tot} = \left[\frac{\Pi_{1,in}}{\omega}\eta_{12} + \frac{\Pi_{2,in}}{\omega}(\eta_1+\eta_{12}) \right]\Delta^{-1} \qquad (8.1.11b)$$

There is a simplification of this result that is very important from the point of view of estimation. Suppose that only system 1 has external excitation. Then $\Pi_{2,in}=0$ and

$$E_{1,tot} = \frac{\Pi_{1,in}}{\omega}(\eta_2+\eta_{21})(\eta_1\eta_2+\eta_2\eta_{12}+\eta_1\eta_{21})^{-1}$$

and, therefore,

$$E_{2,tot} = E_{1,tot}\frac{\eta_{12}}{\eta_2+\eta_{21}} \qquad (8.1.12)$$

This relation allows us to estimate $E_{2,tot}$ if $E_{1,tot}$ is known (by calculation of the response to the environment or by measurements on a similar system.) Only one coupling loss factor need be known since the consistency relation Eq. (8.14) allows the other to be calculated if the mode counts in the band (or average frequency separation of the modes for each system) are known.

Example: 3 Subsystems

Three element systems usually arise when a resonant element (such as a wall) intervenes between two resonant systems of interest. The general equations for this case are

$$\begin{Bmatrix} \eta_1+\eta_{12}+\eta_{13} & -\eta_{21} & -\eta_{31} \\ -\eta_{12} & \eta_2+\eta_{21}+\eta_{23} & -\eta_{32} \\ -\eta_{13} & -\eta_{23} & \eta_3+\eta_{32}+\eta_{31} \end{Bmatrix} \begin{Bmatrix} E_{1,tot} \\ E_{2,tot} \\ E_{3,tot} \end{Bmatrix} = \begin{Bmatrix} \Pi_{1,in}/\omega \\ \Pi_{2,in}/\omega \\ \Pi_{3,in}/\omega \end{Bmatrix}$$
$$(8.1.13)$$

The determinant Δ of the N-matrix is quite complicated and will not be written out here. The matrix inverse to N is found to be

$$
N^{-1} = \frac{1}{\Delta}
\begin{Bmatrix}
\eta_{2t}\eta_{3t} - \eta_{23}\eta_{32} & \eta_{21}\eta_{3t} + \eta_{23}\eta_{31} & \eta_{21}\eta_{32} + \eta_{31}\eta_{2t} \\
\\
\eta_{12}\eta_{3t} + \eta_{13}\eta_{32} & \eta_{1t}\eta_{3t} - \eta_{13}\eta_{31} & \eta_{1t}\eta_{32} + \eta_{31}\eta_{12} \\
\\
\eta_{12}\eta_{23} + \eta_{2t}\eta_{13} & \eta_{1t}\eta_{23} + \eta_{21}\eta_{13} & \eta_{1t}\eta_{2t} - \eta_{12}\eta_{21}
\end{Bmatrix}
\qquad (8.1.14)
$$

where $\eta_{1t} \equiv \eta_1 + \eta_{12} + \eta_{13}$, etc.

To obtain a simple result that illustrates the procedure, but that is <u>not</u> representative of many situations of interest, first assume that only subsystem 1 is externally excited so that $\Pi_{2,in} = \Pi_{3,in} = 0$. Secondly, we set $\eta_{13} = \eta_{31} = 0$ so that the three subsystems form a chain $(1) \rightarrow (2) \rightarrow (3)$. With these assumptions

$$
E_{1,tot} = \Pi_{i,in}(\eta_{2t}\eta_{3t} - \eta_{23}\eta_{32})/\omega\Delta \qquad (8.1.15)
$$

$$
E_{2,tot} = \Pi_{i,in}(\eta_{12}\eta_{3t} + \eta_{13}\eta_{32})/\omega\Delta = E_{1,tot}\frac{\eta_{12}\eta_{3t}}{\eta_{2t}\eta_{3t} - \eta_{23}\eta_{32}}
$$

$$
E_{3,tot} = \Pi_{i,in}(\eta_{12}\eta_{23} + \eta_{2t}\eta_{13})/\omega\Delta = E_{2,tot}\frac{\eta_{23}}{\eta_{3t}}
$$

$$
= E_{1,tot}\frac{\eta_{12}\eta_{23}}{\eta_{2t}\eta_{3t} - \eta_{23}\eta_{32}}
$$

If we assume that $\eta_{23} << \eta_{2t}$ and $\eta_{32} << \eta_{3t}$, (coupling loss small compared to internal damping)

$$E_{3,tot} = E_{2,tot} \frac{\eta_{23}}{\eta_{3t}} = E_{1,tot} \frac{\eta_{12}}{\eta_{2t}} \cdot \frac{\eta_{23}}{\eta_{3t}} \qquad (8.1.16)$$

Clearly, the result in Eq. (8.1.16) is a product of two ratios of the kind shown in Eq. (8.1.12).

8.2   Alternative Form of the Energy Equations

In certain instances, we may know the energy of vibration of the subsystems in the absence of coupling. The uncoupled condition is obtained, as explained in Part I, by causing the response of all other subsystems to vanish. Thus, a boundary between subsystems in which motions are the response variable becomes "fixed". If force (pressure for example) is the response variable, then the boundary becomes "free" when the subsystems are decoupled or "blocked".

The power flow between two subsystems having total blocked energies

$$E_{1,tot}^{(b)} \quad and \quad E_{2,tot}^{(b)}$$

is given by

$$\Pi_{12} = \omega a_{12} E_{1,tot}^{(b)} - \omega a_{21} \, E_{2,tot}^{(b)} \qquad (8.2.1)$$

where the coefficients a_{ij} are different from the η_{ij}'s that appear in Eq. (4.1.3), but satisfy the same consistency relation,

$$N_1 a_{12} = N_2 a_{21} \qquad (8.2.2)$$

The "blocked energy" relation in this form can only be applied to two subsystem problems, so that if system 2 does not have external excitation, one has

$$E_{2,tot} = \frac{\Pi_{12}}{\omega\eta_2} = \frac{a_{12}}{\eta_2} E_{1,tot}^{(b)} \qquad\qquad (8.2.3)$$

since $E_{2,tot}^{(b)} = 0$ if there is not external excitation of system 2. The result in Eq. (8.2.3) is very near to that in Eq. (8.1.12). Evaluation of the coefficients a_{ij} is discussed in Chapter 9.

Fundamentally, the blocked energy of a subsystem is simply a measure of the input power, since

$$E_{tot}^{(b)} = \Pi_{in}/\omega\eta .$$

Thus, any of the relations in paragraph 8.1 that express actual energy of vibration in terms of input power may also be modified to express vibrational energy in terms of blocked energy.

8.3   Parameter Evaluation Using the Energy Equations

The energy equations can also be used in conjunction with experiments to calculate the parameters (coupling loss factors and loss factors) for a system. Eqs. (8.1.5) are linear in these parameters. Thus, if we measure the energy in each subsystem for a known set of input power values $\Pi_{i,in}$ we can generate a set of linear simultaneous equations for the η_i's and η_{ij}'s.

If we have N subsystems, then the total number of parameters in N^2 which consists of N loss factors and $N(N-1)$ coupling loss factors. Of course, we could use the consistency relation Eq. (8.1.4) to reduce the number of required coupling loss factors to $N(N-1)/2$ and an equal number of mode count ratios could also be found.

Any one experiment (for example, setting $\Pi_{2,in} \neq 0$ and $\Pi_{1,in} = \Pi_{3,in} = \cdots = \Pi_{N,in} = 0$ and measuring all the $E_{j,tot}$ values) generates N simultaneous equations for the parameters. Therefore, we must perform N independent measurements (for example, sequentially injecting known power into each of the subsystems and measuring the resulting energies) to obtain the necessary N^2 equations. Of course, any parameter values we may know will reduce the number of measurements accordingly, although we may choose to build some redundancy into the parameter evaluation for greater accuracy or as a check on the procedures.

We should emphasize that the procedure for parameter evaluation just described is not an established method. There has been only partial success for it in the cases for which it has been tried. The difficulties have shown themselves in the form of negative values of parameters. This impossible answer is the result of small errors that occur in each set of measurements that mounts up as one proceeds through the calculations. There is no reported analysis of the sensitivity of the derived parameter values to small errors in measured energy and input power values. Such an analysis is needed to establish the use of the energy equilibrium equations as a useful technique for determining SEA parameter values.

8.4 Useful Approximations and Simplifications

In most SEA calculations, three or four interconnected subsystems may suffice to describe the overall system, but it can easily happen that more subsystems are required. Even though the equations introduced in paragraph 8.1 are readily solved for such numbers of unknowns, it is desirable in many cases to seek ways to abbreviate the calculations.

We may want to simplify the calculations in order to get a "quick look" answer to compare with the more detailed calculation. Or, we may know some of the parameter values only approximately and seek to make sure we are not spending effort on evaluating unneeded parameters. Also, it is sometimes easier to get a better idea of the energy flow process from the simpler calculations so that changes in vibration levels that would result from changing parameter values can be inferred.

The principles of simplification may be listed quite simply. They will not all be applicable or useful in any

particular system, but by using them it is usually possible
to get answers that are of sufficient accuracy. The principles
are as follows:

(1)   Compute the approximate modal energies of the
      directly excited subsystems using $E_{tot}=\Pi_{in}/\omega\eta$.
      The subsystems with the high modal energies E_{tot}/N
      are usually those that will "drive" the other
      subsystems.

(2)   Identify the path(s) from the most energetic sub-
      system to the subsystem of interest. Concentrate
      on those paths that would appear to be dominant
      based on the relative sizes of coupling and
      dissipation loss factors.

(3)   When the loss factor of a subsystem is larger than
      the coupling loss factor connecting it to a more
      energetic system, ignore the coupling loss factors
      in the total damping η_{it}. When the loss factor is
      smaller than the coupling loss factor, assume the
      subsystem in question has the same modal energy
      as its more energetic neighbor.

(4)   If a neighboring subsystem has less modal energy
      than the subsystem being studied and it is not
      "in line" to another subsystem of interest, ignore
      it or at most include the coupling loss factor
      to it as part of the subsystem damping.

The result of these approximations will generally lead
to a "chain" sort of calculation as illustrated by the results
in Eqs. (8.1.12) and (8.1.16). The energy of the adjoining
subsystem is found by taking the product of the energy of the
source subsystem times the ratio of coupling loss factor to
damping for the receiving subsystem. This process is then
repeated for all subsystems along the line.

To illustrate the application of this simplified approach,
consider the diagram shown in Fig. 8.2 that models an airborne
computer. The exterior panels and frame of the computer are
excited directly by the environment, and we are interested in
knowing the dominant path that the energy takes getting to the
circuit boards and what the resulting vibration levels will be.

Suppose that the input power to the computer frame is
known to be $\Pi_{1,in}=0.1$ watts in the 250 Hz octave band (we will
describe ways of computing input power in the following
chapter). Our first task is to compute the energy of system 1.

Noting that η_{12} and η_{14} are both small compared to η_1, we ignore the coupling and compute

$$E_{1,tot} = \Pi_{1,in}/\omega\eta_1 = 6.4\times10^{-3} \text{ joules,} \qquad (8.4.1)$$

and a modal energy of

$$E_1 = E_{1,tot}/N_1 = 6.4\times10^{-4} \text{ joules.} \qquad (8.4.2)$$

Using Eq. (8.1.12) and ignoring coupling losses compared to dissipation, we get

$$E_{2,tot} = E_{1,tot} \frac{\eta_{12}}{\eta_2} = 6.4\times10^{-6} \text{ joules} \qquad (8.4.3)$$

and

$$E_{4,tot} = E_{1,tot} \frac{\eta_{14}}{\eta_4} = 6.4\times10^{-4} \text{ joules,} \qquad (8.4.4)$$

corresponding to modal energies of $E_2 = 6.4\times10^{-8}$ and $E_4 = 3.2\times10^{-5}$ joules respectively.

The energy of the circuit boards due to the air path is given by

$$E_{3,tot}^{(air)} = E_{2,tot} \frac{\eta_{23}}{\eta_3} = 6.4\times10^{-10} \qquad (8.4.5a)$$

while that due to the structural path is

$$E_{3,tot}^{(struct)} = E_{4,tot} \; \frac{\eta_{43}}{\eta_3} = 6.4 \times 10^{-7} \qquad (8.4.5b)$$

Obviously, the structural path is more significant in this example. The modal energy of the circuit board is, therefore:

$$E_3 = E_{3,tot}/N_3 = 1.3 \times 10^{-7} \text{ joules.} \qquad (8.4.6)$$

Since E_3 is greater than E_2, the actual energy flow will be from the circuit board to the air space. That is, the boards are radiating more energy to the air space than they receive from the surrounding air. Of course, this conclusion is dependent on the parameters arbitrarily chosen for the example, but the procedure shown here will usually supply results of sufficient accuracy and provide insight into the physical principles involved.

8.5 Summary

In this chapter, we have shown how energy estimates can be made from the equilibrium relations of SEA. These estimates must use values of SEA parameters, which will be discussed in the following chapter. The estimate is obviously also based on a model of the system. Nevertheless, the work in this chapter will be sufficient to tell us how the system might be changed to reduce the response to an acceptable level.

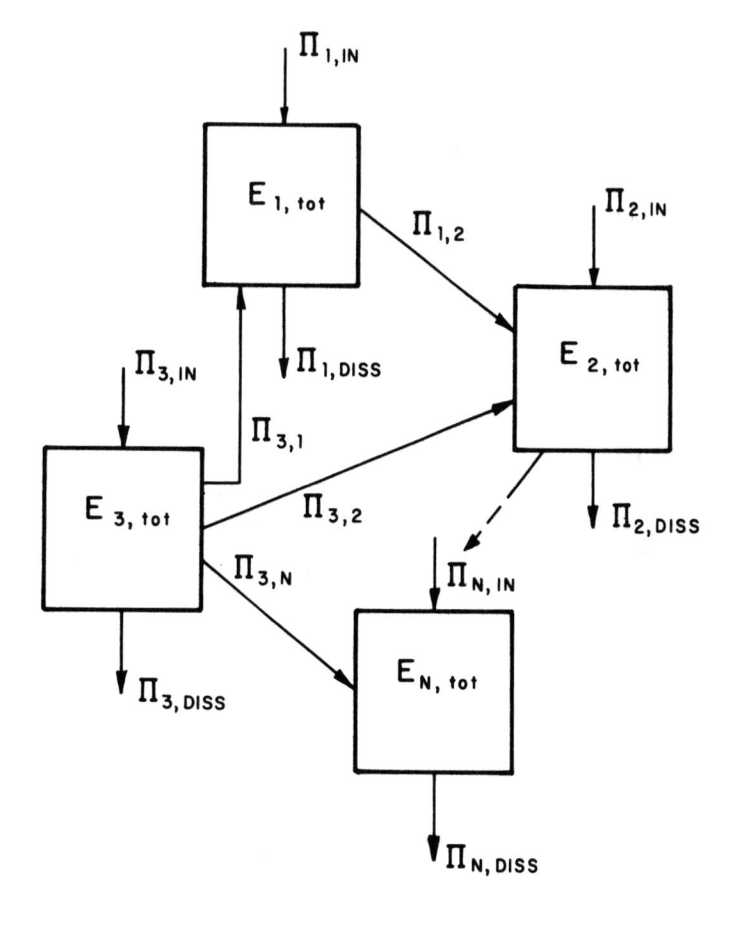

FIG. 8.1

AN SEA MODEL OF SYSTEM CONTAINING N SUBSYSTEMS

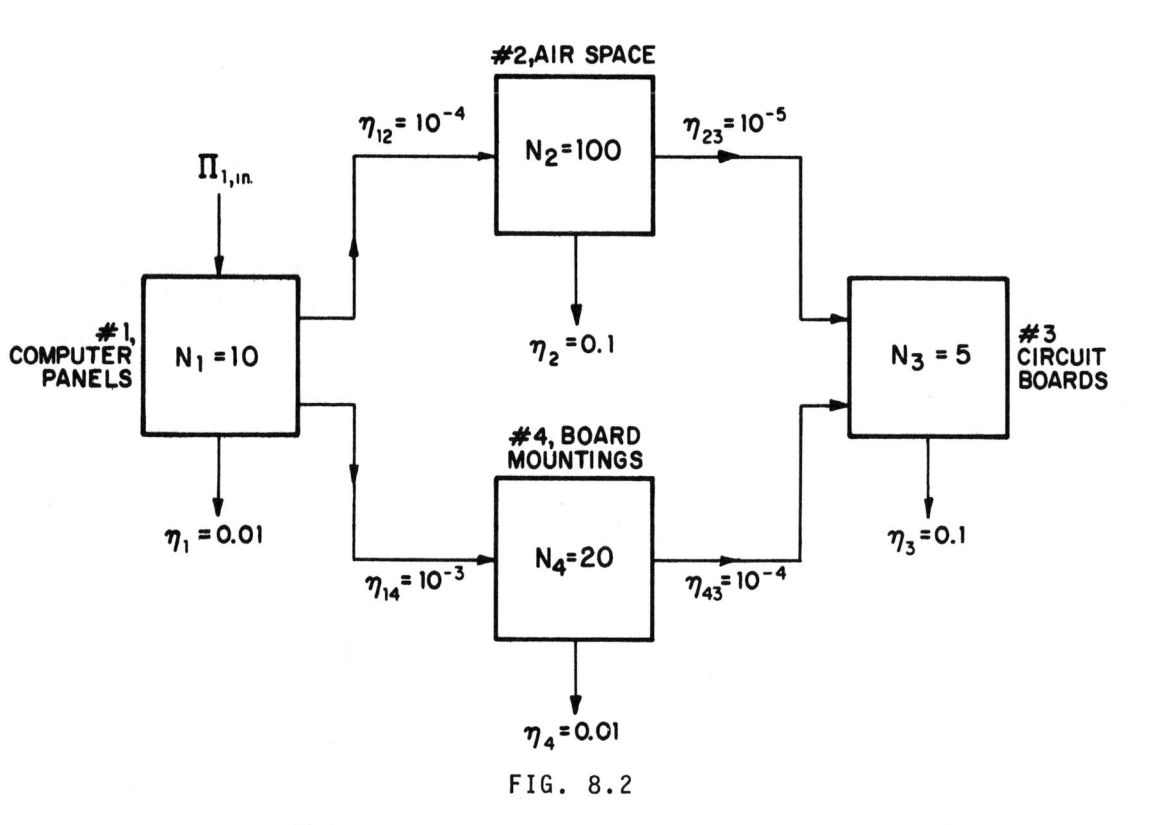

FIG. 8.2

**SEA MODEL OF CIRCUIT BOARD IN COMPUTER FRAME
WITH PARAMETERS FOR THE 250 Hz OCTAVE BAND**

CHAPTER 9. THE MEANING AND USE OF SEA PARAMETERS

9.0 Introduction

In the preceding chapters we have shown how to interpret energy estimates in terms of dynamical response and how the energy estimate itself is computed in terms of the SEA model and its associated parameters. In this chapter we discuss the SEA parameters in more detail, describing what they are from a physical point of view and how one goes about getting values for them.

The parameters govern the power input to each subsystem, the dissipation of energy, energy storage, and energy transfer between subsystems. Section II of this report is a fairly extensive tabulation of SEA parameters, so that we shall not duplicate that work here. Rather, our emphasis is on explaining the parameters and how they are used in the calculations. To obtain formulas or values for the parameters, reference should be made to Section II.

9.1 Dissipation Parameters

The dissipation of stored energy in each subsystem is measured by the parameter η_i, termed the "loss factor." The loss factor is a measure of the ratio of energy dissipated per unit time (one second) to average energy stored

$$\eta = \frac{\Pi_{diss}}{2\pi f E_{stored}} \cdot \qquad (9.1.1)$$

Defined in this way, the loss factor is the reciprocal of the "quality factor" Q used in electrical engineering: $\eta = 1/Q$. The damping parameter commonly used in mechanical engineering is ξ, the ratio of damping to critical damping, and $\eta = 2\xi$.

The loss factor η is occasionally introduced in structural vibration problems as the phase angle of a complex Young's modulus: $E \rightarrow E_o(1-i\eta)$. In other cases, the damping is introduced via a viscous element of value R, in which case, $\eta = R/\omega M$ where M is the subsystem mass. Of course, if the dissipation is actually occurring at the joints or rivet points, neither of these damping descriptions is really descriptive of the physical dissipation process.

It is occasionally asserted that a loss factor description of damping implies a very particular mechanism of dissipation and that erroneous estimates will occur if the actual mechanism is not the assumed ones. In general, response estimates are slightly dependent on the mechanism of damping, but the differences are very small if the loss factor is in the range $\eta=0.1$ or less. Most structures have damping much less than this, so that we need not imply any specific damping mechanism by introducing the loss factor.

In acoustical systems, the dissipation is normally expressed by the rate of decay of sound level after the source of excitation has been turned off. This "decay rate", DR (in dB/sec), is given by

$$DR = 27.3 \, f\eta$$

where f is the center frequency of the band Δf in which the data is taken. Typical values of decay rate for the small acoustical spaces associated with aircraft are of the order of 100 dB/sec or greater. It is often difficult to measure such rapid rates of decay with standard acoustical apparatus. In such a circumstance one may use an oscilloscope for the display.

Damping is probably the single most important parameter in establishing subsystem response because we are dealing with resonant modes. A 10% error in damping will result in a 1 dB error in the response estimate, and a 100% error in damping results in a 3 dB change in estimated response. When coupling factors are larger than the damping, the damping plays a less important role in setting response levels - there is then a tendency for energy equipartition with the adjacent subsystem (energy equipartition means equal modal energies of the subsystems connected by the "large" coupling loss factor).

Damping may be enhanced by addition of "applied damping" treatments. This might consist of a single layer of visco-elastic material adhered to the panel. Such a treatment is

a "free" or "unconstrained" layer, as diagrammed in Fig. 9.1a.
A more complex damping treatment that does not have as large
a weight penalty is the constrained layer, shown in Fig. 9.1b.
An alternative that uses a stiff spacer amplifies the strain
in the viscoelastic layer as shown in Fig. 9.1c. A variant
of this design uses a constrained layer spaced away from the
base panel. Finally, to reduce "drumming" vibration of a
single mode, the damping element can employ resonance to
amplify strain and provide a high degree of damping over a
fairly narrow frequency range as shown in Fig. 9.1d.

9.2 Power Transfer Parameters

   The power flow between subsystems may be described either
in terms of the actual energies of vibration, or in terms
of the energy of vibration that they would have in the ab-
sense of coupling. In the first case, the parameter of
interest is the coupling loss factor, η_{ij}. There is no
general name for the second parameter, but it has been denoted in Part I as a_{ij}. The formulas for power flow from
subsystem i to subsystem j are as follows:

$$\Pi_{ij} = \omega\eta_{ij}E_{i,tot} - \omega\eta_{ji}E_{j,tot} \qquad (9.2.1a)$$

$$= \omega a_{ij}E_{i,tot}^{(b)} - \omega a_{ji}E_{j,tot}^{(b)} \qquad (9.2.1b)$$

Equation (9.2.1a) applies to any situation, but Eq. (9.2.1b)
has only been demonstrated to apply to 2-subsystem situations.

   The principal utility of a_{ij} is that it allows for a
computational algorithm for η_{ij}. This relation was derived
in Part I and will be used in Section II. To illustrate how
this works, however, suppose the two subsystems are joined at
a point and that a point input impedance $Z_{i,in}$ can be defined
for each at the attachment location. Then, the value of
a_{ij} is:

$$a_{ij} = \frac{2\Delta f}{\pi\omega N_i} \frac{R_{i,in} R_{j,in}}{|Z_{i,in} + Z_{j,in}|^2} \qquad (9.2.2)$$

A general relation between the coupling loss factor η_{ij} and the parameter a_{ij} may be found when the jth system becomes "dense", i.e., $N_j f \eta_j / \Delta f \gg 1$. In this circumstance, as shown in Part I [Eq. (4.1.21)]

$$\eta_{ij} = a_{ij} (1-a_{ij}/\eta_i)^{-1} \qquad\qquad (9.2.3)$$

In the case where subsystem i has a single mode, $N_i = 1$, then one can show that [Eq. (3.4.6)],

$$\eta_{ij} = R_{i,in}/\omega M_j \qquad\qquad (9.2.4)$$

where M_j is the total mass of the jth subsystem. When the ith system has many modes, $N_i \gg 1$, one can show that $\eta_{ij} \to a_{ij}$. Thus, the evaluation of the coupling loss factor can be expressed entirely in terms of junction impedances.

The theoretical calculation of coupling loss factors can get quite complicated for one or more of the following reasons:

1. Neither system is "dense"; only a few modes of each of the two subsystems resonate in the band of interest.

2. The junction is not a point, but extends along a line or over an area. More complex impedance functions are necessary to describe the inter-action in this case.

3. The actual interaction may not be "scalar", but "vectorial". This is particularly true when the interacting systems are structures, since a proper description of the interacting forces may include moments and shear and compressional forces. Our purpose in this chapter is to indicate the process by which one obtains the coupling parameter. The detailed methods and available results are covered in Section II of this Part of the report.

We have shown that the coupling loss factor can be related to the junction impedance. It can also be related to other parameters that are well known in certain specialized fields, such as sound transmission between rooms. In that field, the sound transmission is defined by a "transmissibility" τ, which is related to the well known "transmission loss" TL by

$$TL \equiv - 10 \log \tau. \tag{9.2.5}$$

The relation between coupling loss factor and τ is simply

$$\eta_{ij} = \frac{A_w c}{2\omega V_i} \; \tau \tag{9.2.6}$$

where A_w is the area of the wall through which the sound is "leaking", c is the speed of sound and V_i is the volume of room i. Thus, all the data available on TL for various wall constructions becomes a source of information of coupling loss factors between subsystems that are acoustical spaces.

The experimental determination of coupling parameters is usually approached along lines indicated in Chapter 8. A system has power injected into it in a simply way and the resulting response (or subsystem energy) is measured. From this data and a knowledge of the loss factors in the system, the coupling loss factor may be found. This is the manner in which the transmissibility between rooms is found. From Eq. (8.1.12), a measurement of the ratio

$$\frac{E_{2,tot}}{E_{1,tot}} = \frac{\eta_{12}}{\eta_2 + \eta_{21}} \tag{9.2.7}$$

is sufficient to determine η_{12} if η_2 is known and $\eta_{21} << \eta_2$. Thus, if the junction of interest is reproduced between two test structures, structure 1 is excited in some convenient way, and the damping of the receiving structure (#2) is adjusted to satisfy $\eta_2 >> \eta_{21}$, then a simple measure of the vibrational energies of the two structures will determine η_{12}.

9.3 Modal Count of Subsystems

The modal count N_i of a subsystem is the number of modes of that subsystem that resonate in the band Δf under consideration. In some systems the number may be of the order of unity, in others (particularly acoustical subsystems at higher frequencies) the number may be in the thousands. Basically, the mode count is a measure of the number of modes available to accept and store energy.

In SEA work, the modal count is often expressed in terms of a modal density   n   so that

$$N_i = n_i \, \Delta f \ . \qquad\qquad (9.3.1)$$

Since the modal count will vary from one band to another, we may indicate that the modal density will vary with   f,   the center frequency of the band, by writing it $n_i(f)$. Most subsystems have modal densities that vary with frequency.

In Chapter 7, we expressed the modal density $n_i(f)$ in terms of its reciprocal, the average frequency separation between resonant modes,

$$\delta f_i = 1/n_i(f) \qquad\qquad (9.3.2)$$

One may also find the frequency separation expressed in terms of radian frequency

$$\delta \omega_i = 1/n_i(\omega) = 2\pi\delta f_i = 2\pi/n_i(f) \qquad\qquad (9.3.3)$$

and, consequently, $n_i(\omega) = n_i(f)/2\pi$.

Most SEA formulas that are expressed in terms of modal density can be converted to modal count by using Eq. (9.3.1). This is particularly true in dealing with a system that only has one mode in the band of interest Δf. It also applies to cases in which the subsystem may have several modes, but we may wish to concentrate our interest on a single resonant mode. In this case, $N_i=1$ and $n_i(f)=1/\Delta f$ and not $1/\delta f_i$. Note that the fundamental quantity in all cases is the mode count N_i, and the modal density is a derived quantity.

The modal count may be found by both experimental and computational procedures. We shall review both methods in Section II. Here we describe briefly these procedures as a way of determining the most appropriate method in the preliminary design process. Very often several of these alternatives will be possible ways of getting a modal count. One is not faced with this dilemma of choice very often with the other SEA parameters!

The only really verified way of measuring the modal count is to excite the subsystem with a pure tone and observe the response at a second location. The frequency is then swept slowly over the band Δf and the response peaks are counted. Of course, only modes that are non-vanishing at the excitation and observation points will show-up in the response. For this reason, one should select those locations carefully; either at a "corner" location for acoustical systems or along a free edge of a structure.

One will also miss modes if their average spacing δf_i becomes of the same order as their resonance bandwidth $f\eta_i$. The equality

$$\delta f_i = f\eta_i \qquad\qquad (9.3.4)$$

is the condition of modal overlap and marks the frequency range in which one will begin missing modes because they are too close together to be resolved. It is this limitation that the second procedure, the "point conductance" method is proposed to avoid. This method relies on the result that a mean square force $<f^2>$ in the band Δf applied at an "average" point on the subsystem will result in an injected power [see Eq. (2.2.24)],

$$\Pi_{in} = <f^2>/4M_i \delta f_i , \qquad\qquad (9.3.5)$$

and since the mass M_i is known and presumably $<f^2>$ and Π_{in} can be measured, we can find δf_i. This procedure is more complicated than the simpler frequency sweep method and must be considered less well established at present.

There have been many theoretical studies of modal density for both acoustical and structural systems. Generally, the modal density calculations, when tested experimentally, have turned out to be fairly reliable. Modal density is one of the easier parameters to calculate and it seems quite sensible to calculate it if at all possible. In many cases, it will turn out that the appropriate formulas for this calculation are available.

The following is a partial list of system elements for which calculations of modal density exist:

1. Flat plates with various boundary restraints,

2. Flat plates of complex construction including layered plates,

3. Shells and shell segments, including spheres, cones, and cylinders,

4. Acoustical spaces of most shapes, including rectangular, spherical, cylindrical; and volumes representing combinations of these,

5. Various beam and girder shapes including flexural and torsional deformations.

Fortunately, modal count tends to be an extensive property of a system. That is, the total mode count may be estimated by adding the number of modes expected for the various parts of the system. In this way, mode count can be predicted for fairly complex structures.

As an example, consider the estimation of modal count for the equipment shelf shown in Fig. 9.2. Suppose that the two end plates have a diameter of 1 ft and the shelf is 3 ft

long, consisting of four 4 inch webs. This complicated
structure is made up of six plates of varying shapes, but
it is known that the mode count of a plate of thickness h
and area A is given by

$$N = \frac{\sqrt{3}A}{hc_\ell} \; \Delta f \; . \tag{9.3.6}$$

Since h = 3/16 in. and c_ℓ = 17,000 ft/sec, (the speed of
sound in the plate material) are the same for all six plate
segments, the total mode count is found by adding that due
to the parts, or by using the total area of the structure,
which is

$$A_{tot} = (2 \; \frac{\pi D^2}{4}) + 4(3) \; (1/3) = 1.6 + 4 = 5.6 \; ft^2 \tag{9.3.7}$$

to give a mode count

$$N = \frac{(1.7) \; (5.6)}{(0.03) \; (17,000)} \cdot \Delta f = 1.9 \times 10^{-2} \Delta f \tag{9.3.8}$$

Thus, in the 1 kHz octave band (Δf=710Hz) we would expect to
find about 13 resonant modes. The exact spacing and location
of these modes along the frequency axis would depend on details of construction, but the number of modes spread over
this 710 Hz interval would not. Thus, stiffening the structure
by the addition of gusset brackets and the like will perturb
the resonance frequencies, but the modes would still occur
at intervals of about 50 Hz along the frequency axis.

9.4   Input Power Prediction

The input power is one of the quantities that is presumed known in the SEA calculation. However, it is only in
rare instances that the input power will be known directly.
If one is exciting the structure with a shaker and using a

force gauge between the shaker and the structure, it is possible to measure the input power directly. One is more likely to know a mean square force on the structure and some description of the spatial distribution of the load. This is the case of loading by a turbulent flow or acoustical noise pressure field. In this situation, the spatial character of the excitation is defined by a correlation function that will determine how well the spatial shape of the resonant modes will correlate with the excitation.

In a few situations, the excitation may be thought of as highly localized and taken to occur at a point in the system. Eq. (9.3.5) gives an estimate for input power in this instance, assuming the mean square force is known. However, this situation is not as easily realized as it might appear. The difficulty is that the structure has a very low input impedance at a resonance, so that if the shaker has a finite impedance, its force output will drop. This is illustrated in Fig. 9.3. The result is that the effective value of mean square force driving the structure is significantly below the apparent value obtained by multiplying the mean square current times the blocked force-current relation for the exciter.

In the cases of excitation by an acoustical noise field or a turbulent boundary layer, relations have been developed for the input power for a variety of structures that includes flat plates, cylinders, cones and other axisymmetric shapes. The appropriate relations for these cases are presented in Section II. To illustrate how we would use this parameter to calculate input power, we first consider excitation of a flat homogeneous plate by a turbulent boundary layer.

It turns out that the most important determinant of the power injection by a turbulent boundary layer is the ratio of convection speed of the pressure variations (about 80% of the free stream flow speed) to the bending wave speed. For aircraft speeds and aircraft skin panels, the bending speed is usually less than the convection speed. In this case the input power to the structure in the band Δf is known to be [15].

$$\Pi_{in} = \frac{<p^2> \Delta f A_p^2}{R_{in}} \quad \frac{\lambda_p \delta_1}{A_p} \tag{9.4.1}$$

where $<p^2>_{\Delta f}$ is the mean square turbulent pressure measured on
the panel in the band Δf, A_p is the area of the panel,
$R_{in}=4M\delta f$ is the input point resistance of the panel. The
bending wavelength on the plate is λ_p and the displacement
thickness of the boundary layer profile is δ_1.

We can also find the power input to a structure from
a sound field using the expressions for coupling loss factor
already presented. Using

$$\Pi_{in} = \omega\eta_{acoust,struct} E_{acoust,tot} \, , \qquad (9.4.2)$$

the following expression results for the power input from a
sound field

$$\Pi_{in} = \frac{<p^2>_{\Delta f}A_p^2}{R_{in}} \, \sigma_{rad} \, \frac{\lambda^2}{2\pi A_p} \qquad (9.4.3)$$

where $<p^2>_{\Delta f}$ is the mean square pressure measured on the
surface of the panel, λ is the wavelength of sound, and σ_{rad}
is the so-called radiation efficiency of the panel defined
by

$$\sigma_{rad} = \frac{R_{rad}}{\rho c A_p} \, . \qquad (9.4.4)$$

The other parameters are as previously defined. Thus,
on the basis of the above, the designer can predict the power
injected into a panel by a turbulent boundary layer or an
acoustical noise field in terms of the pressure measured
on the panel and other panel properties. Such estimates are
then to be used in the band by band calculation of energy
distribution throughout the system.

9.5   Conclusions

The SEA parameters are naturally occurring quantities in the theory of energy sharing of systems. These parameters may be evaluated by the designer by both analytical and experimental means. In many cases, they are related to more conventional system descriptors. Certainly, this is true for the energy dissipation, expressed by the loss factor. Mode count or modal density are not concepts original with SEA, but they are not widely used parameters in conventional structural dynamics. The coupling loss factor is a parameter that is unique to SEA studies, but even here we find that is often relatable to previously known parameters such as radiation resistance, transmissibility, and junction or point input impedance.

Of course, the particular coupling loss factors and other parameters that we need to find depend fundamentally on the choice of subsystems and their interconnections in the SEA model. This is the topic of the final chapter in this part dealing with the preliminary design process.

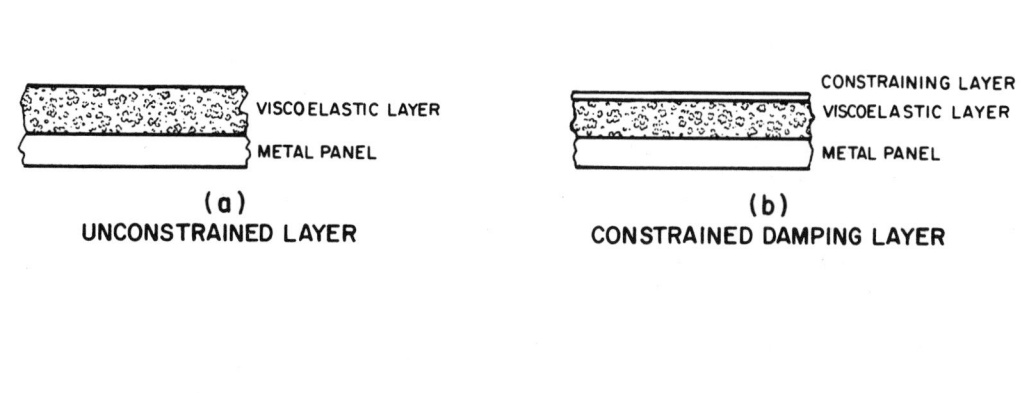
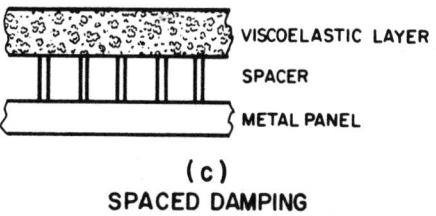
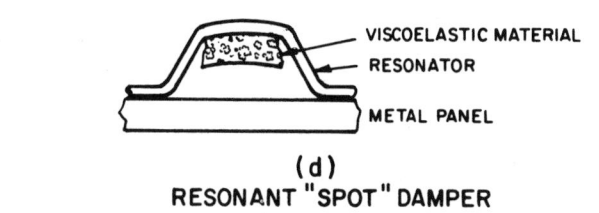

(a)
UNCONSTRAINED LAYER

(b)
CONSTRAINED DAMPING LAYER

(c)
SPACED DAMPING

(d)
RESONANT "SPOT" DAMPER

FIG. 9.1

VARIOUS FORMS OF APPLIED DAMPING

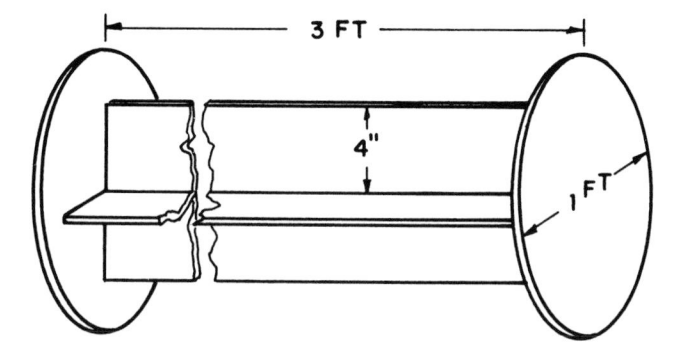

FIG. 9.2

EQUIPMENT SHELF MADE FROM 3/16 IN. ALUMINUM PLATE

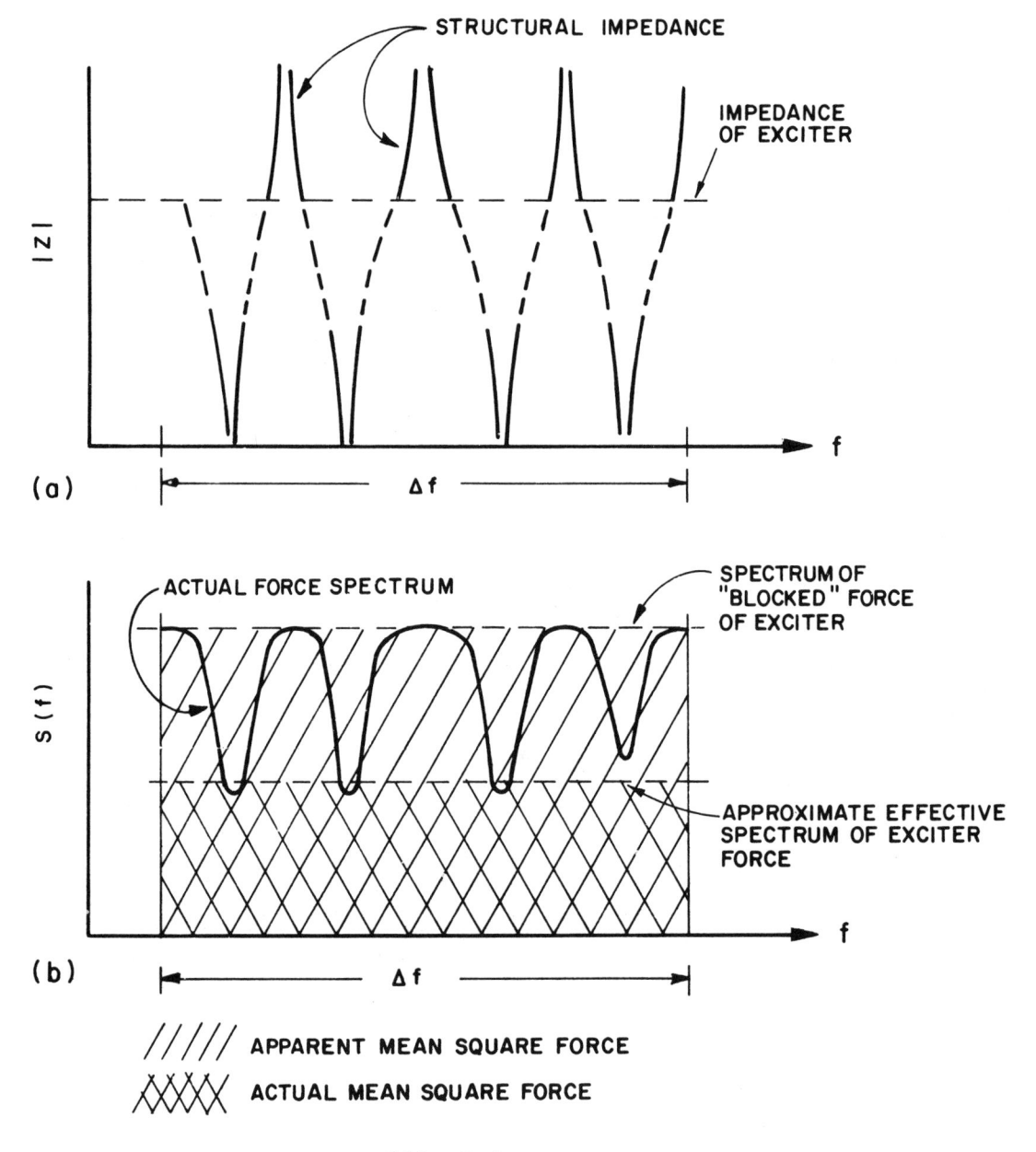

FIG. 9.3

THE IMPEDANCE OF THE STRUCTURE AND THE INTERNAL
IMPEDANCE OF THE EXCITER ARE SHOWN IN (a). THE APPARENT
FORCE SPECTRUM BASED ON THE INPUT CURRENT AND SHAKER
CHARACTERISTICS ARE SHOWN IN (b). ALSO IN (b) ARE SHOWN
THE ACTUAL FORCE SPECTRUM AND ITS "FLAT" EQUIVALENT
BASED ON THE IMPEDANCE DIAGRAMS IN (a)

CHAPTER 10. MODELING THE SYSTEM

10.0 Introduction

We have now worked our way back from the use of the
energy estimate, how that estimate is derived, and what the
parameters are that enter the energy estimate to the most
basic step of all in SEA -- the definition of the SEA model.
This part of the effort is one of the earliest steps in the
use of SEA in a design task. Fortunately, unless the basic
system changes, this may only have to be done once. As
various design modifications are made, perhaps with the goal
of reducing response, the coupling and other parameters will
change and the energy and associated response estimates will
also change.

In a way, however, those aspects of the response estimation
task that we have been discussing are deductive, and in that
sense, straightforward. In the model definition phase, the
designer must extract from the physical system - a fuselage
and attached electronics pod, for example - a model consisting
of groups of resonant interacting modes that will allow an
estimate of response to be made. This is the synthetic part
of the designer's task, and it is less straightforward to
describe in detail just how this is to be accomplished in any
particular case.

In this final chapter of Section I, we examine the procedures for developing SEA models insofar as a largely
synthetic and inductive process can be set down as a set of
procedures. Since a principal motivation for this book
is the application of SEA to high-speed flight vehicles, the
discussion uses examples from this area of engineering, but
the procedures have much broader application. The real
development of synthetic procedures, however, must occur as
the experience of the designer with SEA methods is increased
and, consequently, the important step is to begin to use SEA
to make estimates in the first place.

10.1 Definition of Subsystems

SEA is able to provide estimates of complex system
response because of our ability to group modes together and
deal with them statistically rather than individually and
deterministically. The modes are grouped according to the

following principles:

1. They all resonate in the band Δf_i in the entire SEA model, all parameters are evaluated as averages over the frequency interval Δf also.

2. Modes are grouped by major sections of the system that are to be identified in the final response estimates. Thus, all wing modes, fuselage modes, electronic pod modes, etc., that resonate in the band Δf would be grouped separately.

3. For any section, there may be differing classes of modes that one may wish to identify. For example, a truss or beam may have both torsional and flexural modes that resonate in Δf. These modes may be grouped together if we expect them to be well coupled, or they may be treated as separate subsystems.

A class of modes of a section of a system, resonating in the band Δf, is an SEA subsystem and represented by a "box" in a diagram like that in Fig. 8.1. The expected mode count N_i labels the energy storage capacity of the box. It is important to realize that modes that resonate outside the band Δf (so-called non-resonant modes) are not included in the modal count N_i, which pertains to resonant modes only. These non-resonant modes may play a role in transmitting energy from one sub-system to another, but the energy of vibration that they acquire in doing so is nearly always substantially less than that of the resonant modes. Consequently, we do not count the vibrational energy of non-resonant modes in estimating response.

One of the criteria for modal similarity in defining the subsystem is that the modes have nearly the same damping. Thus, the dissipation of energy by the "box" can be represented by a single loss factor η_i as was done in Chapter 8. A detailed study of the individual modes would show some variation in their damping. Theoretical analyses of permissible variations in damping have not been carried out, but we might assume that the individual modal loss factors could vary over a factor of three or so and the estimates of response based on the average damping would not be too far off. If a mode or group of modes has damping values that differ by a factor of 10 from the mean, than that group should probably be "split off" to form another subsystem.

Another criterion for modal similarity relates to the interaction with the loading environment. Thus, a particular form of excitation, such as acoustical noise, will excite flexural waves on skin panels most directly. In-plane compressional vibration of the skin would be only weakly excited. If the damping and coupling between in-plane and flexural modes were such that equal modal energy between these modes were to be expected, then only a single subsystem representing the skin structure might be necessary. If appreciable difference in modal energy might be expected, or if the distinction between in-plane and flexural or transverse motions is important for some other reason, then the skin structure should be represented by (at least) two subsystems. In this case, the "flexural" subsystem would have an input power Π_{in}, but the in-plane modal subsystem would be excited only by its coupling to the flexural subsystem. There are other considerations as well. The larger that Δf is made, the greater the number of resonant modes for each subsystem and, according to the discussion of Chapter 7, the smaller the variance in the estimate. On the other hand, if the bandwidth is too great, the assumption of uniform loss factor and coupling loss factor for all modes will not be accurate. Also, too broad a bandwidth causes frequency resolution to be lost, which may be important in some applications.

As a final item on the identification and definition of subsystems, it is worthwhile emphasizing that certain elements of a structural and environmental system are not SEA subsystems. For example, a turbulent boundary layer is not an SEA subsystem since it cannot be represented as a set of linear resonators or modes of oscillation. A turbulent flow must be regarded as a source of power, and not a modal subsystem. An acoustical environment may be treated as a power source but in some circumstances (a reverberant test chamber for example) it may be treated as another subsystem.

10.2 Identifying and Evaluating the Coupling Between Subsystems

The identification of the coupling between modal groups that exchange energy can be quite subtle. Certain features of the problem are fairly obvious, and should be dealt with first in the modeling process. For example, consider the system shown in Fig. 10.1, consisting of an exterior shell of an airborne computer that is excited by acoustical noise, and internal frame, and a circuit board mounted into the frame and, of course, an air space within the shell.

According to the procedures indicated in the preceding section, the first task is to divide the system into its major structural and acoustical systems. This division is shown in Fig. 10.2. We note that at least two of the elements have different classes of modes that we may want to treat separately, but for the time being, we shall treat each of the boxes in Fig. 10.2 as an SEA group of "similar" modes.

We now inquire about the energy transfer mechanisms at work in this system. Most obviously, power will be transferred through the mechanical connections between the shell and the frame and the frame and the circuit board. Also, the surface of contact between the enclosed sound field and the shell on one hand and the circuit board on the other. These power flow paths are shown in the diagram in Fig. 10.3.

A "first cut" at finding the circuit board response in this example might be to settle for the system as shown, proceed with evaluation of the parameters (Chapter 9), solve the energy equations for the energy of the circuit boards (Chapter 8) and interpret this energy as strain and acceleration spectra of the components on the boards (Chapter 7). We might then investigate the relative roles of acoustical and vibrational transmission, evaluate stiffening the frame, putting acoustical absorbing material within the shell, or changing the construction of the mountings of the frame into the shell. As each of these changes were made, coupling and damping loss factors would change and the response estimates would likewise change, indicating an increased or decreased vibration of the circuit boards. The basic model configuration would be unchanged, however.

At some point we might want to improve our model by considering effects thus far ignored. One of these is the role of nonresonant modes in the transmission of vibrational energy. The diagram of Fig. 10.3 for example indicates that if the frame were perfectly rigid and there were no resonant modes of the frame in the band Δf, then there would be no energy transferred to the circuit boards by the structural path. This is obviously not so, since a rigid translation of the frame would transfer energy from the shell to the circuit boards. In modal terms, this energy transfer is a result of the nonresonant excitation of modes that have resonance frequencies *above* the band Δf.

In a similar fashion, acoustical excitation of the shell will result in vibration and nonresonant modes of the shell that may be quite effective in exciting resonant acoustical

modes of the cavity within the shell. Since we are treating
the external acoustical fields as a power source in this
example, the effect of nonresonant motion of the shell is to
add an additional power source directly to the cavity. These
two modifications due to nonresonant vibration of subsystems
that connect other subsystems are shown in Fig. 10.4. With
this change in the model, the additional coupling loss factor
η_{13} must be evaluated, as well as the input power $\Pi_{4,in}$.
Obviously, we should expect the calculated circuit board
vibrations to change as a result of these changes, but the
change may not be very great. If the predictions turn out
to be quite insensitive to this modification, then one would
quite likely revert to the model of Fig. 10.3 for systems of
this type.

The preceding discussion illustrates why modeling is a
matter of judgement. The model of Fig. 10.4 is more precise
than that of Fig. 10.3, but it is also more detailed and more
cumbersome. It is only after the revised calculations have
been made with the greater detail included that one can tell
whether or not the extra effort is justified.

10.3   Subsystems Within a Section of the System

In paragraph 10.1, we noted that a section of a vehicle
could contain groups of resonant modes that were sufficiently
dissimilar so that separate subsystems might be necessary to
model the system. In the following we explore this idea
further and illustrate its effects on the model by returning
to Fig. 10.2.

If the two types of shell and frame modes are each
treated as a separate subsystem, then all the subsystems and
their interactions are as shown in Fig. 10.5. Clearly, what
began as a fairly simple model of the circuit board
excitation in Fig. 10.3 has become a very complex model
indeed. Is the complexity necessary? Generally no, but a
blanket answer cannot be given. If the process of refining
the SEA model in going from that shown in Fig. 10.3 to that
shown in Fig. 10.4 were to result in significant changes in
the estimate for board vibration, then the further refine-
ment in going to a model like that of Fig. 10.5 might be
deemed useful. One should always keep in mind the general
results for estimation variance discussed in Chapter 7. If
the refinements in the model causes the modal densities of
certain subsystems to get to be too low, then the variance
may increase so much that the more refined mean value
estimate has little significance.

Most of the changes from Fig. 10.4 represented by Fig. 10.5
are fairly evident but we should comment on the energy
transfer $\Pi_{1a,1b}$ and $\Pi_{2a,2b}$ between the new subsystems. Considering the frame first, we can note that there are two ways
of coupling flexural and torsional motion in a beam. Unless
the cross section of the beam has a high degree of symmetry
(I-beam or box section), there will generally be continuous
coupling between these forms of motion. Likewise, end conditions
on the beams can cause coupling. If a simple beam is clamped
at an angle other than $\pi/2$ to the centerline, flexural waves
will reflect from this boundary as a combination of flexural
and torsional motions.

It will be clear by now that the number of kinds of
structural and acoustical subsystems of interest in SEA is
very great. Many of the parameters needed to analyze
systems comprised of these subsystems are listed and evaluated
in Section II of this report. Even so, it is often necessary
to estimate values for parameters of systems that may not be
in the tables. In these cases, one can often treat the
actual system as "somewhere in between" two limiting cases.
For example, a short tab connecting two structures might be
bracketed by a long beam connection and a rigid connection
between the structures. Also, one can often determine the
coupling loss factor experimentally.

10.4 Discussion

We must keep in mind that although a major emphasis in
this book is the use of SEA in preliminary design, one of
the major advantages of SEA is that the model and the
associated estimates of response can be continually
sharpened and refined as more detail regarding the system is
developed. Thus, at the very early stages, the major
structural sections may be modeled as homogeneous cylinders,
plates, etc. Also, we would probably use broad frequency
bands, octave bands for example, for the frequency resolution.
This would keep the number of parameters down to a reasonable
limit and more realistically reflect our knowledge of the
system. As our knowledge of the details of the system increased, the model could include more modal groups (subsystems) and a finer division along the frequency scale (we
could change to third-octave bands, for example).

None of the other estimation schemes has this capability
of continual adjustment in the procedure to accommodate the
increased knowledge about the system. In a way, we can think

of the SEA model as a communication channel that is able to
accommodate a certain amount of input data (system parameters)
to predict an output (response). A more complex model
represents a channel with greater information handling
capacity, and we can adjust the model to handle the available
information. Thus, SEA is an estimation tool that can be used
throughout a project from preliminary to final design.

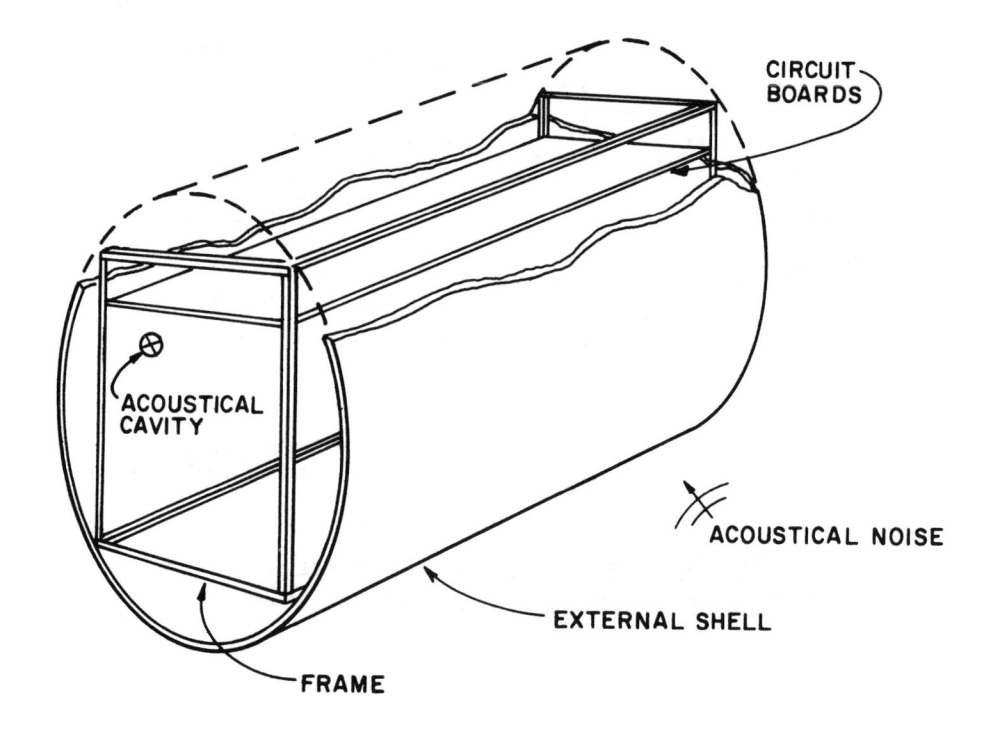

FIG. 10.1

DIAGRAM OF AIRBORNE COMPUTER ASSEMBLY

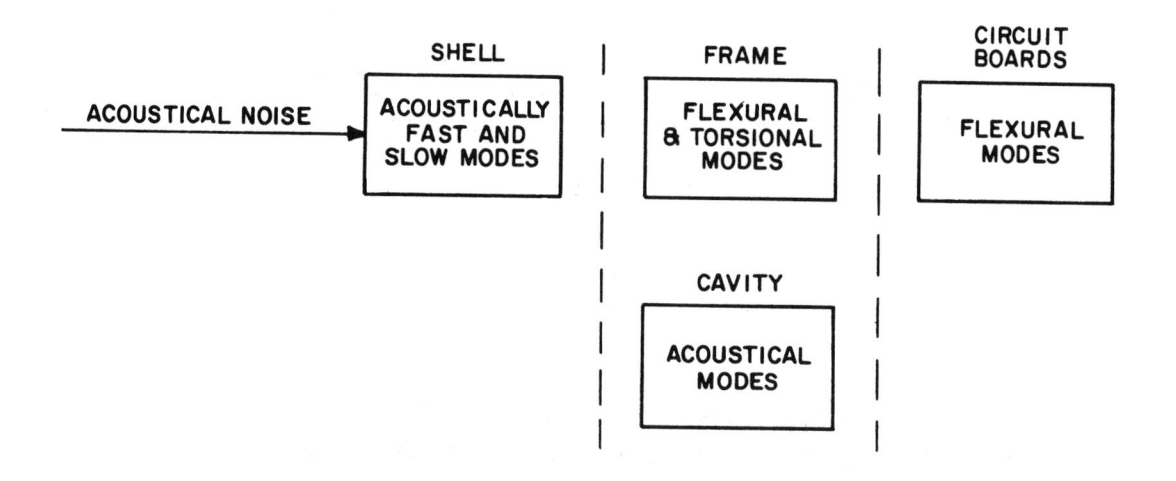

FIG. 10.2

MAJOR SECTIONS OF THE SYSTEM SHOWN IN FIG. 10.1

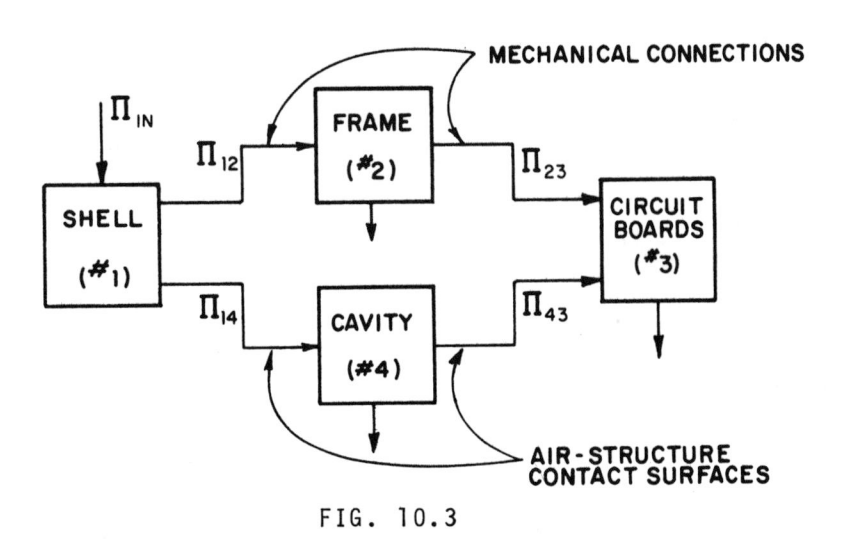

FIG. 10.3

**SYSTEM OF FIG. 10.2 WITH POWER
TRANSFER THROUGH INTERFACES SHOWN**

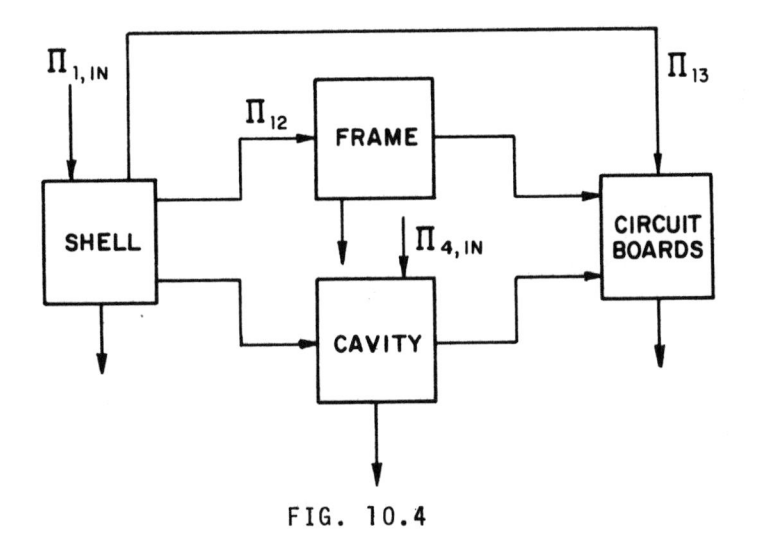

FIG. 10.4

**MODIFICATIONS IN DIAGRAM OF FIG. 10.3
DUE TO NONRESONANT MOTIONS OF FRAME AND SHELL**

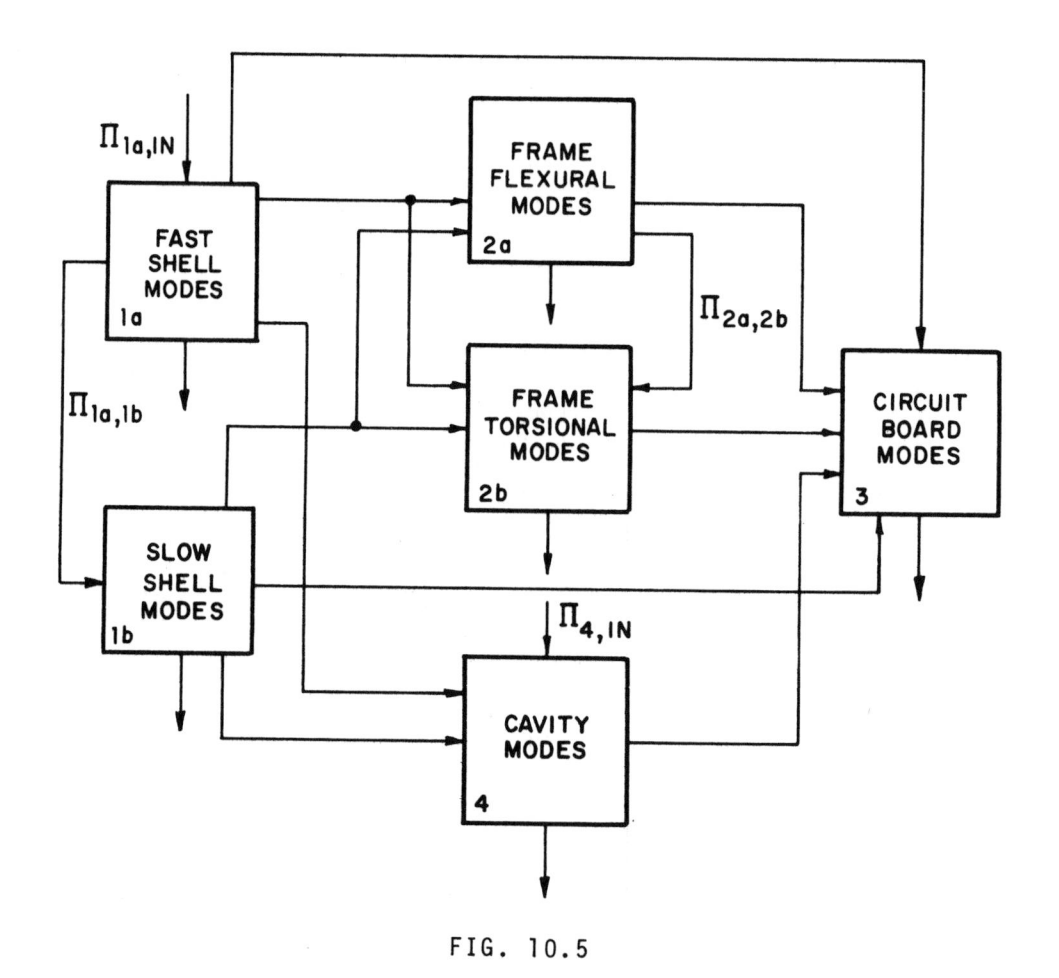

FIG. 10.5

**THE SEA MODEL OF THE SYSTEM OF FIG. 10.1 WITH A GREATER
NUMBER OF MODAL GROUPS AND INTERACTION REPRESENTED**

CHAPTER 11. PARAMETER EVALUATION - THE ENGINEERING BASE OF SEA

11.0 Introduction

In Part I and Section I of Part II of this book, we have concentrated on the theory for and use of SEA in response prediction. The role of the parameters in the various equations for energy and response of the subsystems - parameters such as damping loss factor, mode count and coupling loss factor - has been adequately pointed out. These we term SEA parameters because they enter in all SEA predictions and to a degree, they are used in a way that is unique in the SEA application.

This section is concerned with providing information on the evaluation of SEA parameters. This evaluation may be experimental, theoretical, or a "guesstimate" based on similarity with other situations for which the parameters have been found previously. Indeed, for many practical situations we may expect the guesstimate to be the primary "method" for finding SEA parameters.

Experimental procedures may be "direct" as when, in the case of mode count, one proceeds by counting resonant peaks. The procedure may be "indirect" as when one infers a coupling loss factor from the result of a vibrational response experiment. We present both direct and indirect methods here for all the parameters . The best method in any situation will depend on the range of parameter values involved, as discussed in the paragraphs that follow.

Theoretically derived parameter values are very often used for modal densities or mode count and the coupling loss factor. A few results are available from the literature -- several are quoted here. Situations not covered in the text may be treated from the references which should serve as adequate back-up for new calculations or a quoted result on a similar system that may serve as a useful estimate.

11.1 Who are the SEA Parameters?

The SEA parameters are measures of the dissipation of
vibratory energy, the number of resonant modes in a frequency
band available to store vibratory energy, and the coupling
between the energy containing subsystems.

Chapter 12 describes how damping is measured or estimated
analytically. The mechanisms of damping or dissipation vary
significantly from one system to another. The losses in
mechanical energy may result from molecular effects in polymers, air pumping at joints of built-up structures, or the
motions of dislocations in metals. If the losses are not
too great, however, a single parameter may be used to describe
the dissipation effect in any of them. This parameter is
called the "loss factor", and is the damping descriptor most
often used in SEA.

In Chapter 13 we discuss how one evaluates the mode count;
i.e., the number of resonant modes of a subsystem that are
available for the storage of energy in a frequency band.
The *mode count*, which is the fundamental quantity, may sometimes be estimated as a product of a *modal density* and the
frequency bandwidth for that experiment, Δf. The mode count
will not deviate very much from this estimate if the density
of modes is great enough or the bandwidth for the experiment
is wide enough so that the mode count estimate is at least
10 modes or so.

An important point to remember regarding the modes that
we deal with in SEA is that they should be "similar". This
means that they are of the same mechanical type, that they
are excited in the same way, and that they are coupled to
adjacent systems in similar ways. Suppose that a subsystem
has two types of modes -- say, flexural and torsional, and
that this subsystem is joined into other subsystems in such
a way that we suspect that the energies of the flexural and
torsional modes are not equal. Then the mode count of
flexural and torsional modes should not be added to give
the modal count of the subsystem in that frequency band, but
in fact, two subsystems should be defined, each with its own
mode count.

The coupling loss factor (CLF) is the parameter that
governs the power flow from one subsystem to another. Not
surprisingly, it depends upon the general mechanical parameters of both subsystems to which it refers. Chapter 14
gives a discussion of experimental procedures which may be

used to determine the CLF. The results of calculations of
CLF for a variety of system types -- both acoustical and
structural -- are also presented.

The chances are, however, that the designer who is
attempting to use SEA to obtain response estimates, may not
find his particular subsystem "connection" in the group
represented in Chapter 14, simply because the odds are not
in his favor. Relatively few cases have been worked out in
sufficient detail to allow direct application, and the number
of possible connections between subsystems is very large.
Often, the cases that are known may be an acceptable approximation for the purposes of estimation. The designer will
usually have to set up experiments to evaluate the CLF for
his situation, or use the references as a way of learning the
analytical techniques for a new theoretical derivation.

11.2   How Exactly Can (Must) We Know SEA Parameters?

The requirement for accuracy of the SEA parameters depends
upon the requirement for accuracy in response estimates. The
accuracy required in response estimates depends in turn upon
the use to be made of the estimate. Typical uses include
acoustical noise radiation, malfunction of electronic or
control components and structural fatigue.

The estimation of sound radiation usually involves
accuracy of the order of a few decibels (dB). This is
sufficient to determine the annoyance, reduction in personal
performance, or speech interference effects of noise. The
prediction of troublesome sound levels for any of these
criteria is the least demanding of all estimates of response,
since the form of response estimate is usually the product of
a ratio of loss factors and a ratio of modal counts. Thus,
a 10 percent uncertainty in the parameters could amount to
a 40 percent (or 1.5 dB) uncertainty in the estimate, even if
the response level of the "source" subsystem is known exactly.
Since there may be at least an uncertainty of a couple of
decibels in the "known" level, the uncertainty in the
radiated sound may be at least 3 dB or more.

The prediction of electronic malfunction or structural
fatigue requires much more accurate estimation of response.
For steady state vibration, a 1 dB change in vibration level
can amount to a 100 percent change in expected life. Thus,
the expected uncertainty in the estimate with only a 10 percent
uncertainty in the parameters is greater than desirable.

On the other hand, current estimates -- even those based on
test data from similar systems such as the Mahaffey-Smith
procedure [2] have uncertainties that are at least as large
as 3 dB. In order to predict the life of a structure or
component reliably, the SEA parameters must be known within
a few percent and the response of the "source" subsystem
should be known to an accuracy of less than 0.5 dB. It is
unlikely that such estimation accuracy is achievable,
particularly with the mix of missions characteristic of
current aircraft.

Typically, we may expect an accuracy of our estimates of
modal count and loss factors to be about 10 percent, whether
theoretically or experimentally determined. Thus, a purely
theoretical prediction of structural fatigue or equipment
malfunction cannot be achieved by the use of SEA. On the
other hand, the SEA estimates are at least as accurate as
other procedures, and in addition, the estimates retain
functional dependence on the acoustical and structural parameters. Consequently, one can get an indication of how the
system might be changed to reduce the vibration or noise, and
increase component life.

11.3 How to Use This Section

Part I and Section I of Part II have discussed the basic
theory of SEA and the use of SEA to predict response for
engineering purposes. This section does not attempt to
explain SEA or its computational procedures. Its purpose
is to present methods for finding values of the SEA
parameters -- both theoretically and experimentally -- and to
give some values and formulas that have been found from
earlier work.

Even though the report is a collection of formulas,
graphs and diagrams in topical format, it will be found
advisable to read through each chapter, even though only part
of the chapter may be totally relevant to the problem at hand.
The reason is that there is overlap in nomenclature and
descriptions of experimental techniques between the various
procedures and explanations. Rather than make the text too
wordy with a great deal of repetition, the discussions have
been kept as brief as possible. The result, however, is
that a reading of only the material of direct interest may be
somewhat confusing unless the rest of the chapter up to that
point is read also.

CHAPTER 12. THE DAMPING PARAMETER

12.0   Introduction

In this chapter, we provide methods, formulae and data for the damping or dissipation parameter in SEA. This is commonly measured by the "loss factor" η, but may be known or expressed in a variety of other parameters. The loss factor η is defined as the ratio of energy dissipated per second to the average energy stored in the system.

$$\eta = \frac{1}{2\pi f} \frac{\Pi_{diss}}{E_{tot}} = \frac{T \; \Pi_{diss}}{2\pi \; E_{tot}} \qquad (12.0.1)$$

where $T \; \Pi_{diss}$ is the energy dissipated per cycle of vibration having period $T = 1/f$.

The loss factor can be related to a number of other dissipation parameters that occur in vibration and wave analysis. These factors and their relation to η are shown in Table 12.1. This variety exists because of the importance of dissipation mechanisms in many fields and the natural ways of experimentally determining damping in each field. We shall use the loss factor η for damping almost exclusively in this chapter.

The information on damping is presented in four paragraphs in this chapter. Paragraph 2.1 deals with experimental techniques, which may be applied to the various configurations discussed the later paragraphs. Paragraph 2.2 is concerned with the damping that is naturally available in the material from which the system is constructed. In paragraph 2.3, we discuss damping in built-up structures, which are the kind that interest us in the estimation of response in aerospace applications. Finally, in Paragraph 2.4, we discuss the construction and damping properties of various special add-on treatments that are commercially available.

12.1 Measurement of Damping

Although there is a good deal of tabular and theoretical information available regarding the damping of various structures, the most commonly used method of determining damping is simply to measure it. Fortunately, the accuracy to which the damping is required to be known is not too great, and simple experimental methods can readily achieve them.

There are two basic experimental strategies involved in these measurements: steady-state and decay methods. The steady-state methods are as follows:

a. Frequency response bandwidth for a single mode of oscillation.

b. Frequency response irregularity of a multi-degree-of-freedom system.

c. Measurement of input power.

Technique (a) is the well known "half-power bandwidth" technique. If an experimental arrangement such as that shown in Fig. 12.1 is employed, then a single resonant response like that shown in Fig. 12.2 can usually be found. The loss factor is found from this plot as

$$\eta = \Delta f / f_o \qquad\qquad (12.1.1)$$

In order for the result (12.1.1) to be useful, we must be able to single out a single resonance, which requires that $\Delta f / \delta f \ll 1$, where δf is the average separation between resonant modes of the structure. The ratio $\Delta f / \delta f$ is called the "modal overlap" parameter. When it is large (greater than 2 or 3 for instance), the frequency response curve shown in Fig. 12.2 will not be realized because of the interference between individual resonance curves. In this instance, the response will have the appearance of the curve shown in Fig. 12.3.

If δf_{max} is the measured difference in frequency

between two adjacent maxima in the response curve, then one can demonstrate that the damping loss factor of the system is given by [16]

$$\eta \simeq 0.4 \ <\delta f_{max}>/f \tag{12.1.2}$$

where $<\delta f_{max}>$ is the average value of the separation for a number of adjacent maxima and f is the center frequency of this group of maxima. Parenthetically, we should note that when $\eta f/\delta f > 3$, (conditions of high modal overlap) an individual response maximum does not occur at a modal resonance frequency but is the result of the in phase response of a number of modes at the observation point.

Finally, damping can sometimes be inferred from a measurement of input power to a system. Such an experimental arrangement is shown in Fig. 12.4. A measurement of force and acceleration (or velocity) at the driving point allows one to determine Π_{in}. Measuring the rms acceleration of the structure allows the determination of $<v^2> = <a^2>/4\pi^2 f^2$. The loss factor is then given by

$$\eta = \Pi_{in}/ \ M<v^2>\omega \tag{12.1.3}$$

where M is the mass of the structure.

In addition to the steady state methods, there are transient methods for measurement of damping. These depend on relations between the loss factor and various measures of the rate of decay of a system as detailed in Table 12-1. If one wishes to observe the decay of a single mode of the system, then an arrangement like that shown in Fig. 12.5 may be used. If a single mode is to be studied, then the excitation may be a pure tone. The gate is used to cut-off the excitation, and may also be used to drive the position of the shaker armature out of contact with the structure. Such a provision is useful if the shaker itself provides enough structural damping to affect the measured decay rate.

If the modes are closely spaced so that δf is not much greater than ηf, the abrupt termination of the excitation will cause additional modes to be excited, and a decay curve like

that shown in Fig. 12.6 will result. Such a result can be
avoided by using broader bandwidths of noise excitation that
encompass several modes of vibration, or by using a gating
function that is not so abrupt, but still fast enough not
to interfere with the decay process.

In the case of high modal overlap, the strategy for
damping measurement must change, and this is the case for
steady state measurement also. In this case, even a pure
tone excites several modes, so that one may use several
shakers to simulate the proper generalized force for a given
mode. Such an approach is used for the lower modes of aircraft,
but is not very suitable for higher frequency panel type modes
of substructures. In the latter situation, one simply uses
a band of noise to ensure that many modes of vibration are
excited and decay together. In this case, an average loss
factor for the modal group is obtained.

The sensing of response is made with a microphone or
accelerometer, as appropriate for sound fields or structures.
A short time rms of the signal is taken (averaging time a
few milliseconds or less, depending on the frequency) and
the logarithm taken and the signal displayed. The decay
rate DR in dB/sec is found and the loss factor is found
from

$$DR = 27.3 \ f\eta \ .$$ (12.1.4)

A storage oscilloscope is useful for this measurement because
a graphic level recorder, widely used in sound measurements,
is too slow for many structural decay experiments.

12.2 Damping Values for Materials [17]

Most aerospace structures are constructed from
aluminum alloys which have loss factors in the range from
0.002 to 0.005. Such damping values are generally smaller
than the measured damping of built-up structures by a
sizable factor. Thus, it is generally presumed that the
damping of real structures due to the metal itself is of
little consequence in comparison to the damping due to joints,
rivet contacts, and other similar features.

In Fig. 12.7, we show a graph of values of loss factors for various materials. Metals have an internal loss factor ranging from 10^{-3} to 2×10^{-1}. The larger values are not encountered in aluminum alloys, but may be found in cast iron and special alloys of manganese and copper. Polymer materials have a similar range of loss factors, but are "softer" than metals by a factor of 10 or more. Elastomeric materials are a factor of 10^4 softer then metals, but have higher loss factors than do either metals or polymers, ranging from 0.1 to about 5. Polymers and elastomers have great significance in their role as components of add-on damping treatments, which will be dealt with later on.

The damping provided by the material from which an aero-space vehicle is constructed is so low compared to the values measured for built-up structures that we normally do not con-sider this form of damping to be significant. However, add-on damping treatments do have a high degree of material damping in their polymeric or elastomeric elements, and con-sequently, we are interested in the damping of such materials. In particular, the damping provided by these materials is frequency and temperature sensitive, and aerospace structures are excited over broad frequency ranges and are exposed to large changes in ambient temperature. This subject is well covered by many sources, but it is worthwhile to give a brief review of it here.

Polymeric materials undergo a transition in behavior in terms of their static stiffness from a soft, rubbery (or elastomeric) behavior to a harder, or glassy phase as a certain temperature Θ is passed. The transition is not sharp, but occurs gradually and is therefore, not a "normal" phase transition like that which occurs for example in water between its liquid and solid phases.

In addition, the dynamics of the material undergo a transition in frequency at any temperature about the trans-ition temperature of the material. That is, if the material is in its "rubbery" state at room temperature (and zero-frequency) then if the specimen is cycled quickly enough, the material does not have sufficient time to accommodate, and it acts dynamically as though it were in its more rigid, glassy form.

The time required for the polymer to "slip its bonds" and behave in a rubbery fashion is called the relaxation time for the material, denoted a_Θ. As a result of experimental studies on a variety of materials, the relaxation time a_θ for polymers has been found to follow the relation

$$\log a_\theta = -\frac{8.86(\theta - \theta_0)}{120 + (\theta - \theta_0)} \qquad (12.2.1)$$

where $\theta_0 = \theta_t + 50K$ (expressed in degrees centigrade) and θ_t is the transition temperature mentioned above. Thus, polymeric materials have a dynamic stiffness characterized by a single transition effect that is temperature dependent. The relaxation time gets smaller as the temperature increases, and the frequency of transition increases.

A diagram of the behavior of loss factor and modulus for a rubber material is shown in Figure 12.8 [17]. The tendency for the transition frequency f_c to rise as the temperatures is increased is clearly evident from the behavior of the maximum in the loss factor function and the line of inflection in the stiffness modulus. A great deal of data has been collected for various polymer materials and is available in standard reference works [17].

From the above discussion, it might be inferred that polymers in the glassy state are not good damping compounds. Generally, this is true although such materials may be used occasionally where some damping is desirable. Nevertheless, high modulus damping materials are desirable, particularly, in their application as unconstrained layers. To achieve such behavior an elastomer, such as polyvinyl chloride (PVC) is loaded with a stiff filler material. This filler has two important effects - it increases the overall rigidity of the material and it amplifies the strain in regions near the filler particles. Thus, the stiffness and damping of the combination are both increased [18].

The mechanisms of damping in metals include thermal conduction, grain boundary motion, molecular site transition, and dislocation oscillations. At larger amplitudes, non-linear effects such as plastic flow also occur. Because of the complexity of mechanisms, but more importantly, because of the wide range of activation energies for these processes, there is little possibility of developing such simple functions as that shown in Fig. 12.8 to illustrate the internal damping of metals. Accordingly, the best we can do in this instance is to refer to reference material in which values of material damping of various metals are tabulated [17].

12.3 Damping of Built-Up Structures

We note above that aerospace structures that are made up of aluminum alloy sheets, ribs, stringers and rivets will under test display a loss factor of the order of 0.02, whereas the loss factor of the metal itself is likely to be of the order of 0.002. Although the vehicle designer has very little control over the damping obtained this way, it is nevertheless useful to review what is known regarding the damping of built-up structures.

It is reasonably well established that the increased damping is due to the riveted joints of the structure. There are at least two theories regarding the source of this increased damping. One of these [19] assumes that the damping is due to surface slip and plastic deformation of the contacting asperities of the overlapping surfaces. The resulting damping is nonlinear. The damping in this case should increase as the level of vibration increases.

The second theory assumes that the dissipation is due to viscous flow in the region between the metal surfaces along the riveted joints. If a liquid is present, such as oil, this mechanism is fairly obvious - as the metal surfaces vibrate, the gap changes its depth in a cyclic fashion. To stay in contact, therefore, the liquid must flow in and out of the narrow gap, and viscous dissipation is substantial. But it turns out that even if no liquid is present, the flow of air in and out of the gap is capable of dissipating sufficient energy to account for the observed damping [20]. The obvious implication of this mechanism is that the damping will be reduced as the air pressure is reduced and so, we might expect less damping as altitude is increased.

The theory of damping by gas-pumping is rather complicated, but we can summarize the results here. The formal expression for the loss factor has only been found for high frequencies (when the flexural wavelength on the plate is of the order of a rivet spacing or less), but we may presume that the general parameter dependence will likely apply to lower frequencies also. The loss factor for the case in which the plate vibrates and the attached beam is assumed stationary is [20]

$$\eta = \frac{A_b c^2 \gamma P}{16\pi^3 f^3 A_p m_p \kappa_p h c_\ell} \quad H(\theta) \qquad (12.3.1)$$

where

A_b =   area of beam-plate overlap

A_p =   area of structural panel

m_p =   surface density of panel

κ_p =   radius of gyration of plate cross section

γ =   ratio of specific heats of gas

c_ℓ =   longitudinal wave speed in plate material

h   =   gap thickness

p   =   gas pressure

c   =   speed of sound in gas

The function $H(\Theta)$ is a measure of the ease of flow of the gas within the gap and $\Theta = h/\delta$ is the ratio of the gap depth to the length parameter

$$\delta = 2 \ (\nu/\omega)^{\frac{1}{2}}$$

where ν is the kinematic viscosity of the gas. A graph of this function is presented in Fig. 12.9.

A comparison of the prediction according to Eq. (12.3.1) and laboratory studies of the damping of a single beam riveted to a plate is shown in Fig. 12.10. In this the plate is 1/64 inch thick and the attached beam is 1/4 x 1 x 17 inches. The correspondence between the theoretical and measured values of damping is impressive. In particular, the known tendency for the observed value of damping to have a broad range of values near 0.01 for built-up structures is supported by this data and the calculations.

The results on air pumping are presented here because this is practically the only theory that gives reasonable predictions of damping in riveted, aerospace type structures. We do not wish to infer that this is a closed matter, however.

Certainly there are many cases in other areas of acoustical
and vibration engineering in which the damping in built-up
structures is not due principally to this mechanism.

12.4 The Damping of Add-On Systems

Since the energy of vibration tends to be inversely
proportional to damping an increase of damping is frequently
desired to reduce vibration amplitudes. Since the "natural"
loss factor will be of the order of 0.01, we must increase
the damping so that the loss factor of the structure becomes
of the order of 0.1 in order to have an appreciable effect
on the response. To achieve such a high level of damping,
however, one must usually resort to some kind of "add-on"
damping treatment. The performance of such treatments is the
subject of this section.

The most widely used add-on damping systems are

(1)   a free or unconstrained layer of damping material,
        applied either by troweling, spraying, or in the
        form of tiles;

(2)   a constrained layer of damping material, in the
        form of a tape with foil and elastomeric adhesive,
        or with the elastomer and constraining sheet applie
        separately;

(3)   spaced damping consisting of a spacing structure
        and either a free or constrained layer of damping.

(4)   A resonant damper designed to produce high damping
        for a particular mode of vibration.

Diagrams of these various damping systems are shown in Chapter
9 and in the present chapter also.

Free (Unconstrained) Layer [21]. It is convenient to
examine results for the unconstrained and constrained treatments separately, both with and without a spacing layer.
Some typical free layer configurations are shown in Fig. 12.11
The base panel is "1", the elastomer layer is "3" and, when
present, the spacer layer is "2". The loss factor of this
composite is (assuming stiffness of the damping layer is
small compared to the structural stiffness) [21]

$$\eta = \eta_3 \, \frac{E_3}{E_1} \, \frac{h_3 w_3 H_{31}^2}{I_1} \tag{12.4.1}$$

where

η_3 = loss factor of layer 3, discussed in paragraph 12.2

E_3 = Young's modulus of layer 3

E_1 = Young's modulus of layer 1

h_3 = thickness of layer 3

w_3 = width of layer (take as unity when treatment covers material)

H_{31} = distance between neutral axes of layers 1 and 3

I_1 = moment of inertia of layer 1.

The interpretation of Eq. (12.4.1) is straightforward and revealing. The first factor is a ratio of material parameters only, and we note that $\eta_3 E_3 / \eta E_1$ is the relative dissipation in the damping layer to the apparent dissipation in the base structure for equal strains. The desirability of combining "high" stiffness and loss factor in the damping material is obvious from this expression. The second factor is geometric - it is a ratio of the moment of inertia of the damping layer to that of the base structure. It represents the square of a strain amplification factor produced by the geometry of the configuration. Thus, the purpose of the spacing layer "2" is to increase the strain in the damping layer and consequently obtain more damping.

As a practial matter, it is possible to increase the damping by a factor of 10 or so by spacing the material away from the structure. At higher frequencies this "gain" will be less, because the spacing layer will tend to shear and not stretch the damping layer as shown in Fig. 12.11. A typical curve of loss factor achieved with a spacing structure is shown in Fig. 12.12. The frequency and temperature dependence of η will be that of η_3, as discussed in paragraph 2.2.

Constrained Layers. The use of a constraining layer above the damping layer is another device to amplify strain in the damping layer. In this instance, the constraining layer augments shear in the damping material rather than elongation as in spaced free layer damping. A comparison of constrained layer and free layer distortions is shown in Fig. 12.13. In aerospace applications, the damping is thin and also soft compared to the base structure and the constraining layer. In this circumstances a fairly complicated formula results for the loss factor of the composite structure, which is best presented graphically.

Fig. 12.14 shows the dependence of the composite loss factor on the "shear parameter" χ, a parameter that is essentially inversely proportional to frequency. The general formula for χ is

$$\chi = \frac{G_3 w_3}{h_3} \left(\frac{\lambda_b}{2}\right)^2 \left(\frac{K_1 + K_2}{K_1 K_2}\right) \tag{12.4.2}$$

where $K_i = E_i h_i$ is the extensional stiffness of the layer, λ_b is the bending wavelength, G_3 is shear modulus of the damping layer (damping is due to shear in this configuration, not to extension as in Eq. (12.4.1) and Fig. 12.11) w_3 is the width of the damping layer (coverage is not necessarily complete) and h_3 is the thickness of the damping layer. Subscripts correspond to the diagram in Fig. 12.15. Since

$$\lambda_b \sim 1/f^{\frac{1}{2}} \ ,$$

one has $\chi \sim f^{-1}$.

The optimal value of the shear parameter χ is given by

$$\chi_{opt} = (1+Y)^{-\frac{1}{2}} (1+\beta^2)^{-\frac{1}{2}} \tag{12.4.3}$$

where Y is the "structural parameter". The value of Y for various configurations of interest is shown in Fig. 12.16. The parameter β is the loss factor for shear in the damping material and is usually very close in the value to η_3. The maximum composite loss factor η_{max} is given by

$$\eta_{max} = \beta Y \left[2 + Y + 2/\chi_{opt}\right]^{-1}. \qquad (12.4.4)$$

This expression is graphed in Fig. 12.17 as a function of damping layer loss factor for various configurations (structural parameters).

By combining the information in Figs. 12.14, 12.16, and 12.17, along with material damping information from paragraph 12.2, the damping of any constrained layer system may be estimated. Constrained layers are preferred for aerospace application because they tend to have less weight for a given damping. This is demonstrated in Fig. 12.18, which shows a comparison of damping versus added weight for free and constrained layers. It is evident that as long as the acceptable weight increase is less than 10 per cent, one would achieve much more damping from the constrained layer. The penalty that one pays for this performance is the narrow frequency range of performance. That is, the constrained layer treatment must be more closely "tuned" to the desired frequency range.

TABLE 12.1

COMPARISON OF COMMONLY USED DAMPING MEASURES

Dissipation Descriptor	Symbol	Units	Relation to η
Loss Factor	η	-	η
Quality Factor	Q	-	$1/\eta$
Critical Damping Ratio	ξ	-	$\frac{1}{2}\eta$
Reverberation Time	T_R	seconds	$2.2/f\eta$
Decay Rate	DR	dB/sec	$27.3 f\eta$
Logarithmic Decrement	δ	nepers/sec^{-1}	$\pi f\eta$
Wave Attenuation*	m	nepers/m	$\pi f\eta/c_g$
Mechanical Resistance**	R	newt-sec/m	$2\pi f\eta M$
Damping Bandwidth (Half-Power)	BW	hertz	$f\eta$
Imaginary Part of Modulus $E_r + iE_i$	E_i	newt/m^2	$E_r \eta$
Acoustical Absorption Coefficient***	α	-	$(8\pi f V/cA)\eta$

*c_g is group velocity for system in meters/sec.

**M is the system mass

***A is area of walls, V is room volume, c is speed of sound.

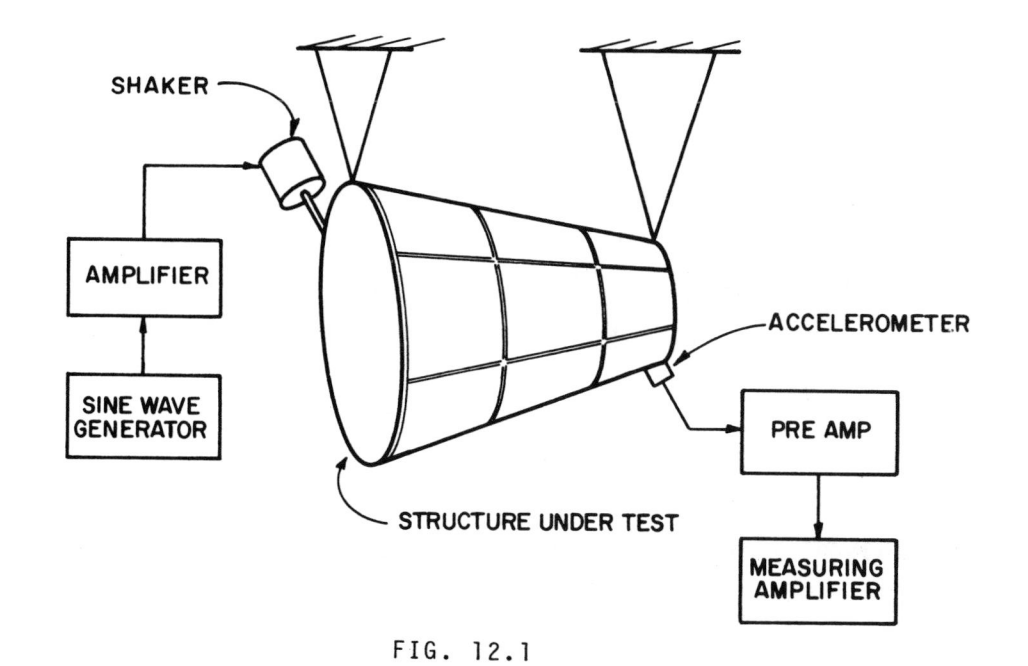

FIG. 12.1

EXPERIMENTAL ARRANGEMENT FOR MEASURING FREQUENCY
RESPONSE OF A STRUCTURE

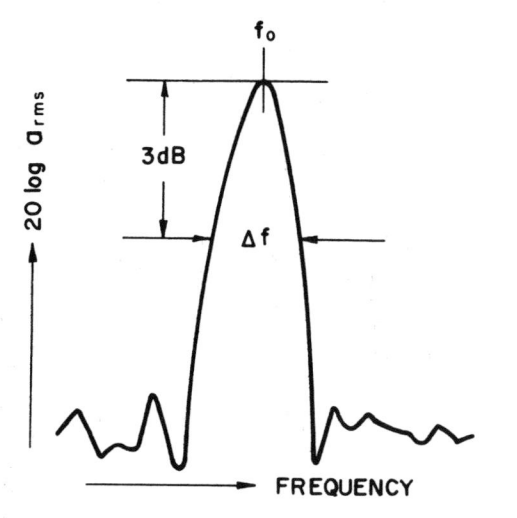

FIG. 12.2

RESPONSE CURVE FOR A SINGLE MODE

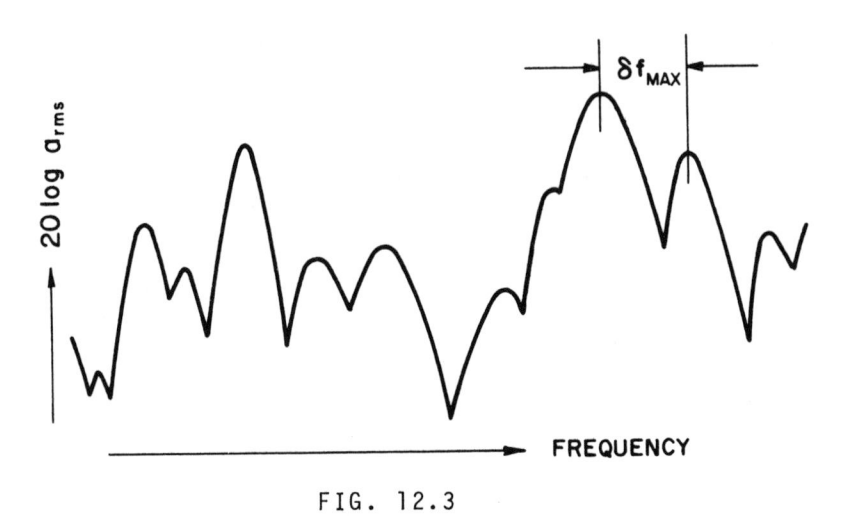

FIG. 12.3

RESPONSE CURVE FOR A SYSTEM WITH HIGH MODAL OVERLAP

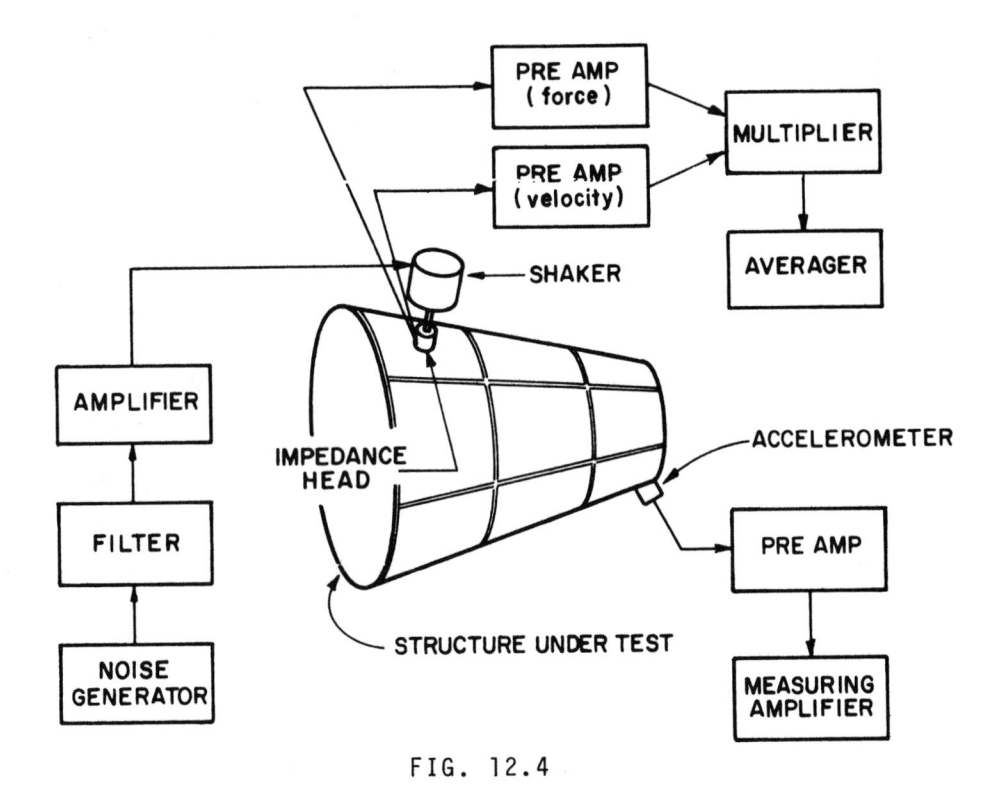

FIG. 12.4

ARRANGEMENT FOR DETERMINING DAMPING FROM INPUT
POWER AND AVERAGE RESPONSE

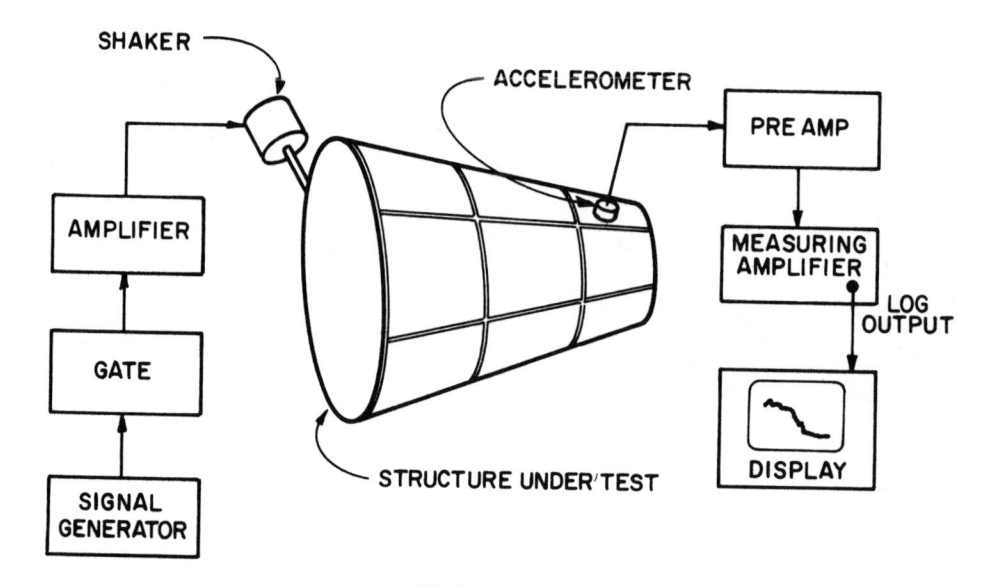

FIG. 12.5

ARRANGEMENT FOR MEASURING DECAY RATE OF STRUCTURAL VIBRATIONS

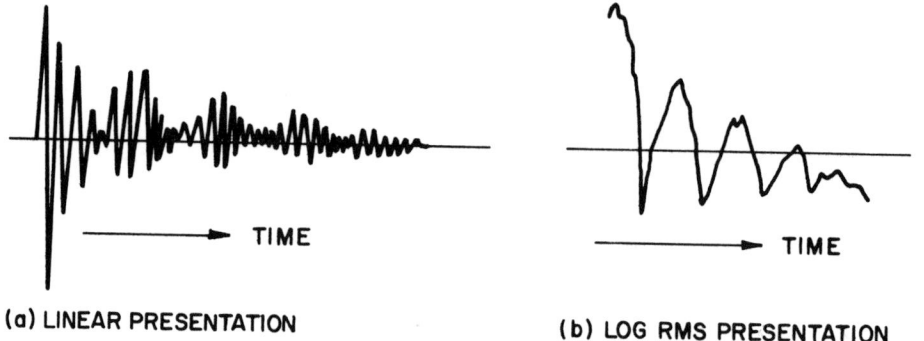

(a) LINEAR PRESENTATION   (b) LOG RMS PRESENTATION

FIG. 12.6

SIMULTANEOUS DECAY OF TWO MODES HAVING SLIGHTLY DIFFERENT NATURAL FREQUENCIES

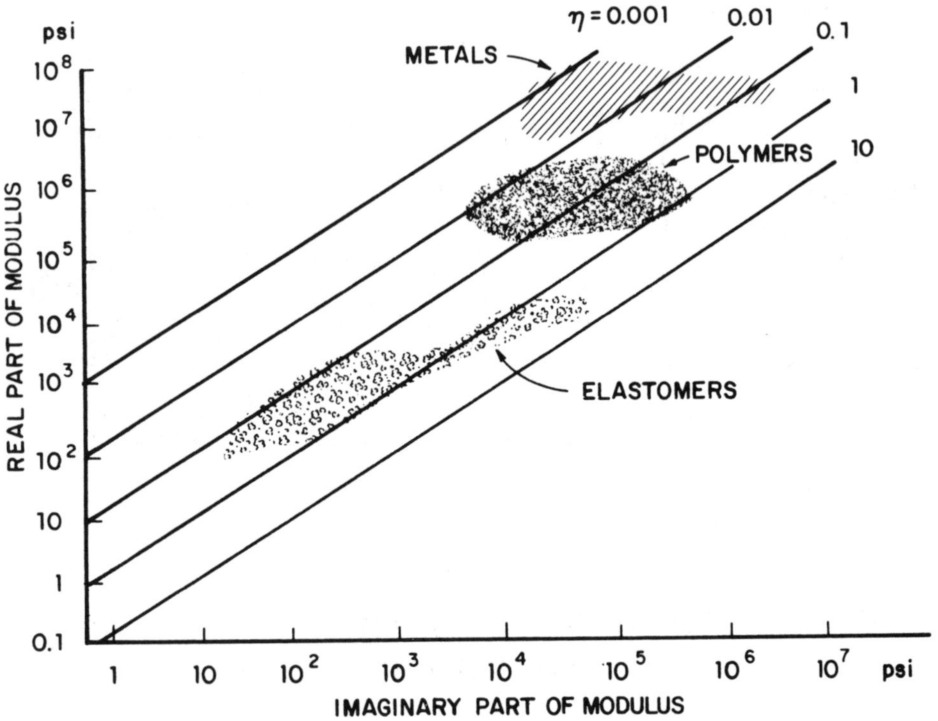

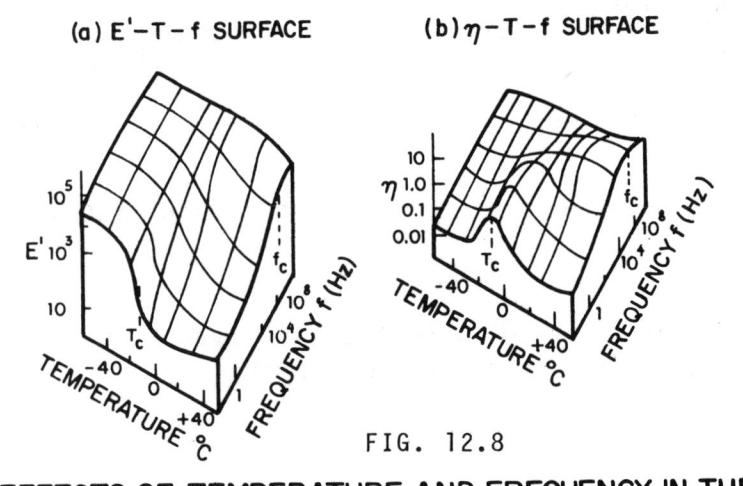

FIG. 12.7

LOSS FACTORS AND ELASTIC MODULI FOR VARIOUS CLASSES OF MATERIALS (ADAPTED FROM REF 3)

(a) $E'-T-f$ SURFACE   (b) $\eta-T-f$ SURFACE

FIG. 12.8

EFFECTS OF TEMPERATURE AND FREQUENCY IN THE TRANSITION REGION ON THE STORAGE MODULUS AND LOSS COEFFICIENT OF BUNA N RUBBER

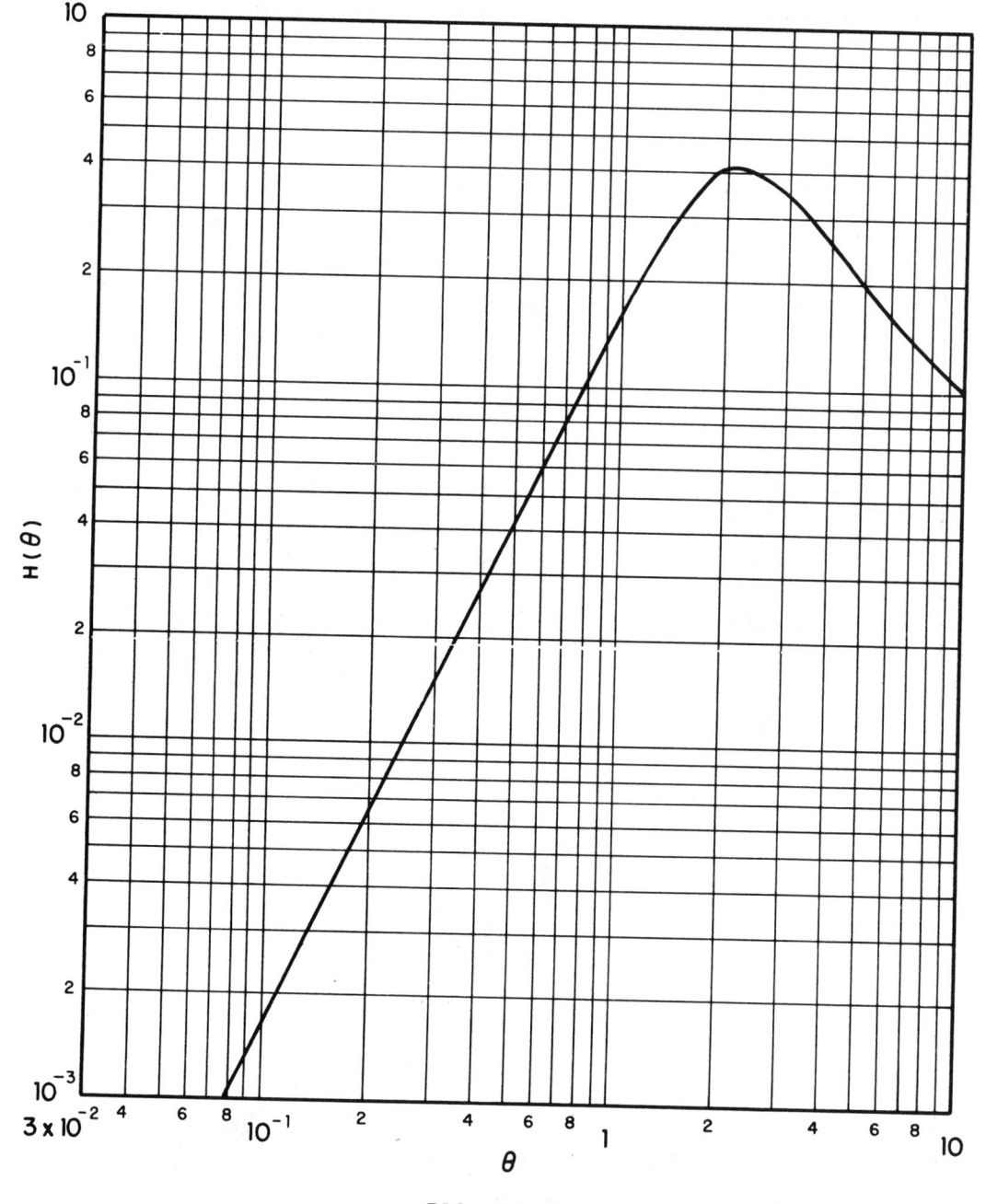

FIG. 12.9

DEPENDENCE OF LOSS FACTOR ON RATIO OF GAP BETWEEN
PLATE AND BEAM TO VISCOUS BOUNDARY LAYER THICKNESS

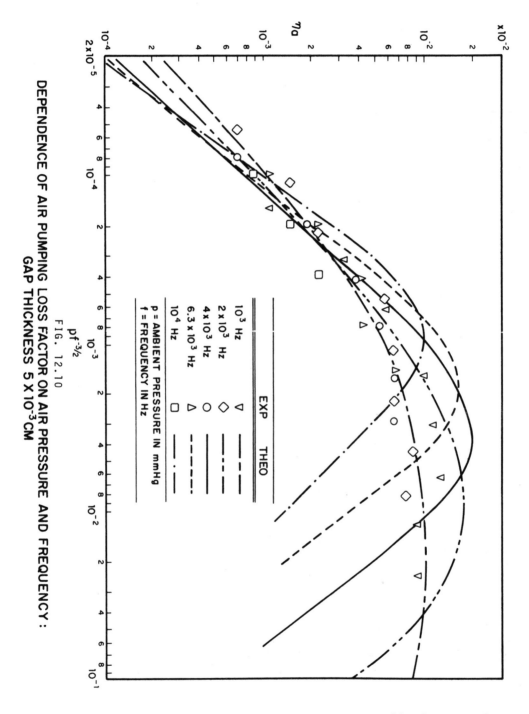

DEPENDENCE OF AIR PUMPING LOSS FACTOR ON AIR PRESSURE AND FREQUENCY:
GAP THICKNESS 5 X 10⁻³CM

FIG. 12.10

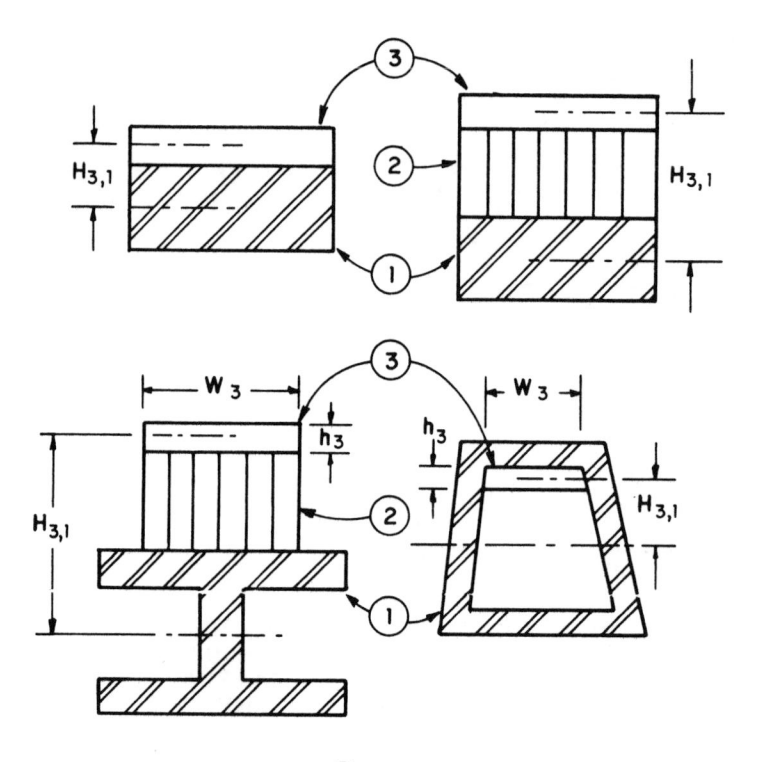

① PRIMARY STRUCTURE
② SPACER
③ DAMPING MATERIAL

FIG. 12.11

TYPICAL FREE-LAYER CONFIGURATIONS

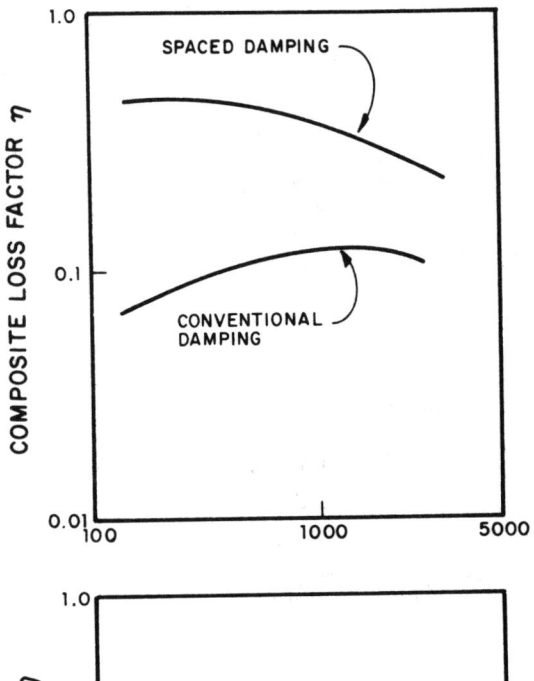

(a) COMPOSITE LOSS FACTOR VS.
FREQUENCY FOR SPACED
DAMPING AND CONVENTIONAL
DAMPING APPLIED TO A 12'x2"x0.25"
SOLID STEEL BEAM. DAMPING
MATERIALS AND COVERAGE THE
SAME IN BOTH CASES.

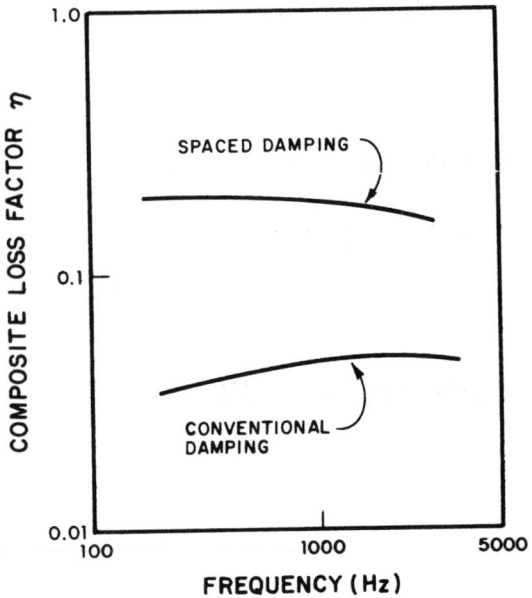

(b) COMPOSITE LOSS FACTOR VS.
FREQUENCY FOR SPACED
DAMPING AND CONVENTIONAL
DAMPING APPLIED TO A 0"x1/8"
SQUARE TUBULAR BEAM.
DAMPING MATERIALS AND
COVERAGE THE SAME IN
BOTH CASES

FIG. 12.12

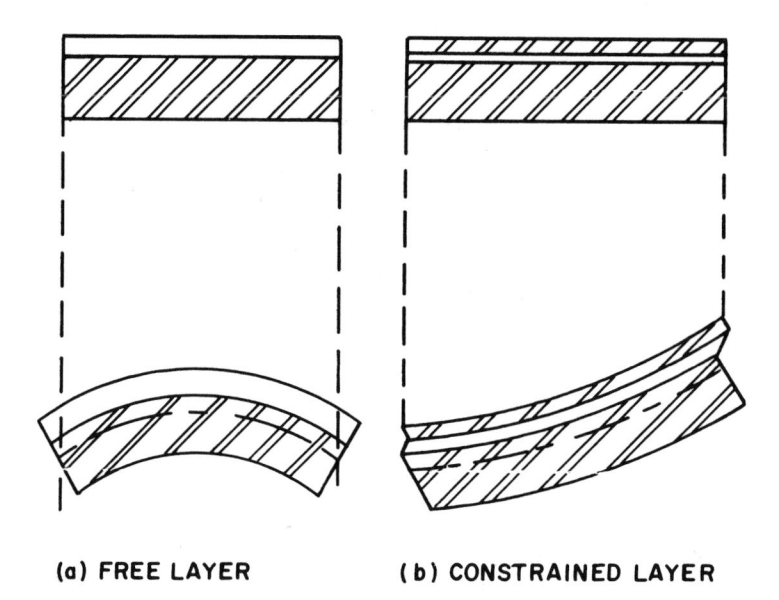

(a) FREE LAYER   **(b) CONSTRAINED LAYER**

FIG. 12.13

SYSTEMS WITH VISCOELASTIC LAYERS

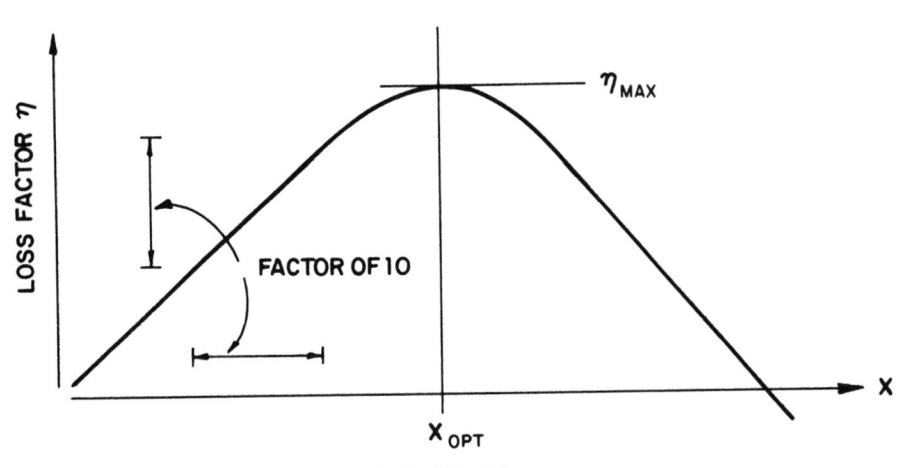

FIG. 12.14

COMPOSITE LOSS FACTOR OF CONSTRAINED DAMPING LAYER ON BASE STRUCTURE

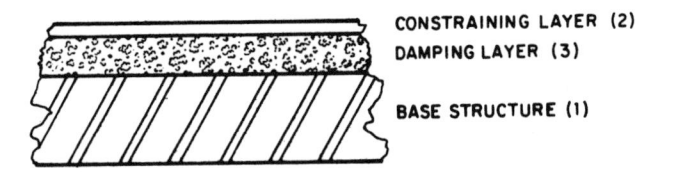

FIG. 12.15

CONFIGURATION OF CONSTRAINED LAYER COMPOSITE

STRUCTURE		STRUCTURAL PARAMETER γ	STRUCTURE		STRUCTURAL PARAMETER γ
PLATES WITH THIN TAPE		0.1	PLATES WITH THICKNESS RATIO OF 2 TO 1		2.0
HIGHLY ASYMMETRIC COMPOSITES	$\gamma = \dfrac{h_1}{2h_2}$	$\gamma = \dfrac{h_1}{2h_2}$	IDENTICAL PLATES OR THIN-CORE SANDWICH		3.0
I-BEAMS WITH EQUAL FLANGES AND DEPTH RATIO OF 2 TO 1		0.75	SYMMETRIC FINNED COMPOSITES		> 3
IDENTICAL I-BEAMS		1.0	THICK-CORE SANDWICH		> 3

FIG. 12.16

TYPICAL VALUES OF STRUCTURAL PARAMETER

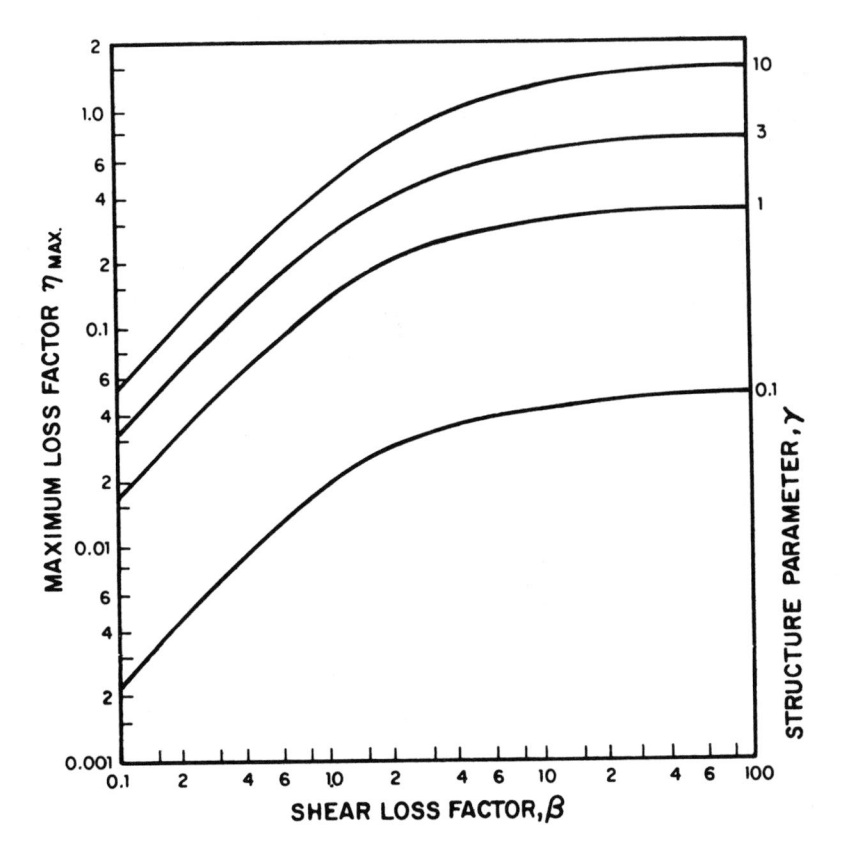

FIG. 12.17

**MAXIMUM DAMPING OF
CONSTRAINED LAYERS VS. MATERIAL LOSS FACTOR**

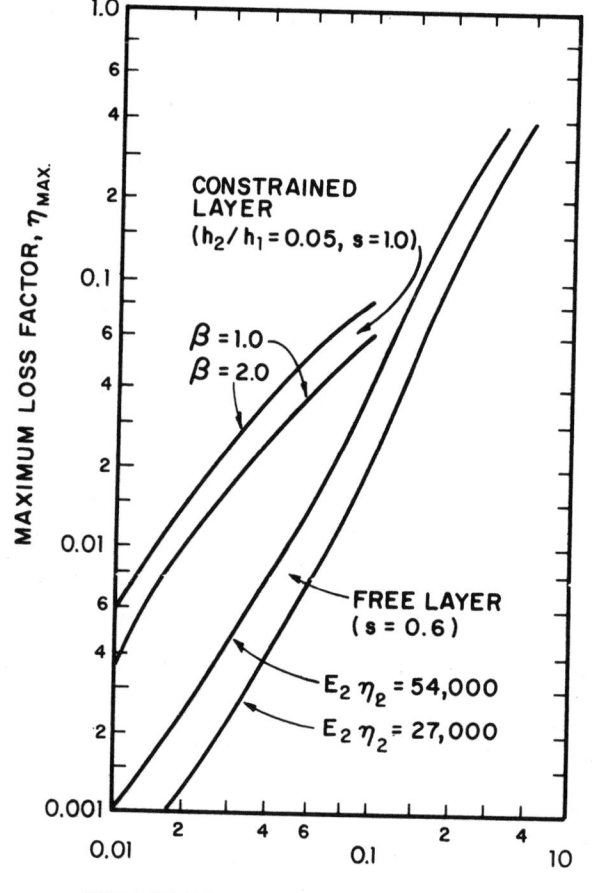

FIG. 12.18

COMPARISON OF FREE AND CONSTRAINED – LAYER TREATMENTS
ON STEEL PLATES. SUBSCRIPTS 1 AND 2 REFER TO PLATE
AND DAMPING MATERIALS RESPECTIVELY

CHAPTER 13. EVALUATING THE MODE COUNT

13.0 Introduction

The mode count is important in response prediction because it tells us how many resonant modes are available to store energy in the subsystem being studied. In Part I and Section I of Part II, we have seen that this parameter enters most of the equations for response, if not directly, then indirectly through the consistency relation for the coupling loss factors. Mode count also enters relations between modal response and subsystem response in a frequency band containing many modes.

When the mode count is experimentally determined, it can enter the predictions directly, but when we determine it theoretically, we are more likely to calculate the modal density, the number of modes per unit frequency interval. Indeed, we more often find the modal density appearing in the theoretical expressions rather than the mode count. Although the modal density is a simpler quantity to calculate, we should not lose sight of the fact that it is the expected mode count that is fundamental, and that systems that have small mode count may still be modeled as SEA systems even though the concept of modal density is inappropriate.

The mode count (or modal density) is the SEA parameter that is generally the easiest to determine. In frequency regions in which there are few modes and the modal density may be difficult to calculate, a simple measurement on the system (or a model of it) can often be made. In frequency regions in which the modes are very dense, and it is difficult to resolve them experimentally, one can usually make an adequate theoretical estimate. Thus, by a combination of experimental and calculational procedures, one can usually determine the mode count to acceptable accuracy.

Division of the problem into regimes of sparse and dense modal distribution is reminiscent of the situation with regard to damping. In fact, it turns out to be true for all the parameters, including coupling loss factors as well, that different functional forms or different experimental methods for their evaluation are required, depending on the density of modes.

In this chapter, we first discuss the measurement of mode count and subsequently, the theoretical formulae for important cases of acoustical spaces and structural subsystems.

13.1 Experimental Procedures for Determining Mode Count

The simplest and most common set-up for finding mode count is that shown in Fig. 13.1. The idea is simply to excite the system with a pure tone, usually at a point, and observe the response at a second point as the excitation frequency is slowly swept over the band of interest. For the mechanical structure in Fig. 13.1, the excitation is with a shaker and response is measured by an accelerometer. If the system is an acoustical system, excitation would be with a small loud-speaker or horn driver and response would be measured with a microphone.

Even though simple, this procedure can given erroneous or incomplete results unless certain precautions are taken. To make certain that as many modes as possible are excited; the excitation should be near a free boundary for a structure or a rigid wall for a sound field, since all the modes have anti-nodes along such boundaries. To further enhance the likeli-hood of all modes being excited and sensed, several source and receiver locations should be chosen. If the graphs are overlaid, modes that barely respond on one sweep will respond more on another. Even if some of the modes are missed, the estimates are only sensitive to the relative error in mode count. Thus, if two out of twenty modes are missed, only a 10% error in the mode count will result. This error would normally result in a 1 dB error in a response prediction.

Some other features of the experimental set up should be mentioned. For example, the use of a logarithmic presentation on the chart is preferable to a linear presentation. With logarithmic output, modes that respond with small amplitude are much more likely to be seen. Also, the sweep must be slow enough so that two modes are not simultaneously excited and beat against each other, or so that a mode that is decaying does not beat with the non-resonant response at the shaker frequency. Such beats will cause additional peaks on the chart which might be mistaken for modal resonances.

The procedure just described works as long as the separation between modes δf is at least three times the bandwidth of a mode, $\pi/2 \ \eta f$ or

$$\pi \eta f / 2 \delta f \le 0.3. \qquad\qquad (13.1.1)$$

If the average modal separate δf is too small, then too many modes will be missed by the sine-sweep test. In this event, an alternate, but less proven technique may be useful.

The alternate procedure relies upon the result that, as shown in Part I, the average input power due to a force $L(t)$ in a band Δf is [see Eq. (2.2.24)]

$$\Pi_{in} = <L^2>_t / 4M\delta f \qquad\qquad (13.1.2)$$

and the mean square response is, therefore,

$$<v^2> = \Pi_{in} / 2\pi f \eta M = <L^2>_t / 8\pi f \eta M^2 \delta f \qquad\qquad (13.1.3)$$

Thus, if $<v^2>$ is measured for a known $<L^2>$ and η, one can infer δf. In this procedure, the difficulty is to control the mean square force, as discussed in Chapter 12. Even if one cannot precisely determine $<L^2>$, the relative response of the system in different bands can determine relative values of δf. Then if the exact value of δf is known in a few bands, it can be determined in all the bands.

13.2 Mode Counts of Acoustical Subsystems

The average frequency separation between modes in a one-dimensional acoustical system of length ℓ and sound speed c is

$$\delta f = c / 2\ell \qquad\qquad (13.2.1)$$

This formula applies to any cylindrical space in which one-

dimension is significantly greater than the other two and the
wavelength of sound is greater than the greatest cross-
dimension ℓ_c:

$$\lambda > 2\ell_c \text{ or } f_c < c/2\ell_c \, . \qquad\qquad (13.2.2)$$

The modal separation for a thin, flat acoustical space
of area A and perimeter P is given by

$$1/\delta f = 2\pi f A/c^2 + P/c \qquad\qquad (13.2.3)$$

This formula is valid as long as the wavelength of sound is
greater than twice the depth ℓ_D of the space

$$\lambda > 2\ell_D \text{ or } f < c/2\ell_D \qquad\qquad (13.2.4)$$

The modal separation for a three-dimensional space of
volume V, surface area A and total edge length ℓ is given by

$$1/\delta f = 4\pi f^2 V/c^3 + \pi f A/2c^2 + \ell/8c \qquad\qquad (13.2.5)$$

These formulas and the cases they relate to are shown in
Figs. 13.2, 13.3 and 13.4. Generally, these results are
adequate to supply good estimates of mode count for subsystems or elements that have standing sound waves as the
energy storage mechanism.

13.3 Flat Structures

The simplest structure that we can write the modal density
for is the homogeneous flat panel of area A

$$\delta f = hc_\ell/\sqrt{3} \; A \tag{13.3.1}$$

where h is the thickness of the panel cross section and c_ℓ is the longitudinal wavespeed in the panel material. If the material is steel, aluminum or glass, then

$$c_\ell = 17,000 \text{ ft/sec}$$

and if h is expressed in ft, one has

$$\delta f = h(ft) \times 10^4/A(ft^2) \tag{13.3.2}$$

If the panel is not homogeneous, but is of a sandwich or other layered construction, a similar but more complicated relation can be given. The wave velocity c_ϕ must be known as a function of frequency $\omega = 2\pi f$. It is then substituted into

$$\delta f = \frac{c_\phi^2}{\omega A} \left[1 - \frac{\omega}{c_\phi} \frac{dc_\phi}{d\omega} \right]^{-1} \tag{13.3.3}$$

For example, it often happens for such structures that the wave speed increases as a power of the frequency

$$c_\phi = B \, \omega^n \tag{13.3.4}$$

in which case

$$\delta f = \frac{B^2 \omega^{2n-1}}{A} (1-n)^{-1} \tag{13.3.5}$$

When n = 0 (acoustical case, Figure 13.3), $\delta f \sim 1/f$ as found
in Equation (13.2.3). When n = 1/2 (homogeneous plate), then
δf = const., as found in Eq. (13.3.1). For other constructions, a dynamical analysis of the panel construction
must be carried out to determine c_ϕ.

In some cases, a plate will be orthotropic, i.e.,
stiffer along one direction than the other, as shown in Fig. 13.5.
If these two directions are labelled "1" and "2", the modal
density $n(f) = 1/\delta f$ is in effect an average value between
isotropic plates of the two stiffnesses

$$1/\delta f = \frac{1}{2} \omega A \left\{ \frac{1}{c_1^2} \left(1 - \frac{\omega}{c_1} \frac{dc_1}{d\omega}\right) + \frac{1}{c_2^2} \left(1 - \frac{\omega}{c_2} \frac{dc_2}{d\omega}\right) \right\}$$

(13.3.6)

or, if the construction is such that $c_1, c_2 \sim \omega^n$, we have

$$1/\delta f = \frac{1}{2} \omega A (1-n) \left\{ \frac{1}{c_1^2} + \frac{1}{c_2^2} \right\}$$

(13.3.7)

where c_1 and c_2 are the wavespeeds in the two principal
directions. Obviously if $c_1 = c_2$, we revert to Eq. (13.3.5).

13.4 Mode Count of Curved Structures

Since most aerospace structures consist of complete or
segments of curved shells, the effect of curvature on mode
count of structures is of great significance. Although a
fair amount of work has been done on such structures, we are
not able to write simple formulas for the mode count in all
instances. In this paragraph we provide some of the available
results that are of greatest interest in the present context.

Circular Cylinders. The modal density of a circular
cylinder may be considered a variation of the modal density of

a flat structure of the same area and construction (homo-geneous, layered, etc.) but modified to the account of the effects of curvature. The form of the modal density is shown in Fig. 13.6. The characteristic frequency that separates the high frequency flat plate behavior from low frequency cylinder behavior is the so-called ring frequency f_{ring}, defined by

$$f_{ring} = c_\ell/2\pi a \qquad\qquad (13.4.1)$$

where c_ℓ is the longitudinal wavespeed introduced in paragraphs 3.3 and   a   is the radius of curvature of the cylinder.

Using results from Szechenyi [22] we can present simple curve-fitting formulas for the modal density, which are as follows:

$$\frac{\delta f(\text{flat plate})}{\delta f(\text{cylinder})} = (\frac{f}{f_{ring}})^{\frac{1}{2}}; \quad \frac{f}{f_{ring}} \leq 0.5$$

$$= 1.4 \frac{f}{f_{ring}}; \quad 0.5 < \frac{f}{f_{ring}} < 0.8$$

$$= 0.8 \frac{0.1}{F-1/F} \{F \cos [\frac{1.75}{F^2} (\frac{f_{ring}}{f})^2]$$

$$- \frac{1}{F} \cos [1.75 F^2 (\frac{f_{ring}}{f})^2]\};$$

$$\frac{f}{f_{ring}} > 0.8$$

$$(13.4.2)$$

where the quantity F represents the band limits over which
the average is taken from

$$F \frac{f}{f_{ring}}$$

to

$$\frac{1}{F} \frac{f}{f_{ring}}$$

Thus, F = 1.222 for a 1/3 octave band and 1.414 for an octave
band.

   Doubly Curved Shells. Occasionally a shell or a segment
of a shell will be in the form of a doubly curved surface, as
shown in Fig. 13.7. A torus is an example of such a shell.
Torodial sections are frequently used to join a cylinder with
a spherical cap, for example. Suppose the shell has area A,
longitudinal wavespeed c_ℓ, and cross-sectional radius of
gyration κ. Then if R_1 and R_2 are the two principal radii
of curvature, we can define two ring frequencies

$$f_{r1} = \frac{c_\ell}{2\pi R_1} \; ; \quad f_{r2} = \frac{c_\ell}{2\pi R_2} \tag{13.4.3}$$

and we arbitrarily assume $R_2 > R_1$ so that $f_{r2} < f_{r1}$.

   With this hypothesis, Wilkinson [23] has computed the
modal densities for a doubly curved shallow shell.

$$f < f_{r2} < f_{r1}; \quad n(f) = 1/\delta = 0. \tag{13.4.4}$$

There are no resonant modes in this frequency regime. In
the frequency range "between" the two ring frequencies,
$f_{r2} < f < f_{r1}$

$$\frac{\delta f\,(\text{flat plate})}{\delta f\,(\text{shell})} = \frac{\sqrt{2}}{\pi}\ \frac{f^{\frac{3}{2}}}{(f^2-f_{r1}^2)^{\frac{1}{2}}(f_{r1}-f_{r2})^{\frac{1}{2}}}\ F(\pi/2,\kappa)$$

$$(13.4.5)$$

and in the frequency range above the two ring frequencies $f_{r2} < f_{r1} < f$

$$\frac{\delta f\,(\text{flat plate})}{\delta f\,(\text{shell})} = \frac{2}{\pi}\ \frac{f^2 F(\pi/2,\ 1/\kappa)}{(f^2-f_{r1}^2)^{\frac{1}{2}}(f+f_{r1})^{\frac{1}{2}}(f-f_{r2})^{\frac{1}{2}}}$$

$$(13.4.6)$$

where $\kappa = (f+f_{r1})^{\frac{1}{2}}\,(f-f_{r2})^{\frac{1}{2}}/\{2f(f_{r1}-f_{r2})\}^{\frac{1}{2}}$,and $F(\pi/2,\xi)$ is the elliptic integral of the first kind, defined by [24]

$$F(\pi/2,\xi) = \frac{\pi}{2}\int_{0}^{\pi/2}(1-\xi^2\sin^2 t)^{-\frac{1}{2}}\ dt. \qquad (13.4.7)$$

When $|\xi| < 1$ this can be expressed in the series [24]

$$F\left(\frac{\pi}{2},\xi\right) = \frac{\pi}{2}\sum_{n=0}^{\infty}\frac{\Gamma(n+\frac{1}{2})\,(2n)!}{\Gamma(\frac{1}{2})\,(n!)^3}\,2^{-2n}\xi^{2n} \qquad (13.4.8)$$

When $f \gg f_{r1}$, then $|\xi| \gg 1$ and Eq. (13.4.6) becomes

$$\frac{\delta f\,(\text{flat plate})}{\delta f\,(\text{shell})} = 1 \qquad\qquad (13.4.9)$$

Thus, the modal density again becomes equal to that of a flat plate of equal area at high frequencies and short wavelengths.

Two interesting special cases can be derived from the relations (13.4.4, 5, 6). If $R_2 \to \infty$ and $f_{r2} \to 0$, the shell modal density becomes for $f < f_r^2$.

$$\frac{\delta f \text{(flat plate)}}{\delta f \text{(cylinder)}} = \frac{\sqrt{2}}{\pi} \frac{f^{\frac{3}{2}} F[\pi/2, \sqrt{(f+f_r)/2f_r}]}{(f^2-f_r^2)^{\frac{1}{2}} f_r^{\frac{1}{2}}} \quad (13.4.2')$$

and for $f > f_r$.

$$\frac{\delta f \text{(flat plate)}}{\delta f \text{(cylinder)}} = \frac{2}{\pi} \frac{f^2 F[\pi/2, \sqrt{2f_r/(f+f_r)}]}{(f^2-f_r^2)^{\frac{1}{2}}(f+f_r)^{\frac{1}{2}}f_r^{\frac{1}{2}}} \quad (13.4.2'')$$

On the other hand, $R_1 = R_2$ for a spherical cap, and the middle range vanishes to give, for $f < f_r$

$$\frac{\delta f \text{(flat plate)}}{\delta f \text{(sphere)}} = 0 \quad (13.4.10)$$

and for $f > f_r$

$$\frac{\delta f \text{(flat plate)}}{\delta f \text{(sphere)}} = \frac{2f^2}{\pi(f^2-f_r^2)} F(\frac{\pi}{2},0) = \frac{f^2}{f^2-f_r^2} \quad . \quad (13.4.11)$$

For a sphere, therefore, the modal density approaches its asymptotic flat plate value very quickly above the ring frequency and there is a total depletion of modes below the ring frequency.

Conical Shells and Shells of Varying Radius of Curvature. A large number of structural shells of aerospace interest are sections of surfaces of revolution in which the curvature varies along a coordinate. It is possible to develop a general theory for the mode shapes and resonant frequencies for such shells as recently reported by Pierce [25] and Germogenova [26].

When the radius of curvature increases linearly along the axial coordinate, the shell is a cone. It is not possible to find a general expression for the modal density for these structures, but as shown in Chapter 15, it is possible to find the mode count for particular situations.

Although general formulas for mode count are not available for shells of revolution, many situations can be worked out to a satisfactory degree of accuracy by following the procedures in the references.

13.5 One-Dimensional Structure

A commonly encountered aerospace structure is the one-dimensional beam, stringer, or frame member. Although the one-dimensional geometry tends to simplify the mode count prediction, the dynamics of such structures can be rather complex. A simple beam of rectangular or circular cross section has separate propagation modes for flexural, torsional, and longitudinal wave propagation. However, channel or hat-section stringer or frame has coupling between these motions so that the actual vibrational modes combine the "pure" wave types.

Fortunately, the additive properties of mode count simplifies the process of prediction. We may calculate mode counts on the hypothesis that the coupling does not occur and then add the mode counts for the pure modes to obtain our estimate for the mixed motions when the geometry of the beam cross-section or end conditions are such that we expect such coupling to be important.

The average modal separation δf of a one-dimensional system of length ℓ and phase speed $c_\phi(\omega)$ is given by

$$n(f) = 1/\delta f = \frac{2\ell}{c_\phi}\left[1 - \frac{\omega}{c_\phi}\,\frac{dc_\phi}{d\omega}\right] . \tag{13.5.1}$$

In the case of longitudinal, the phase speed c_ϕ is independent of frequency:

$$1/\delta f(\text{long'l}) = 2\ell/c_\ell; \quad c_\ell = \sqrt{E/\rho} \qquad (13.5.2)$$

Where E is the Young's modulus and ρ is the lineal density of the beam. In the case of torsional waves

$$1/\delta f(\text{torsional}) = 2\ell/c_t; \quad c_t = c_s \sqrt{J/\kappa_\phi^2 A} \qquad (13.5.3)$$

where J is the moment of rigidity for the cross-section in question, κ_ϕ is the radius of gyration about the c.g. of the cross-section, A is the area of the cross-section of the beam, and c_s is the speed of shear waves in the material. The quantity J for various cross-sections of interest may be found in references on the strength of materials.

The bending wave speed is generally dispersive, as it is in plates. Again, if $c_\phi \sim \omega^n$, then

$$1/\delta f(\text{flexural}) = \frac{2\ell}{c_\phi}(1-n) \qquad (13.5.4)$$

which is ℓ/c_ϕ for $n = 1/2$. For more complex cross-sectional shapes, the more general formula (13.5.1) must be used.

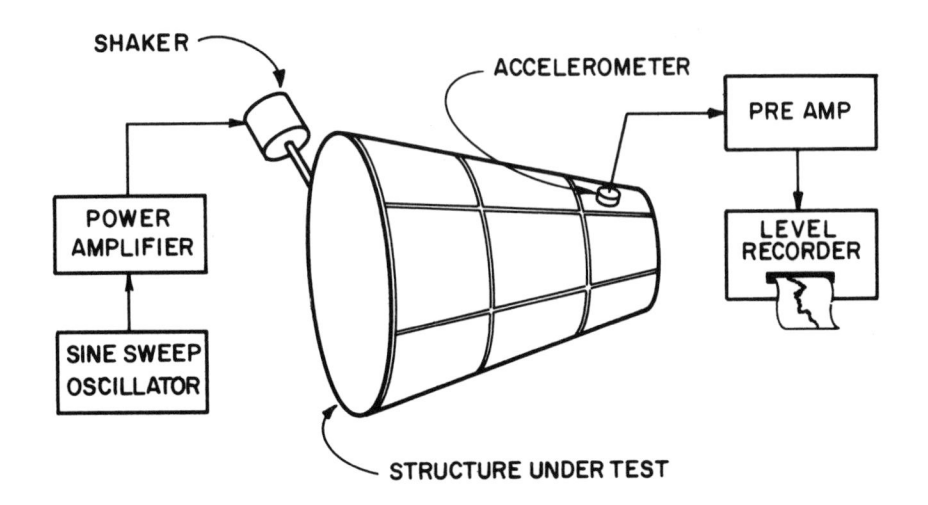

FIG. 13.1

ARRANGEMENT FOR FINDING MODE COUNT
BY SINE SWEEP TECHNIQUE

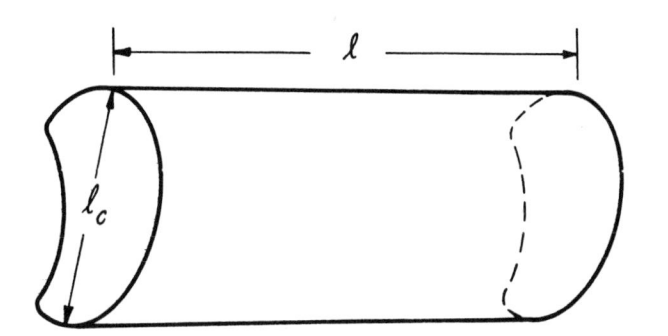

FIG. 13.2

ACOUSTICAL SUBSYSTEM ; 1-DIMENSIONAL CYLINDER

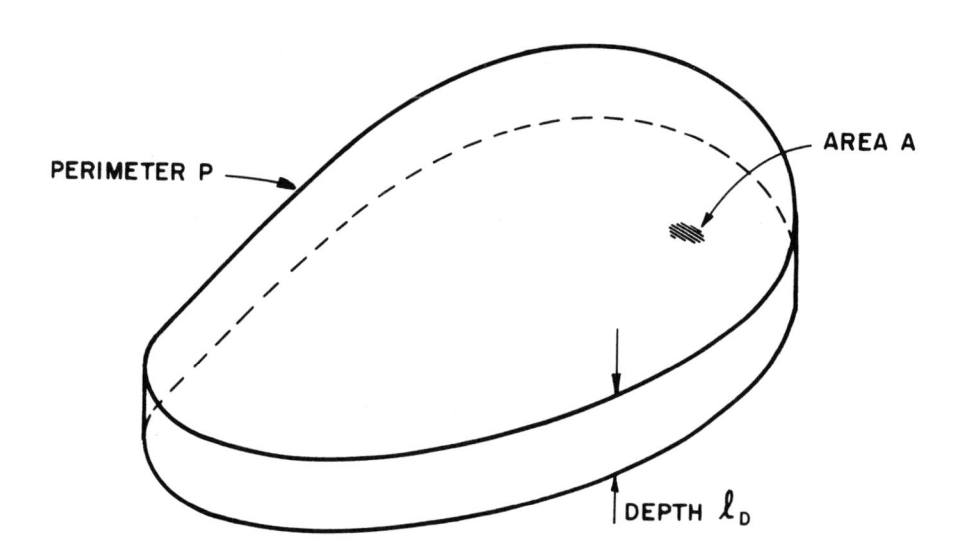

FIG. 13.3

ACOUSTICAL SUBSYSTEM; 2-DIMENSIONAL CAVITY

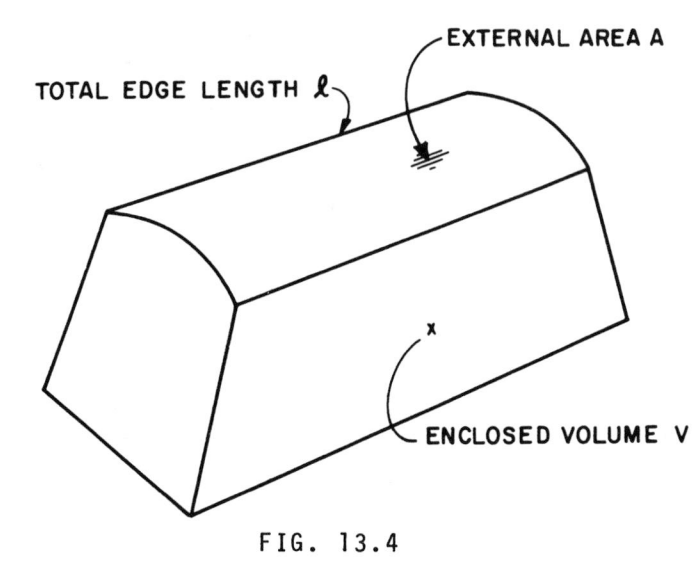

FIG. 13.4

ACOUSTICAL SUBSYSTEM : 3-DIMENSIONAL

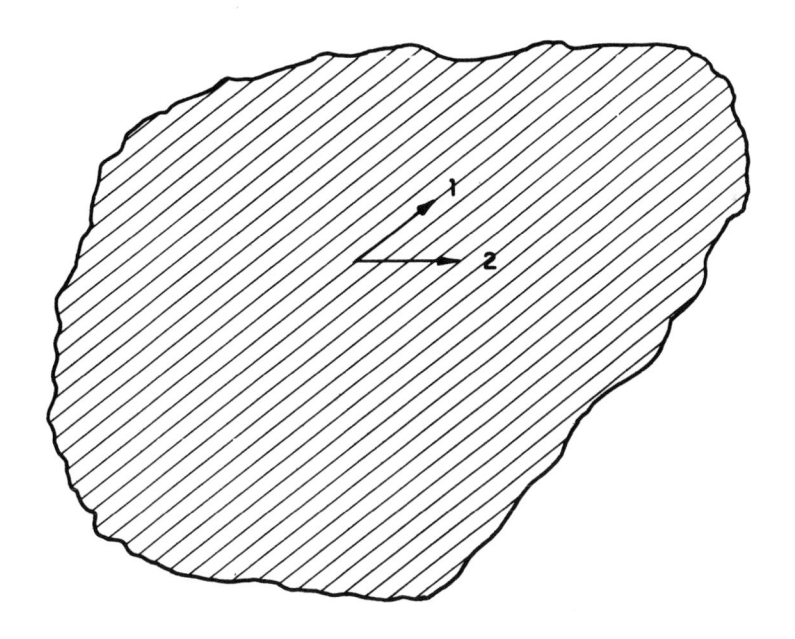

FIG. 13.5

ANISOTROPIC PLATE : PHASE SPEED c_1
IN 1-DIRECTION, c_2 IN 2-DIRECTION

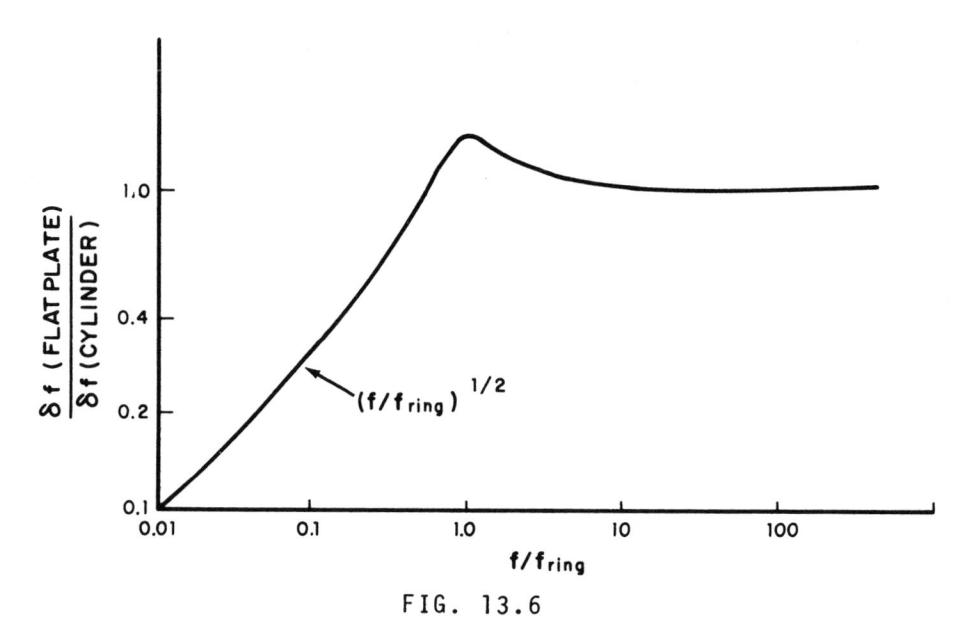

FIG. 13.6

MODAL DENSITY OF CYLINDER

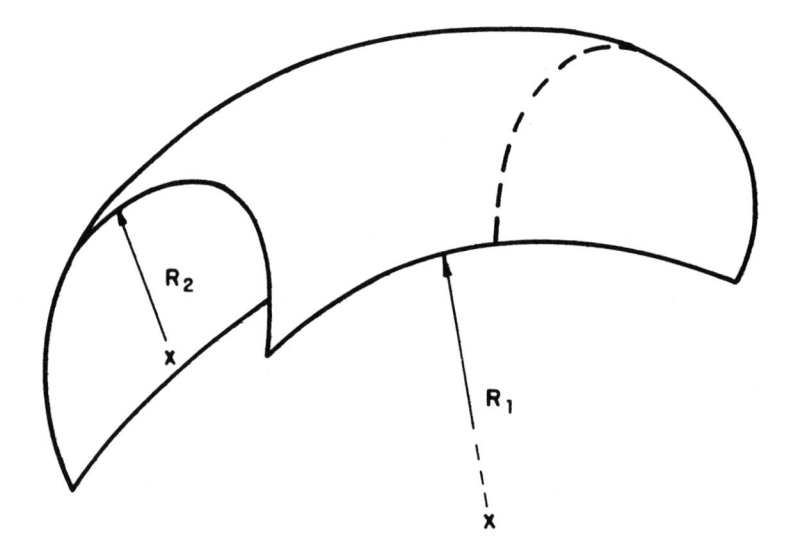

FIG. 13.7

**SECTION OF A TORUS, A SHELL WITH TWO CONSTANT
RADII OF CURVATURE**

CHAPTER 14. EVALUATING COUPLING LOSS FACTORS

14.0 Introduction

This chapter is concerned with presenting known values of the coupling loss factor (CLF) and methods for obtaining the CLF. In contrast to modal densities and damping, parameters that are well known outside the context of SEA, the coupling loss factor is uniquely associated with SEA. As we have seen in Part I and Part II, Section I, however, it is sometimes possible to relate the CLF to parameters (such as structural input impedances) that may have been evaluated for other purposes.

There are a variety of ways that the CLF may enter a calculation - as a "radiation resistance", as a ratio of power flow to stored energy, or as a frequency average over junction impedance functions. In addition, there is a variety of acoustical and mechanical systems that are of interest in aerospace applications that may be joined together. This richness of possibilities makes it very difficult to give an exhaustive listing of CLF values that will cover all potential cases of interest. Nevertheless, we present here known CLF data on some systems of interest and indicate wherever possible how information on other systems may be generated either by experiment or calculation.

The chapter is organized by the kinds of subsystems that are in contact with each other. Thus, we begin with coupling between two acoustical spaces. In this case, it happens that there is a great deal of experimental and theoretical data available on the CLF, or more precisely, a related parameter, the transmission loss. The second category of coupled systems is that in which one is structural and the other is a sound field. In this case, the coupling loss factor is related to the radiation resistance of the structure, for which there is a fair amount of information in the literature, some of which is experimental, but most is theoretical.

The final paragraph of the chapter is concerned with structure-structure interactions. These interactions are the most difficult for several reasons. First, experiments on such interaction are difficult to carry out and the number of reported results is fairly small. Second, the

kinds of motion may be quite complex for some structures and
the "impedances" may be matrix quantities. Third, the formal
derivation of CLF values for these cases can be extremely
complicated, so that theoretical results, while conceivable,
may require such extensive analyses that we are reluctant
to do the work, particularly in a preliminary design situation.

14.1 Coupling Between Acoustical Spaces

In acoustical spaces, the response is measured by the
m.s. pressure in a band, $<p^2>$. If the dimensions of the
space (call it subsystem 1) are greater than an acoustical
wavelength, then the power incident upon a wall of A_w is

$$\Pi_{inc} = <p_1^2> \; A_w/4\rho c \qquad\qquad (14.1.1)$$

where ρ is the density of air and c is the speed of sound.
The transmissibility of the wall is defined as

$$\tau = \Pi_{trans}/\Pi_{inc} = 4\rho c \; \Pi_{trans} \; <p_1^2> \; A_w \qquad (14.1.2)$$

where Π_{trans} is the power transmitted through the wall into
the receiving acoustical space.

The transmitted power can be related to the CLF since,
in the absence of flow back from the receiving system (subsystem 2) to the "source" cavity (subsystem 1), an
assumption that is made in most acoustical transmission
studies, one may also write

$$\Pi_{trans} = \omega\eta_{12}E_1 = 2\pi f\eta_{12} \; <p_1^2> \; V_1/\rho c^2 \qquad (14.1.3)$$

and, consequently, comparing (14.1.2) and (14.1.3), we can make the association:

$$\eta_{12} = \tau c A_w / 8\pi f V_1 \tag{14.1.4}$$

All of the voluminous data on transmission loss (TL),

$$TL \equiv -10 \log \tau \tag{14.1.5}$$

becomes, therefore, a source of information on the coupling loss factor between acoustical spaces [27].

Unfortunately, as large as this data base is, it is primarily oriented to building constructions (gypsum board, brick, plaster, etc.) and relatively little data by comparison is available for aerospace constructions. Also, building constructions tend to be formed from flat panels so that the curved geometries characteristic of aerospace construction are not well represented. For these reasons, it will often be necessary to determine η_{12} experimentally for systems representative of aerospace configurations.

Experimental Determination of CLF for Acoustical Spaces. The experimental procedure for finding η_{12} is to excite one of the cavities with a source of sound as shown in Fig. 14.1 (usually the loudspeaker is driven by a band of noise) and to infer τ from the m.s. pressure in the two spaces. Since the transmitted sound power into cavity 2 must be dissipated there, one has

$$\Pi_{trans} = \frac{\langle p_2^2 \rangle V_2}{\rho c^2} \, 2\pi f \eta_2 \tag{14.1.5}$$

where η_2 is the loss factor and V_2 is the volume of cavity 2. This parameter is usually measured by a decay rate experiment of the kind discussed in Chapter 12, using the relation

$$DR \text{ (dB/sec)} = 27.3 \, f\eta_2 , \tag{14.1.6}$$

Using (14.1.5), (14.1.2) and (14.1.4), we then obtain

$$\tau = \frac{<p_2^2>}{<p_1^2>} \quad \frac{8\pi fV_2}{cA_w} \eta_2 = \eta_{12} \quad \frac{8\pi fV_1}{cA_w} \qquad (14.1.7)$$

or,

$$10 \log \eta_{12} = L_{p_2} - L_{p_1} + 10 \log \eta_2 + 10 \log V_2/V_1$$

$$(14.1.8)$$

where, of course, L_p is the sound pressure level. Thus, a simple measurement of sound pressure levels in the cavities and of decay rate of the receiving space is sufficient to determine the CLF. One should be careful, however, that the difference in sound pressure level $L_{p_1} - L_{p_2}$ should be at least 10 dB to ensure that the assumption that all the transmitted power is dissipated in the receiving space is satisfied.

Data on CLF for Aerospace Structures. The transmission loss of a thin flat aluminum panel has been measured and calculated by Crocker [28] whose data is shown in Fig. 14.2. A flat aluminum structure has a TL minimum at the critical frequency

$$f_c = 12,500/h \text{ (mm)} \qquad (14.1.9)$$

where h is the thickness of the panel in millimeters. For frequencies an octave or more below f_c, the TL is well approximated by the so-called "mass law" formula [29]

$$TL_M = 20 \log f + 20 \log W - 33 \qquad (14.1.10)$$

where f is the frequency in hertz and W is the mass density of the panel in lb/ft^2. Since aerospace structures are fairly thin, we are usually interested in the frequency range well below f_c and this mass law formula will give an acceptable approximation for TL.

Eq. (14.1.10) is derived from the non-resonant transmissibility of the panel, but one must sometimes include a resonant transmissibility, particularly if the structure has many reinforcing ribs or frames. The resonant transmissibility is given by

$$\tau_{res} = \frac{2}{\pi} \frac{\rho^2 c}{m_s^2} \frac{V_1}{A_w} \frac{\sigma_{rad}^2}{\eta_w} \qquad (14.1.11)$$

where σ_{rad} is the radiation efficiency of the structure, m_s is the mass per unit area of the structure (wall) and η_w is its loss factor. The parameter $\sigma_{rad} = R_{rad}/\rho c A_w$, and R_{rad} is evaluated from data presented in paragraph 14.2.

For frequencies above the critical frequency, the transmission loss (or coupling loss factor) depends on the damping of the wall η_w in the form [30]

$$TL(f > f_c) = TL_M + 10 \log \eta_w + 10 \log \left(\frac{f}{f_c} - 1\right) + 3 \qquad (14.1.12)$$

where TL_M is given by Eq. (14.1.10) and η_w is to be determined by the methods of Chapter 12. Since most aerospace panels have a fairly high critical frequency, we are not ordinarily interested in the condition $f > f_c$. On the other hand, lightweight sandwich constructions used in fatigue resistant designs may have a much lower critical frequency. For such constructions

$$f_c \simeq 7000/t \quad (mm) \qquad (14.1.13)$$

where t is the total thickness of the aluminum sandwich in
mm. Thus, an 8 mm thick sandwich will have a critical
frequency of 900 Hz. We may well be concerned with finding
the CLF in frequency bands up to 4 or 5 kHz, in which case
we would wish to use Eq. (14.1.12).

The relations (14.1.10) and (14.1.12) apply an octave
or so away from $f = f_c$. To "fill in" the range $f \approx f_c$, a
transition curve like that appearing in the data shown in
Fig. 14.2 can be used. The resulting curve will usually
provide an estimate of TL that is accurate enough for
estimation purposes.

<u>Cylindrical Structures.</u> Sound transmission through
cylindrical structures is complicated by an additional
frequency parameter, the ring frequency

$$f_r = c_\ell / 2\pi \dot{a} \tag{14.1.14}$$

where a is the radius of the cylinder. As noted in Chapter 13,
above this frequency the cylinder acts as a flat plate, so
that the flat plate CLF formulas apply. At and below f_r,
membrane stiffness effects cause changes in the transmission
loss, as may be seen in the experimental data shown in Fig.
14.3. Deviations from the prediction according to Eq.
(14.1.10) and the data at and below f_r may be taken as
representative of the effect of the ring frequency on TL. [22].
Near f_r, the TL is reduced 2 to 3 dB and below f_r the TL is
increased by 3 to 4 dB as compared to flat plate values.

14.2 Coupling Between Structures and Acoustical Spaces

The CLF between a structure and a sound field is most
simply expressed by the average radiation resistance of the
structure interacting with an infinite space. This is an
acceptable approximation when the wavelength of sound is a
third or less a typical dimension of the cavity. At lower
frequencies, a mode-to-mode CLF may have to be developed.

When the radiation resistance of the structure is a
suitable measure, the CLF between the structure and the
acoustical cavity, η_{sa}, is given by

$$\eta_{sa} = R_{rad}/\omega M_s \qquad\qquad (14.2.1)$$

where M_s is the mass of the structure. This radiation resistance R_{rad} has been calculated for several structural systems of interest in aerospace applications. The coupling from the acoustical space to the structure can be found from the consistency relation,

$$\eta_{as} = \eta_{sa} \, N_s/N_a \qquad\qquad (14.2.2)$$

where N_s and N_a are mode counts for the structure and the acoustical space respectively, as determined in Chapter 13.

Radiation from Finite Flat Panels. The radiation resistance of one side of a flat panel of area A_p and perimeter P is given by [31]

$$R_{rad} = \rho c A_p \left\{ \frac{\lambda_c P}{\pi A_p} \, \frac{2}{\pi} \, \sin^{-1} \left(\frac{f}{f_c} \right)^{\frac{1}{2}} \right\} \beta \qquad (f < f_c)$$

$$= \rho c A_p \{1 - f_c/f \}^{-\frac{1}{2}} \qquad\qquad (f > f_c) \, (14.2.3)$$

where f_c is the critical frequency given in Eqs. (14.1.9) and (14.1.13) and λ_c is the wavelength of sound at the critical frequency. The parameter β is related to edge fixation. If the edges are simple supports, $\beta = 1$; if they are clamped $\beta = 2$. Usually these values will bracket more realistic mounting conditions, for which one may use $\beta = \sqrt{2}$. This formula applies at frequencies on octave or more above the fundamental panel resonance frequency.

Radiation from a Support on the Panel. When a supporting member such as a stringer or frame is attached to the panel, the radiation at frequencies below the critical frequency is increased. The amount of increase depends on the stiffness of the stiffener and its lineal mass. The increment in radiation resistance is given by

$$\Delta R_{rad} = \rho c A_p \left\{ \frac{4\lambda_c L}{\pi^2 A_p} \left(\frac{f}{f_c}\right)^{\frac{1}{2}} D\left(\frac{f}{f_c}\right) \right\} \qquad (14.2.4)$$

where L is the length of the stiffener and D is a function that depends on the lineal mass m of the stiffener and the mass per unit area of the panel m_s according to the parameter

$$M_o = 2 m/\lambda_c m_s \quad . \qquad (14.2.5)$$

We have graphed $10 \log D(f/f_c)$ in Fig. 14.4. The plot in this Figure assumes that the critical frequency of the beam is $0.01f_c$, the critical frequency of the panel. For aerospace structures, this would mean a beam critical frequency of 100 Hz or so, a reasonable value for most constructions.

    Radiation Resistance of Cylinders. The radiation resistance of a cylinder of area A_c above the critical frequency f_c is the same as for a flat structure above f_c, and is given by Eq. (14.2.3)

$$R_{rad} = \rho c A_c (1-f_c/f)^{-\frac{1}{2}}; \quad f > f_c \qquad (14.2.6)$$

    As noted earlier, the critical frequency of aerospace structural panels is usually well above the ring frequency. In this event, the structure behaves as a flat plate in the frequency range $f_R < f < f_c$ and the following radiation resistance formula applies:

$$R_{rad} = \rho c A_c \left\{ \frac{4\lambda_c}{\pi^2 A_c} \left(2\pi R + L\right) \sin^{-1}\left(\frac{f}{f_c}\right)^{\frac{1}{2}} \right\} \beta;$$

$$\left(f_R < f < f_c \right) \qquad (14.2.7)$$

where R is the radius of the cylinder and L is the
length of any stiffeners on the cylinder. The parameter β
relates to fixation of the supports in the same way that it
did in Eq. (14.2.3). [Note that if the critical frequency
is less than the ring frequency (a situation that can occur
for equipment pods) than this range corresponds to
$f_c < f < f_R$ and Eq. (14.2.6) applies.]

When $f_r < f_c$ and $f < f_r$, we can write the following
expression for the radiation resistance of a cylinder [22]

$$R_{rad} = \frac{\sqrt{3}\rho c A_c}{2\beta} \left(\frac{f}{f_r}\right)^{\frac{3}{2}} \frac{f_r}{f_c} \qquad (14.2.8)$$

where

$$\beta = 2.5 \sqrt{\frac{f}{f_R}} \quad \text{for} \quad f < 0.5 \; f_r$$

$$\beta = 3.6 \; \frac{f}{f_r} \quad \text{for} \; 0.5 \; f_r < f < 0.8 \; f_r.$$

Using Eqs. (14.2.6, 7, 8), one can estimate the coupling of
the curved surface of a cylinder with either an internal or
external sound field. The radiation resistance of the flat
ends of the cylinder may be estimated from Eqs. (14.2, 3, 4).

Departure from Simplified Results at Low Frequencies .
As noted in Chapter 12, the mode count in frequency bands is
reduced as one lowers the frequency of interest, particularly
if proportional bandwidth filters are used. The result is
that modes have less of a chance of "overlapping" each
other and the coupling between systems shows substantial
variations. Under these circumstances, one may have to calculate mode-to-mode values of CLF. There are not many
instances for which this has been done, but there is work by
Fahy [32, 33] on such coupling between the vibration of
structures and the acoustical modes of the contained fluid.

The frequency below which the simplified values of radiation resistance presented above may be in error has been estimated for the cases of the coupling between a wall of a box-like enclosure and the enclosed fluid. If the panel has thickness h and dimensions a x b and the box has dimensions a x b x c, then the limiting frequency f_ℓ is found from [32]

$$f_\ell = f_c [c^4/2\pi^2 (\eta_s + \eta_a)(a+b)(abc)\ f_c^4 (\Delta f/f)]^{2/7} \qquad (14.2.9)$$

where f_c is given by Eq. (14.1.9) and (14.1.10), η_a and η_s are the loss factors of the acoustical space and the structure respectively, and Δf is the bandwidth in which the data is taken. Normally such data is taken in constant percentage bands so that $\Delta f/f$ is a known constant.

The limiting frequency for cylindrical structures differs from that of flat structures because of the increased importance of membrane stresses. The expression for f_ℓ in this instance is given by [21]

$$f_\ell \simeq f_r [c^2/\pi^2 \ell a^2 (\eta_a + \eta_s) f_r^3 (\Delta f/f)]^{\frac{1}{2}} \qquad (14.2.10)$$

where ℓ is the length of the cylinder, a is its radius, and f_r is the ring frequency given in Eq. (14.1.14). The formulas in (14.2.9) and (14.2.10) are designed to be conservative, and the actual departures from predictions based on the CLF formulas of paragraph 14.2 may well occur at an octave or so lower than f_ℓ as predicted by these formulas.

In the range $f < f_\ell$, the calculation of CLF between the sound field and the structure becomes very complex and highly specialized to the geometry and structural details. The references should be consulted for details on the computational procedures to be followed in such cases.

14.3 Coupling Between Structural Subsystems

The most common interface between subsystems in flight

vehicles is in the form of a mechanical connection between the structural elements. The number of possible subsystems to be connected together (cylinders, plates, cones, beams, etc.) and the variety of kinds of connectors (rivet lines, brackets, welds, etc.) results in quite a large array of structural connections of possible interest to the designer. In this section, we quote results from various theoretical and experimental studies of the coupling loss factor for structures. We also indicate how some of the existing results might be extended to cover other cases of interest.

The first system to be discussed is a structural beam (stringer or frame) that separates two panels. Since the beam will reflect some of the flexural wave energy incident on it and transmit the rest, one can define a transmissibility τ for the wave energy in the same way that such a parameter was introduced for the transmission between two acoustical spaces. If the mechanical power incident on one side of a beam or length L is Π_{inc} and the power transmitted to the second side of the beam is Π_{trans} the transmissibility

$$\tau \equiv \Pi_{trans}/\Pi_{inc} \qquad\qquad (14.3.1)$$

is related to the coupling loss factor η_{12} (where the source side is "1" and the receiving side is "2") by the relation

$$\eta_{12} = \frac{c_g L}{2\eta^2 f A_1} \ \tau \ . \qquad\qquad (14.3.2)$$

Most measurements and calculations of transmission of vibrational energy are designed to evaluate τ, from which η_{12} may be inferred.

Experimental Methods for Finding η_{12}. There are three principles that may be applied in the measurement of the coupling loss factor for structural connections.

(1) The receiving system may be damped sufficiently so that very little of the energy received by it will "return" to junction. Alternately, one may say that $\eta_2 \gg \eta_1 + \eta_{12}$. In this event, the

apparent damping of subsystem "1" will be $\eta_1 + \eta_{12}$. If η_1 can be found separately (estimated, measured by "clamping" subsystem "2", etc.) then η_{12} can also be determined.

(2)   The receiving system is very heavily damped at its boundaries and a steady state excitation is applied to system 1. The mean squared velocity of system "2" is a measure of the transmitted power and the mean squared velocity of system 1 measures the power incident on the boundary. The experiment determines τ, from which η_{12} may be found in accordance with Eq. (14.3.2)

(3)   The receiving system is sufficiently damped so that its modal energy E_2 is appreciably less than the modal energy of the source structure, E_1, but the damping is not so great as in cases (1) and (2). In this circumstance, the vibration of subsystem "2" is almost directly proportional to the value of η_{12}.

As an example of case (1) above, consider the experimental set-up shown in Fig. 14.5 [34]. In this experiment, the damping of plate 1 is measured by a decay rate experiment before attachment to plate number 2. The results are plotted as η_1. The damping of plate 2 is increased as much as possible and plate 1 is connected to it in the desired way. The new loss factor η is also plotted and for this situation is equal to

$$\eta = \eta_{12} + \eta_1 \,. \qquad\qquad (14.3.3)$$

Thus by taking the difference between η and η_1, the CLF is found. The CLF has also been interpreted as a transmissibility in Fig. 14.5 according to Eq. (14.3.2). A comparison with the theoretical value of τ_{th} is shown in this figure.

The transmissibility as measured in case (2) above has been found experimentally for beam structures by Heckl [35]. A typical set-up is shown in Fig. 14.6. The source section has no applied damping, the receiving section is heavily damped by a "wedge" of material. The excitation is with a shaker (shown as F in the drawing). The change in velocity level across the beam

$$\Delta L = 10 \log \langle v_1^2 \rangle / \langle v_2^2 \rangle \qquad (14.3.4)$$

is related to the transmissibility

$$\tau = 2 \langle v_2^2 \rangle / \langle v_1^2 \rangle \qquad (14.3.5)$$

where again plate 1 is the source and plate 2 is the receiver.

Finally, one can determine η_{12} from steady state response measurements when the receiving system is not so heavily damped as in cases (1) and (2). The damping must be high enough so that the modal energy of the receiving system is appreciably below that of the directly excited system. The steady state response of the indirectly excited structure $\langle v_2^2 \rangle$ is found in terms of the response $\langle v_1^2 \rangle$ of the directly excited system, as follows

$$\frac{M_2 \langle v_2^2 \rangle}{N_2} = \frac{M_1 \langle v_1^2 \rangle}{N_1} \quad \frac{\eta_{21}}{\eta_2 + \eta_{21}} \qquad (14.3.6)$$

The errors using this method of determining η_{21} tend to be rather high, so that the equation is a better estimator of response from a knowledge of the parameters than it is for estimating parameters when the response levels are known.

Transmission Through Plate Junctions. Theoretical formulas and data are available for panels joined along lines, either by a beam, a framing member, or by a simple bend. Theoretical deviations for line connections between plates are fairly complicated, particularly when the bending and torsional rigidity of the beam must be included. Some of the results are presented here, the background information necessary to make calculations for other systems may be found in the references.

Two plates joined at right angles were studied by Lyon and Eichler [34]. The configuration is shown in Fig. 14.7. The coupling has simple theoretical values for plates of equal thickness and also when one plate is much stiffer than

the other. The expressions for the transmissibility from
the semi-infinite plate no. 1 to the infinite plate no. 2
are

$$\tau = 8/27 \text{ (plates of equal stiffness)}$$

$$\tau = D_1/D_2 \text{ (plate 2 much stiffer than plate 1)}$$

$$(14.3.7)$$

where $D = m_p \kappa^2 c_\ell^2$ is the bending rigidity of each plate. For
homogeneous plates, $D \sim h^3$, so that the second relation above
is satisfied if $h_1 \leq h_2/2$. A comparison between theoretical
and experimental results for such a junction may be seen in
Fig. 14.5.

   The transmission from one panel to another through a
reinforcing beam has been studied by Heckl [35]. If the
panels are of equal thickness, then the transmissibility may
be well approximated by

$$\tau = 2 \frac{\kappa_p c_p m_p}{c_{fB} m_B} \left(1 + 64 \frac{m_p c_{fp}}{\omega m B}\right) \qquad (14.3.8)$$

where κ_p is the radius of gyration of the plate cross-section,
c_p is the longitudinal wave speed in the plate material, m_p
is the mass per unit area of the plate, m_B is the mass per
unit length of the beam, c_{fp} and c_{fB} are flexural wave speeds
for the plate and the beam respectively. A comparison between
theoretical and experimental results for a simple plate-beam
combination is shown in Fig. 14.6.

   In addition to these theoretical results, a number of
more complicated systems have been studied experimentally by
Ungar, et al. [36].

   Junctions of Panels and Beams. Coupling loss factors
have been found in a few cases for beams which connect to
plates, and these results are presented below. In general,
however, such results are sparse and one will likely be

forced to calculate the coupling loss factor. Procedures
for such calculations using junction impedances are discussed later in this section.

If a beam is cantilevered to a plate as shown in
Fig. 14.8, the CLF may be expressed in terms of junction
moment impedances [34]

$$\eta_{bp} = (2\rho_b c_b \kappa_b S_b)^2 \; (\omega M_b)^{-1} \, \text{Re}(Z_p^{-1}) \; |Z_p/(Z_p+Z_b)|^2$$

$$(14.3.9)$$

where Z_p and Z_b are moment impedances of an infinite plate
and a semi-infinite beam respectively, given by

$$Z_b = \rho_b c_b^2 S_b \kappa_b^2 \kappa_{fb} (1+i)/\omega \qquad\qquad (14.3.10)$$

$$Z_p = \omega (1-i\Gamma)/16\rho_s \kappa_p^2 c_p^2 \; . \qquad\qquad (14.3.11)$$

When the plate and beam are of equal thickness and unconstructed
of the same material, this simplifies considerably to become

$$\eta_{bp} = w/4\ell \qquad\qquad (14.3.12)$$

where w is the beam width and ℓ is its length.

Another important situation is for a beam that joins
the edge of a plate, as shown in Fig. (14.9). In this case,
the CLF can again be found from junction impedances but the
impedance functions are matrices since both moment and transverse force cause energy transfer at the boundary. The
formulas are not presented here because of their complexity,
but a comparison of theoretical and experimental results for
1 x 1/16 in. beam connected to a 1/16 in. thick plate is shown

in Fig. 14.10. When the axis of the beam is perpendicular
to the line of the plate edge, then there is theoretically
no coupling between flexural and torsional motions of the
beam. For such a situation, torsional and flexural motions
are decoupled. The total coupling loss factor is a
weighted average of the coupling loss factors for flexural
and torsional wave types,

$$\eta_{bp} = \delta f_b (\eta_{bp}^{flex}/\delta f_{flex} + \eta_{bp}^{tors}/\delta f_{tors}) \qquad (14.3.13)$$

where δf_{flex} and δ_{tors} are the average frequency spacings
between flexural and torsional modes respectively and

$$\delta f_b = (1/\delta f_{flex} + 1/\delta f_{tors})^{-1} \qquad (14.3.14)$$

is the average frequency spacing for all mode types.

Junction Impedances and Coupling Loss Factor. As we
have seen, junction impedances play an important part in the
evaluation of the coupling loss factor, particularly when
the coupling is concentrated over a small area of the plate.
Several authors have published tables of these junction
impedances. The reader should refer to the publications for
junction impedance formulas and data [37].

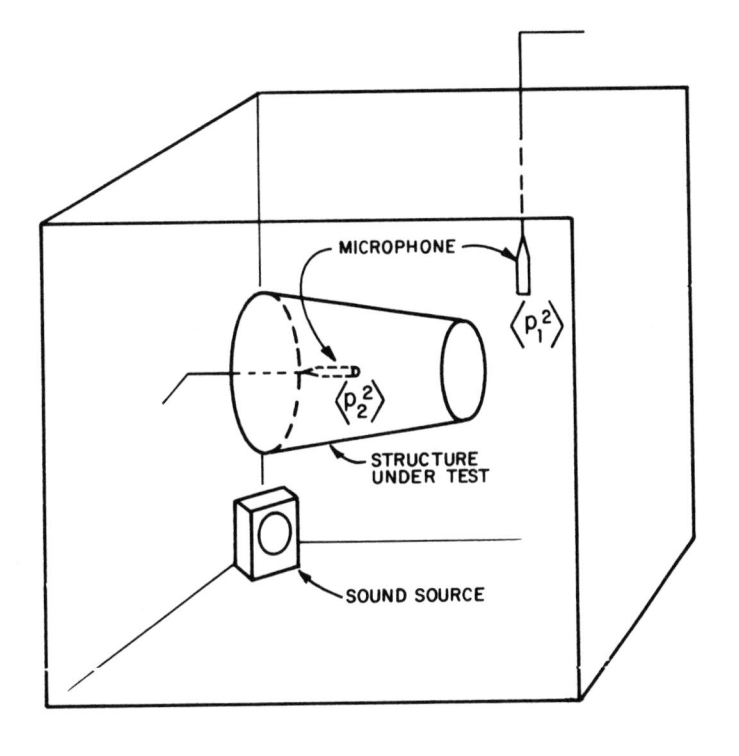

FIG. 14.1

**SET UP FOR MEASURING ACOUSTICAL TRANSMISSIBILITY
OF AN AEROSPACE STRUCTURE**

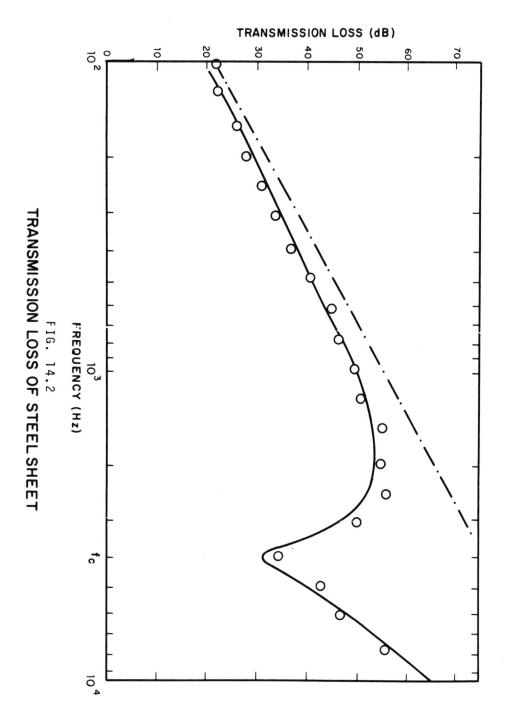

FIG. 14.2
TRANSMISSION LOSS OF STEEL SHEET

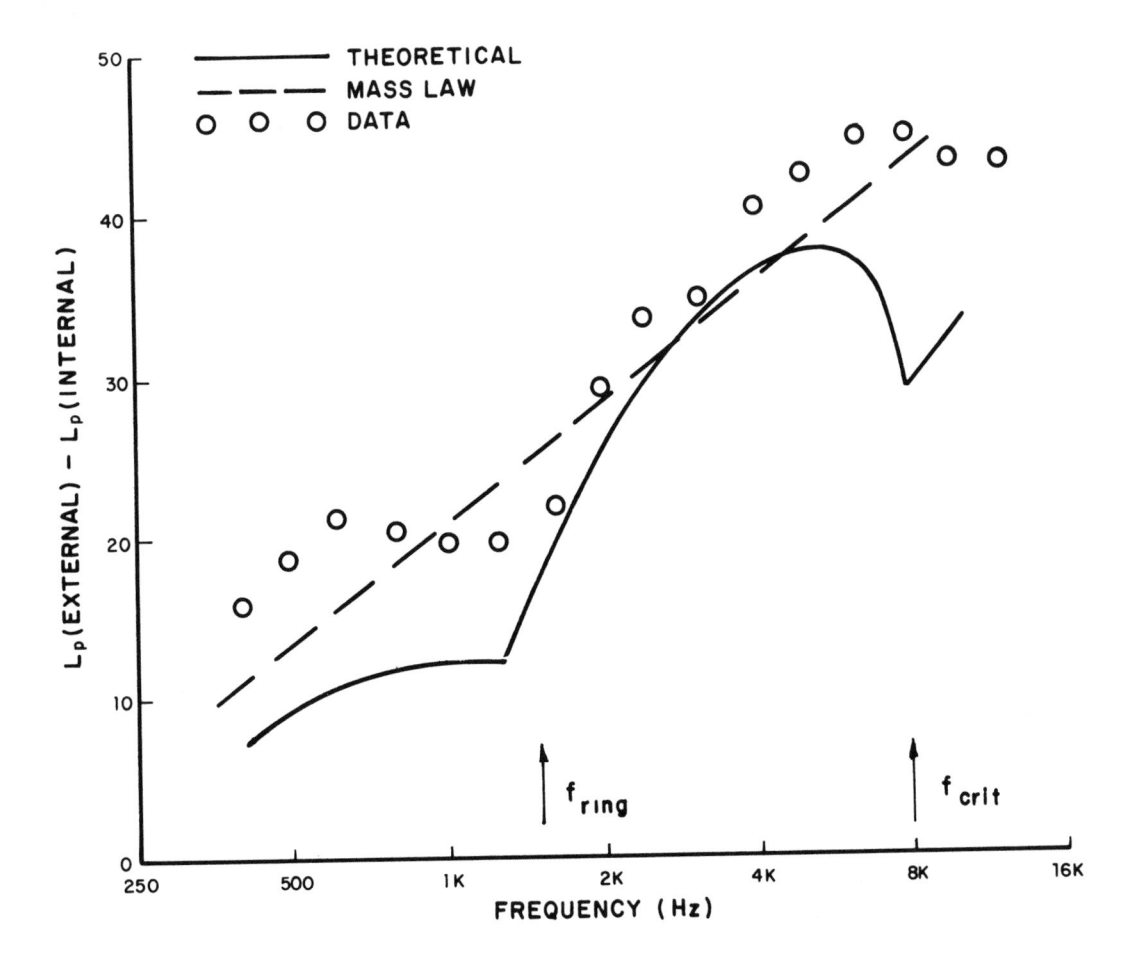

FIG. 14.3

NOISE REDUCTION OF CYLINDER STRUCTURE ;
THEORY AND EXPERIMENT

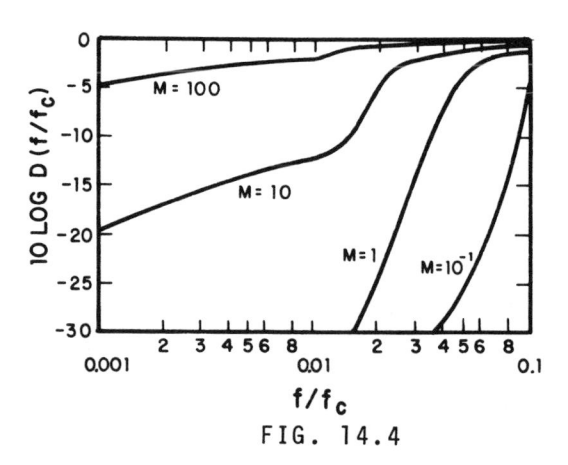

FIG. 14.4

**EFFECT ON RADIATION RESISTANCE
OF FINITE MASS OF BEAM ON A PANEL**

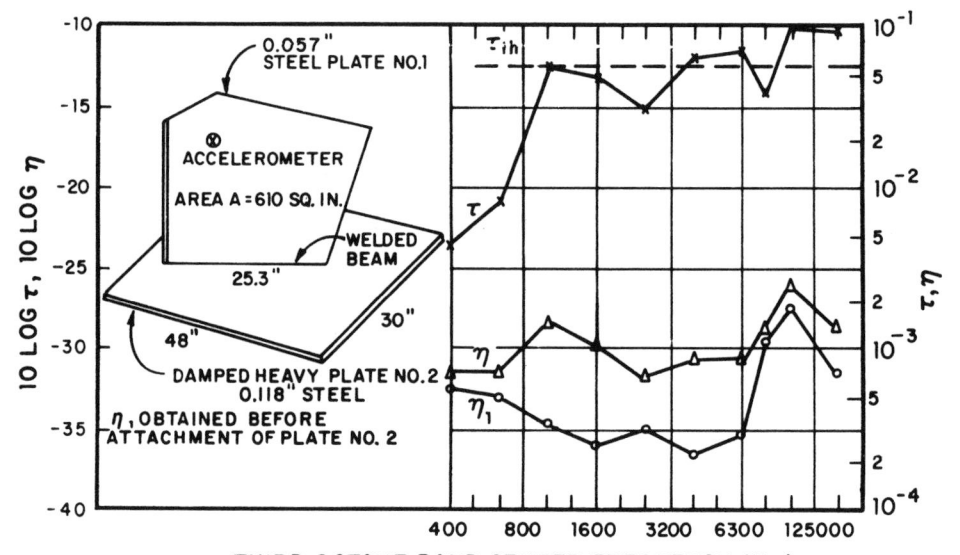

FIG. 14.5

**EXPERIMENTAL DETERMINATION OF
η_{12} (AND τ) FOR TWO CONNECTED PLATES**

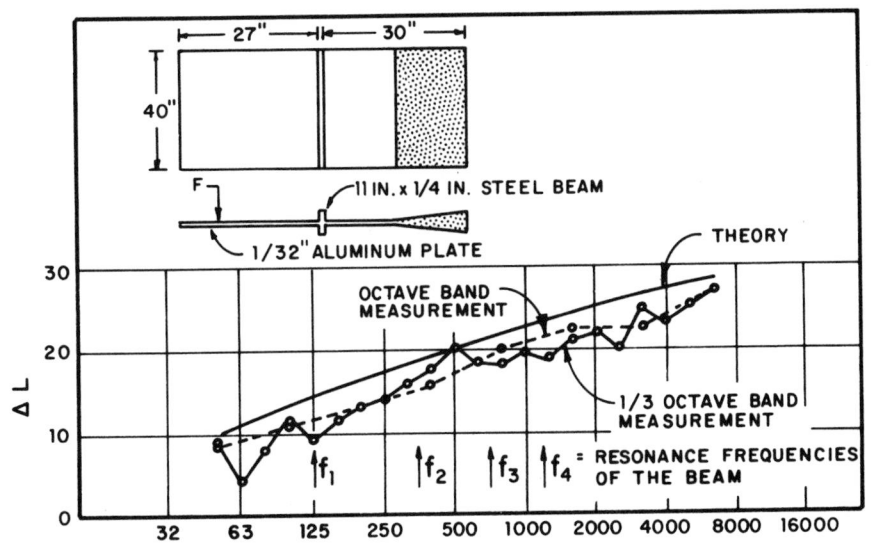

FIG. 14.6

EXPERIMENTAL DETERMINATION OF TRANSMISSIBILITY
OF A BEAM ON A PLATE

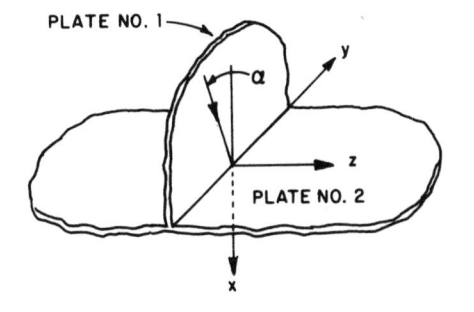

FIG. 14.7

TWO PLATES, RIGIDLY CONNECTED ALONG
A LINEAR JUNCTION

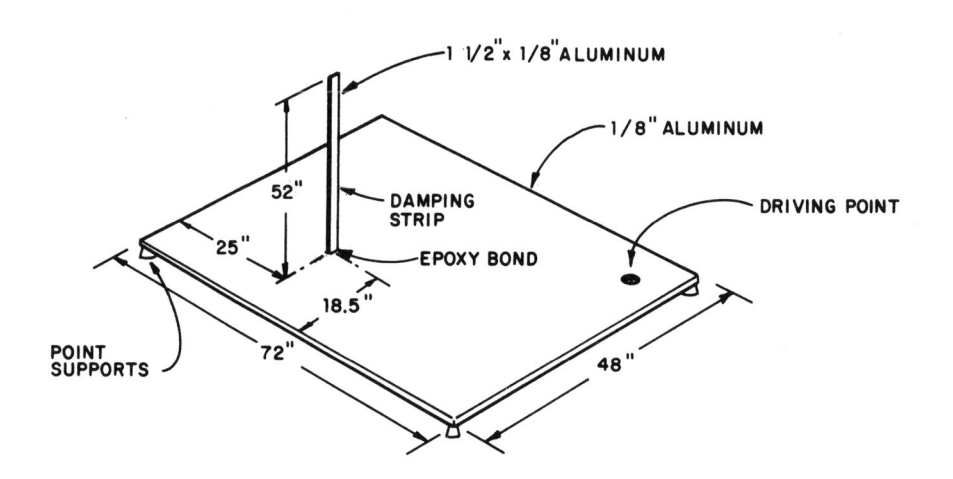

FIG. 14.8

BEAM CANTILEVERED TO A PLATE

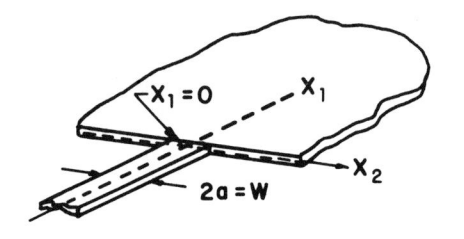

FIG. 14.9

BEAM CONNECTED TO EDGE OF A PLATE

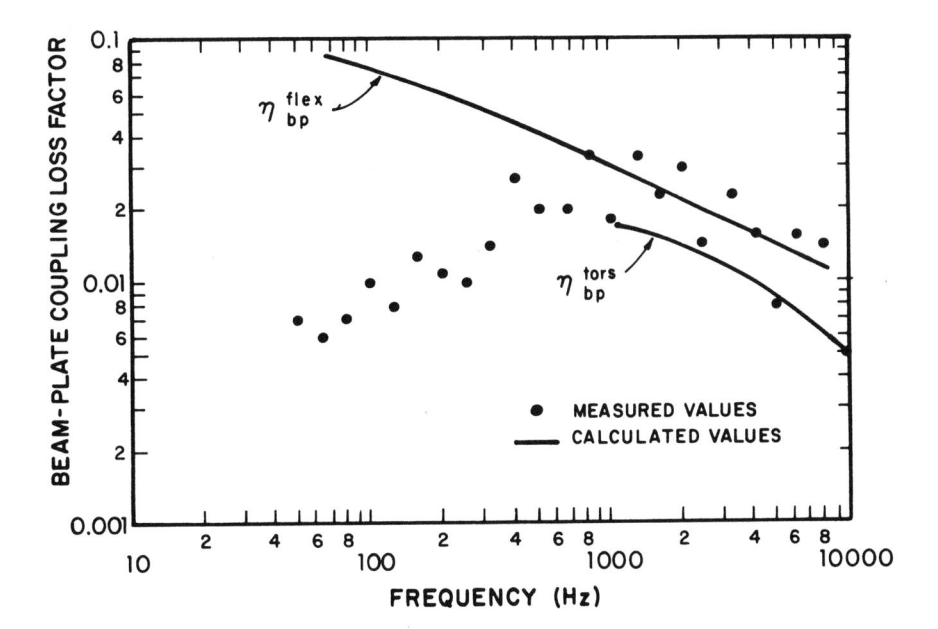

FIG. 14.10

COUPLING LOSS FACTORS FOR A BEAM
CONNECTED TO THE EDGE OF A PLATE :
THEORY AND EXPERIMENT

SECTION III - EXAMPLE OF RESPONSE ESTIMATION

by Huw G. Davies

CHAPTER 15. VIBRATION OF A REENTRY VEHICLE

15.0 Introduction

This chapter is intended to demonstrate the application of SEA to a specific problem in response estimation. We shall discuss the problem in some detail, and from a variety of aspects. Much of the work in this part is taken from a report by Manning.

The vehicle which is sketched in Figure 15.1 is about 12 ft long with a maximum diameter of 39 in. We shall be concerned here with the vibration of the shell of the vehicle caused by high-speed turbulent-boundary-layer flow, and with the transmission of vibration from the shell to equipment mounted on the upper instrument shelf.

15.1 Modeling the Vehicle

15.1.1 Low Frequency Model of the Vehicle

For very low frequencies (less than 50 Hz in the present case) the vibration of the vehicle is dominated by large scale flexural modes of the whole vehicle. The vehicle thus behaves as a free-free beam of rather complicated cross-section. The numbers of such flexural modes in any third octave band below 50 Hz is small. The modal density of the skin panel flexural modes is also small. Thus the usual techniques of vibration analysis at low frequencies may be used and no advantage is gained from the SEA approach at these frequencies. The analysis techniques required have been described in Appendix VIII of Ref. 39, for example, and we shall not review those techniques here.

The vibration prediction techniques just mentioned become cumbersome when many modes contribute to the vibration level. The transition region in this instance is about 100 Hz, third-octave bands above this frequency have more than two resonant flexural skin modes.   (The modal densities in-volved are discussed later.)   The techniques of SEA are thus

applicable above this frequency. In the remainder of this
Part we shall consider only vibration at frequencies above
100 Hz.

15.1.2 High Frequency Model

High speed turbulent flow over the surface of the reentry vehicle generates high frequency vibration of the skin.
This vibration is in turn transmitted to the various components carried within the vehicle. Our prediction of the
density of resonant skin modes of the vehicle shows that
there are many such resonant modes in all the third octave
bands above, say, 100 Hz. The techniques of SEA are thus
useful in predicting vibration levels of the skin and of the
interior components in this frequency range. In what follows
we shall be concerned solely with this frequency range. (The
term 'low frequencies' will refer to the lower part of this
range, that is to say, from 100 to 1000 Hz, in the remainder
of this chapter).

As indicated above, we shall restrict attention here to
the transmission of vibration from the vehicle skin to the
upper instrument shelf shown in Fig. 15.1, and to the subsequent estimation of vibration levels on this shelf. The
configuration is clearly similar to that in the example discussed in Chapter 10, and the modeling techniques are thus
similar also. We could treat the skin, the stiffener   the
ring connector, the instrument shelf, and the interior
acoustic space as separate systems, each described by one or
more groups of similar modes. This is clearly a very complicated model requiring a large amount of computation.
Experience suggests that a simpler model may be adequate to
describe the main features of the vibration.

We may note here the work of Manning [7] on the transmission of vibration to a shroud-enclosed spacecraft. Some
features of the SEA model used by Manning could be applied
to the present system. For example, his theoretical predictions of the coupling loss factors from cylindrical shells
to ring stiffeners could be used. However, although
features of the geometry are similar, the scale of the
fixture studied by Manning is considerably larger than the
present fixture. Modal densities of components thus are considerably reduced in the present system, so that the use of
multimodal subsystems in our SEA model for these components
may not be justified. Clearly then, a reasonable SEA model

can be formed for any case only after some preliminary
estimation of such parameters as the modal densities of some
of the components of the fixture being studied.

We begin with a two subsystem model. Since the random
pressure on the vehicle skin generates mainly transverse
motion, the group of transverse modes of the skin itself
should describe the skin vibration. The second group of modes
that we choose represents the vibration of the instrument
shelf. Coupling between the two groups of modes is provided
by the ring connector. This two component system requires
the evaluation of a minimum number of SEA parameters. Even so,
the evaluation of these parameters in this example turns out
to be quite involved.

A more complicated model may be used if it appears to be
necessary. We note, however, that an advantage of the SEA
approach is that groups of modes we may treat separately in
a more complicated model will often show up, for example, as
an additional dissipative mechanism in the simple model. This
added dissipation would be included implicitly in an experimentally obtained value of the loss factor used in the
simpler model.

Our model has been chosen deliberately to be simpler than
the model used by Manning [7]. This is so for a variety of
reasons. As pointed out above we expect that the main
features of the vibration can be explained on the basis of a
simple model. To a degree we are limited to this simple
model by the restricted amount of data available for the prediction of coupling loss factors. We have already noted that
there are very few modes of vibration of the ring connector
between the shell and the shelf in the frequency range of
interest; we would not be justified in treating this ring as
a separate subsystem in our SEA model. Finally, since the
ring connector provides a fairly strong structural link with
the skin, we are justified in neglecting the transmission
of vibration from skin to shelf via an airborne path.

15.2 Modal Density

15.2.1 Modal Density of Vehicle Skin

The vehicle skin is treated as a truncated conical-shell
of uniform thickness h, semi-vertex angle ϕ, and slant

length L as shown in Fig. 15.2. The shell has a maximum
radius a_1 and a minimum radius a_2. The coordinates x and θ
describe the surface of the cone.

We follow the approximate description of the cone
vibration given by Manning, et al. [38]. We assume that a
small section   S   of the cone with mean radius   a   has
the same vibration properties as a section   C   on a cylinder
of the same material, the same thickness   h,   and the same
radius   a(x).   In particular, we assume that for resonant
vibration at frequency ω, the local wavenumbers (k_x, k_θ) on
section   S   of the cone are the same as the corresponding
wavenumbers on the cylinder section C. (This is in the spirit
of the treatments of Refs. 25, 26).

For a given mode, the wavenumbers are given by

$$k_\theta = n/a(x), \qquad k_x = \frac{m\pi}{L}.$$

We note that these wavenumbers must be functions of the coordinate   x.   Also, if we write

$$k_p = \left(\frac{\omega}{\kappa c_\ell}\right)^{\frac{1}{2}}$$

as the wavenumber for resonant vibration at frequency ω on
a flat plate of the same material and the same thickness as
the cone, resonant motion corresponding to the local values
of the wavenumbers can only occur if

$$k_\theta(x) = \frac{n}{a(x)} < k_p.$$

The corresponding local k_x wavenumber is subsequently obtained
in this case from the frequency equation of a cylinder [3].

$$\left(\frac{\omega a^2}{c_\ell}\right) = (\kappa a)^2 (k_x^2 + k_\theta^2)^2 \quad + \frac{k_x^2}{k_x^2 + k_\theta^2} \qquad (15.1)$$

where $a(x)$ represents the local radius of the cone. Examination of this equation shows that for a given (m,n) mode resonant at frequency ω three types of solutions are possible. The three cases are sketched in Fig. 15.2 (b).

Case I:

 $k_\theta(x) < k_p$ everywhere on the cone. That is, for the given number n, numbers m can be found such that $k_x = m\pi/L$ satisfies Eq. (15.1). The corresponding mode shape at resonance extends over the entire length of the cone.

Case II:

 $k_\theta(x) < k_p$ for $x_0 < x < x_\ell$, and $k_\theta(x) > k_p$ for $x_\ell > x > x_b$. For this case there are no solutions of Eq. (15.1) for values of $x < x_\ell$. Clearly x_ℓ is given by $n/a(x_\ell) = k_p$. The resonant motion of the corresponding mode is restricted to values of $x > x_\ell$, that is, to the larger end of the cone.

Case III:

 $k_\theta(x) > k_p$ everywhere on the cone. No resonant motion is possible for this case.

No simple analytic results exist for the modal density of a cylindrical shell. Hence no simple analytic result can be obtained for the modal density of the cone by our technique of matching cone sections to equivalent cylinder sections. Empirical curves for a cylinder are given in Fig. 15.3 (from Ref. 22). For the cone, the modal density for resonant modes at a given position can be found approximately by using the appropriate value of the radius $a(x)$ in the non-dimensional parameter

$$\nu_0 = 2\pi f_0 a(x)/c_\ell$$

in Fig. 15.3. The parameter f_0 here is the center frequency

of the band being considered. In keeping with the three cases
discussed above, we note that for a given frequency the local
modal density tends to increase towards the larger end of the
cone. Of course, at frequencies well above the ring frequency
associated with the radius of the smaller end of the cone all
the equivalent cylinders have the same modal density as a
flat plate of equal area. The modal density of the cone is then
the same as that of a flat plate of equal area and the modal
density becomes independent of frequency.

15.2.2   Modal Density of Instrument Shelf

The instrument shelf is a fairly stiff assembly. It has
been determined experimentally that there are nine resonant
modes below 2000 Hz. The modes are not uniformly spaced, and
their precise location in frequency is not known. We shall
discuss later some ways of overcoming this lack of data.

One approximate theoretical estimate may be obtained by
treating the instrument shelf as an equivalent flat plate.
This is the same as adding the modal densities of each flange
of the shelf structure. Suitable values would be an area
$A = 4$ ft^2 and $\kappa c_\ell = 200$ ft^2/sec. This leads to a model
density $n_\eta(\omega) = A/4\pi\kappa c_\ell = 1/200\pi$, giving, for example, 2.3
modes in the third octave band at 1000 Hz. Although it uses
reasonable parameter values, this estimate is somewhat high
when compared to the experimental values found at lower frequencies. We shall, however, use this estimate in one of
the vibration prediction schemes discussed below. We expect
that the estimate will be more accurate at high frequencies.

The experimental estimate of the numbers of modes at low
frequencies (100 to 1000 Hz) suggests that some third octave
bands will contain resonant modes of the shelf, others will
not. A reasonable model of the shelf at low frequencies
would thus be a one degree of freedom system attached by mass-
less moment arms to the vehicle skin. This model should give
reasonable predictions on the shelf vibration at least in the
vehicle axial direction, assuming that the ring connector acts
as a rigid body. Clearly, this model is most applicable at
low frequencies. We shall discuss the parameter values involved
here in the next section.

15.3 Coupling Loss Factor

The coupling loss factor required in the present SEA
model is that which describes energy transmission between
the two groups of modes describing the vehicle skin and the
instrument shelf subsystems.

Since the coupling loss factor is often the most difficult
SEA parameter to predict accurately we shall discuss its
evaluation in some detail. Two highly simplified models of
the instrument shelf lead to two theoretical predictions. In
addition, some of the data may be interpreted in terms of these
two models, thus giving two experimental predictions. Finally,
in later parts of our discussion we shall show how the data may be
used to predict vibration levels on the shelf without explicitly
evaluating the coupling loss factor.

15.3.1 Approximate Theoretical Prediction #1 (high frequencies)

The instrument shelf is a fairly complicated assembly; a
detailed analysis does not seem feasible. However, as far as
the coupling loss factor is concerned, the essential features
of the vibration of the shelf may be obtained by treating the
shelf as an equivalent flat plate. A further simplification
is to treat the skin-shelf transmission as a simple plate-
plate transmission. This is certainly justifiable at higher
frequencies when wave motion on the cylinder is controlled
solely by the material bending stiffness.

The coupling loss factor may then be taken from Chapter 14
(see also Ref. 10):

$$\eta_{21} = \frac{2}{\pi} \frac{L}{k_2 A_2} \tau ,$$

where η_{21} represents the coupling loss factor from shelf to
skin, $L = 2\pi a$ is the joint length, k_2 is the wavenumber
on the equivalent plate of area A_2 and τ is the transmissibility
factor. Since the material and thickness of the skin and of
the equivalent plate are, at least, similar, we taken $\tau = 8/27$.
By substituting in suitable values

$$[L = 2\pi(1.2)\text{ft.}, \quad A_2 \simeq 4\text{ft}^2, \quad \kappa c_\ell = 200 \text{ ft}^2/\text{sec}]$$

we find

$$\eta_{21} \simeq 5 \, \omega^{-\frac{1}{2}} \simeq 2 \, f^{-\frac{1}{2}} \tag{15.2}$$

15.3.2. Experimental Prediction #1

15.3.2.1 Description of the Experiment

An evaluation of the coupling loss factor may be obtained from the experimental data. Because of the complexity of the structural properties of the ring connector, an experimental measure of the coupling loss factor certainly is warranted. The ideal experiment would be conducted on the vehicle itself. This was not feasible in the experiments. A model of the instrument shelf mounted in a cylindrical rather than a conical shell was used.

In a vibration experiment of this type, one of the properties one would like to model accurately is the source impedance of the vehicle shell. This impedance depends in a complicated way on the geometric and material properties of the shell. The cylindrical shell used by Manning had the same surface density and bending wavespeed as the actual vehicle skin, and thus provided a realistic test model.

The model used by Manning is shown in Fig. 15.4. The steel shell has a diameter of 29 in. and thickness of 5/16 in. The instrument shelf is attached to the cylindrical shell by an L-shaped ring, a connection that is structurally similar to that used in the actual vehicle.

The experimental structure was excited by two 50 pound shakers located as shown in Fig. 15.4. Third octave bands of noise covering the frequency range 50 Hz to 10 kHz were used. No attempt was made to excite the fixture at the levels predicted during actual flights since the goal in these tests

was to obtain vibration levels on the instrument shelf
relative to those on the cylindrical shell.

Acceleration levels measured in one-third octave bands
were measured at a number of points on the fixture. Three
groups of measurement positions were used: on the instrument shelf, on the shell at the point of connection to the
shelf, and in the reverberant field of the shell, that is,
away from the ends, the connector, and the shakers. All
levels were measured relative to a single reference acceleration level measured at the ring connector. The data is shown
in Figs. 15.5, 15.6, and 15.7. Figs. 15.5 and 15.6 show
the variation of vibration level in the reverberant field of
the shell and on the ring connector, respectively. Fig. 15.7
shows the transmissibility from the test shell to various
points on the instrument shelf. The transmissibility gives
the acceleration level at a point on the shelf relative to
the average acceleration level in the reverberant field of
the test shell.

15.3.2.2 Determination of η_{12} From the Data (Model #1)

In this paragraph we discuss how estimates of the
coupling loss factor may be obtained from the vibration data
based on model #1, the equivalent plate model. As we shall
see, although the data is useful, it is insufficient for
an accurate prediction of the coupling loss factor without
some further assumptions. Educated guesses are necessary
for some parameters. The interpretation of the data based
on the equivalent plate model is expected to provide reasonable estimates only at high frequencies.

We consider here a two component system, the skin and the
shelf, with only one system, the skin, being directly excited.
Eq. (14.3.6) applies to the relative total energies of each
system in this situation.

$$E_{2,\text{tot}} = E_{1,\text{tot}} \frac{\eta_{12}}{\eta_2 + \eta_{21}} \ , \qquad\qquad (15.3)$$

where 1 and 2 refer to the skin and to the shelf, respectively.
We have also the symmetry relation

$$N_1 \eta_{12} = N_2 \eta_{21}. \tag{15.4}$$

In the tests discussed above the transmissibility was expressed in terms of the average acceleration levels of the skin and of the shelf. In terms of these levels, Eq. (15.3) may be written

$$\frac{M_2 \langle a_2^2 \rangle}{M_1 \langle a_2^2 \rangle} = \frac{N_2}{N_1} \frac{\eta_{21}}{\eta_2 + \eta_{21}}, \tag{15.5}$$

or

$$\eta_{21} = \eta_2 \left(\frac{N_2 M_1 \langle a_1^2 \rangle}{N_1 M_2 \langle a_1^2 \rangle} - 1 \right)^{-1} \tag{15.6}$$

Eq. 15.6 gives an estimate of η_{21} provided that the quantities on the right hand side are known. Now, $\langle a_1^2 \rangle / \langle a_2^2 \rangle$ has been measured, M_1 and N_1 are known quite accurately, and M_2 may be estimated with reasonable accuracy. We have seen that N_2 is not easy to estimate, and η_2 was not measured.

The test data thus does not seem particularly useful for determining the coupling loss factor. However, several comments on this attempted evaluation can be made:

1. While η_{21} may not be directly evaluated in this case, the data obtained on the transmissibility may still be used in our SEA approach to predict vibration levels on the actual instrument shelf. We shall discuss this evaluation later.

2. This type of test, while on occasion the only possible way of obtaining a value of the coupling loss factor, can be inherently inaccurate. We note in

particular the case of very small η_2 ($\eta_2 << \eta_{21}$).
In this case, equipartition of energy among the
modes of the shelf and the skin occurs. η_{21} does
not control the relative levels in this case, and
so, the response values are quite insensitive to
the value of η_{21}.

3. The above comments notwithstanding, we may, with
suitable assumptions (sometimes guesses!) evaluate η_{21}
if it proves necessary to do so. It is instructive
to do so here in order to compare the values obtained
with our theoretical predictions.

The following table gives some representative values of
the numbers of modes in third octave bands for the text
fixture skin, and for the equivalent plate representing the
shelf (see Paragraph 15.2.2).

TABLE 15.1

MODE COUNTS FOR SHELL AND SHELF STRUCTURES

1/3 Octave Band Center Frequency Hz	$N_1 = n_1 \Delta\omega$	$N_2 = n_2 \Delta\omega$	N_1/N_2
125	.8	.3	2.6
250	2	.6	3
500	6	1.2	5
1,000	18	2.3	8
2,000	64	4.6	14
4,000	109	9.2	12
8,000	208	18.4	11

For the values chosen in our model we have $M_1/M_2 \simeq 4$. The
data from Ref. 1 provides values of $<a_1^2>/<a_2^2>$. Since this

ratio always corresponds to at least 10 dB in the frequency range of interest, Eq. (15.6) may be approximated to

$$\eta_{21} \simeq \eta_2 \; \frac{N_1 M_2 \langle a_2^2 \rangle}{N_2 M_1 \langle a_1^2 \rangle}$$

or

$$10 \log \eta_{21} \sim 10 \log \eta_2 + T + 10 \log N_1/N_2 - 6 \qquad (15.8)$$

where T represents the transmissibility to be obtained from Fig. 15.7.

A comparison of the predictions obtained from Eqs. (15.2) and (15.8) is shown in Table 15.2.

TABLE 15.2

VALUES OF COUPLING LOSS FACTOR

1/3 Octave Band Hz	Eq. 15.2 $10 \log \eta_{21}$	Eq. 15.8 $10 \log \eta_{21}/\eta_2$
125	− 7.5	− 11
250	− 9	− 19
500	− 10.5	− 23
1000	− 12	− 27
2000	− 13.5	− 7
4000	− 15	− 5
8000	− 16.5	− 8

For a relatively well damped shelf, a guess of $\eta_2 \sim 0.1$ would be realistic. Even with this large value, the two estimates differ by up to 15 dB at frequencies below 2000 Hz.

Several factors contribute to this discrepancy. Perhaps the most drastic assumption is that of the multi-modal nature of the response of the instrument shelf in all bands. This feature enters into both estimates. We note, however, that at frequencies above 1000 Hz where there are indeed a number of modes in each frequency band the agreement between the two estimates (with $\eta_2 = 0.1$) is reasonable. The equivalent plate model may thus be appropriate in the 2, 4 and 8 kHz bands.

15.3.3   Alternate Prediction of η_{21}

15.3.3.1   Theoretical Prediction #2 (low frequencies, 100 to 1000 Hz)

Our previous assumption of the multimodal nature of the response of the instrument shelf proved to be unrealistic at low frequencies, particularly for the frequency bands where we know from experimental data that the shelf response is not multimodal. A more realistic low frequency model is as follows: In each frequency band of interest, we assume that the shelf can be treated as a lumped parameter system with one degree of freedom, that is, as a single resonator. Assume that the resonator is free to vibrate along the axial direction of the vehicle, and is caused to vibrate by moment arms attached to the skin of the vehicle. A diagram of this model for the test fixture in Fig. 15.4 is shown in Fig. 15.8.

This model clearly predicts motion of the shelf in the axial direction only. Now, from the data of Ref. 38 we may note the following: at lower and middle frequencies, the reverberant field acceleration level of the cylinder is typically 20 dB higher than the average level at the ring connector, showing that the ring has a considerable stiffening effect as far as transverse motion of the shell is concerned. We are thus perhaps justified in considering only the moment transmitted by the L-shaped ring connector. We may subsequently estimate the acceleration levels in the cross-sectional plane by equating them to those at the ring connector.

This is an overestimate, however, as one can see from paragraph 15.3.3.3.

Because of the axial symmetry of our model, we may treat the shell as a beam of thickness h and width $2\pi a$, acted on by a concentrated moment. Fig. 15.8a shows a schematic drawing of the model. We assume that a value of η_{21} can be obtained by considering the power input to an infinite beam when the instrument shelf is excited by a force F. From Fig. 15.8b we see that the power supplied to the undamped resonator representing the shelf is all transferred to the beam, so that power balance requires that

$$\Pi_{in} = \omega \eta_{21}\ E_{2,tot} \tag{15.9}$$

where $E_{2,tot}$ is the total energy of the oscillator (shelf). The input impedance for the force F is

$$Z_{tot} = i\ \omega\ M_2 + i\ \frac{K}{\omega} + \frac{Z_{beam}}{k^2 a^2}\ ,$$

where

$$Z_{beam} = 2E\ \frac{I}{\omega}\ k^3 (1-i)$$

is the point input impedance for an infinite beam (Section II, see also Ref. 41). It follows from equation 15.1.3.3.1 that

$$\eta_{21} = R_e (Z_{tot})/\omega M_2$$

$$= \frac{1}{2\pi}\ \frac{M_1}{M_2}\ (\frac{\kappa c_\ell}{\omega})^{\frac{3}{2}}\ \frac{1}{a^3}$$

For the values of the parameters used, this result
predicts a value for the coupling loss factor of

$$\eta_{21} \simeq 20 \ f^{-\frac{3}{2}}$$ (15.10)

and hence

$$\eta_{12} \sim 20 \ f^{-\frac{3}{2}} \ (n_1 \Delta\omega)$$ (15.11)

where $(n_1 \Delta\omega)$ is the number of resonant skin modes in the
frequency band.

15.3.3.2 Experimental Prediction #2

The vibration data can also be interpreted in terms
of this second model. We merely replace equation 15.6 by the
similar equation

$$\eta_{21} = \eta_2 \left(\frac{M_2 <a_1^2>}{n_1 \Delta\omega M_2 <a_2^2>} - 1 \right)^{-1}$$ (15.12)

where $<a_2^2>$ now refers to only the axial acceleration of the
shelf.

We note that if an approximation similar to that leading
to equation (15.8) is made, the two models while predicting
different values of η_{21} for a given frequency band predict
the same value of η_{12} for that band.

15.3.3.3 Transverse Vibration of Shelf for Modal #2

Our low frequency model predicts only vibration in the
axial direction. It was suggested above that an overestimate
could be obtained by equating the shelf vibration levels in the
direction perpendicular to the vehicle axis to the vibration

levels on the skin at the ring connector.

Values of the vibration level at the ring connector can be obtained by allowing the mass in Fig. 15.8(b) to move transversely. For only transverse motion we find the coupling loss factor (by the same method as in paragraph 15.3.3.2 to be

$$\eta_{21} \sim 4 \, f^{-\frac{1}{2}} \, . \qquad\qquad (15.13)$$

15.3.3.4 Comparison of Coupling Loss Factor Predictions

The theoretical and experimental predictions of η_{12} are compared in Fig. 15.9. We have estimated possible values of the shelf loss factor η_2 as shown. The experimental predictions are based on the chosen values of η_2 (the data available in fact predicts only $(10 \, \log \eta_{12}/\eta_2)$.

We have used extremely simple models for our theoretical predictions. Nevertheless, the agreement between theory and experiment is fair. In particular, the difference in the low and high frequency behaviour is readily seen.

This fairly successful use of crude models of complicated structures demonstrates the strength of SEA. Although, for example, our lumped-mass-on-a-beam model seems very different from the actual reentry vehicle, the two systems are sufficiently similar that useful results can be obtained on a statistical basis by treating both systems as two members of the same ensemble of similar systems. We require merely that such gross parameters as, for example, the total mass be preserved.

15.3.3.5 Summary of Couping Loss Factor Values

It remains for us to choose appropriate values of η_{12} and η_{21} for each frequency range from among the many predictions just discussed. Fig. 15.9 indicates that there is a change in the apparent behavior at low and high frequencies. We therefore propose using the low frequency model for frequencies at and below 1000 Hz, and the high frequencies at and above 2000 Hz.

If purely theoretical predictions are required, then η_{12} and η_{21} are given as follows:

$f_o \leq 1000Hz$:

$$\eta_{21} = 20 \ f^{-\frac{3}{2}} \tag{15.10}$$

$$\eta_{12} = \frac{1}{N_1} \ \eta_{21} \tag{15.14}$$

$f_o \geq 2000Hz$:

$$\eta_{21} = 2 \ f^{-\frac{1}{2}} \tag{15.2}$$

$$\eta_{12} = \frac{N_1}{N_2} \ \eta_{21} \tag{15.15}$$

The change-over frequency point has been chosen in fact from comparisons with experiments. However, a theoretical basis may be given for this changeover if we say that the high frequency model applies when there are more than two resonant modes of the shelf in a third-octave band.

Because of the complexity of the fixture, we expect our experimental predictions to be more accurate than the purely theoretical predictions. It follows that the best approach is to use values of $10 \log (\eta_{12}/\eta_2)$ obtained from Fig. 15.8 with corresponding values of $10 \log(\eta_{21}/\eta_2)$ obtained by using Eqs. (15.14) and (15.15) in the appropriate frequency range.

15.4 Prediction of the Vehicle Skin Vibration Levels.

In this paragraph, we discuss the response of a conical shell to a turbulent-boundary-layer pressure field. This work together with the mean square pressure shown in Figure 15.11 can be used to predict the actual acceleration levels of the vehicle shell.

15.4.1 Response of a Conical Shell to a Turbulent Boundary Layer (TBL) Pressure Field

15.4.1.1 General Outline of the Calculation

Fig. 15.10a indicates an important feature of a TBL pressure field. At a frequency ω, the fluctuating wall pressure possesses wave number components k_x that are very nearly equal to ω/U_c, U_c being the mean convection velocity of the pressure field. At frequency ω, a section S of the cone at radius a [Fig. 15.10b] accepts power from the pressure field, if there is a local vibration pattern on section S that is resonant at frequency ω and also has a local wave-number component $k_x = \omega/U_c$. Such a wavenumber vector at resonance is shown by a heavy dot in Fig. 15.10a. (If the frequency ω is lower than the local ring frequency c_ℓ/a for section S, two such wavenumber vectors may be possible.

From the known results for a cylindrical shell, we can calculate this wavenumber vector, as well as the associated power input from the pressure field to section S (see Ref. 43). Also, from the calculated value fo the wavenumber component k corresponding to the heavy dot in Fig. 15.10a and from the known radius a at section S, we determine the number n of the cone mode to which this particular resonant motion over the section S belongs

$$n = k_\theta a \qquad\qquad (15.16)$$

The spatial extent of this parent mode is also easily determined. If $k_\theta a/k_p a_2 < 1$, Case I of paragraph 15.2.3.1 applies and the parent mode extends over the entire surface of the cone. Otherwise, Case II applies. The non-zero region of the parent mode for Case II is sketched in Fig. 15.10(a). The radius a_o at the boundary AB of this region is determined from the conditions:

$$\frac{k_\theta a}{k_p a_o} = 1. \qquad\qquad\qquad (15.17)$$

Now we assume that the power transmitted by the pressure field to the section S is retained entirely by the parent mode and is manifested as uniform coherent vibration of this mode.

This procedure then enables us to estimate the average mean-squared acceleration (in any given frequency band), over the extent of the parent mode, that results from the interaction between the TBL pressure field and the section S. By making the assumption that there is negligible coherence between vibration of different modes, we obtain the total acceleration-level at a particular location on the cone surface (in a frequency band) by merely summing up contributions from all the sections S of the conical shell.

The cone vibration would be contributed partly by modes that exclude the smaller end of the cone from their vibration patterns. The larger end of the cone, however, is always included. In other words, the power accepted by the various cone sections from the TBL pressure field is distributed in a preferential direction: namely, towards the larger end of the cone. Consequently, the vibration levels on the cone surface are expected to increase from the smaller to the larger end of the cone.

15.4.1.2 Response Estimate for the Conical Shell

15.4.1.2.1 Structural Model

The conical shell is modeled as an isotropic conical

shell 110 in. long, with maximum and minimum diameters of 6.4 and 39 in. The surface mass density is taken to be 15 lb/ft^2. Based on the experimentally determined values of the bending wave speeds in the two layered and multi-layered sample bars, we take the quantity κc_ℓ to be 90 ft^2/sec. The longitudinal wave speed c_ℓ by itself enters in the description of the extensional or membrane stress-controlled vibration. We assume c_ℓ to be 10,000 ft/sec. The chosen values of κ and c_ℓ are then consistent for a shell thickness of about 3/8 in. The structural loss factor η_1 is taken to have a value of 0.025 at all frequencies. This estimate is based on the value of η_1 deduced at 2000 Hz from shock transmission studies.

15.4.1.2.2 Flow Model

The free stream velocity U of the flow is taken to be 21,000 ft/sec. The mean convection velocity U_c of the TBL pressure field is taken as 0.6 U. The dynamic head q is 50 atm, and the root-mean-square pressure fluctuation p_h on the cone surface is 0.02 q. The boundary layer displacement thickness is taken to increase linearly from the smaller end to the larger end of the cone, with a mean value of 0.2 in. This choice is equivalent to taking the displacement thickness $\delta^* = 0.0026x$, x being the distance along the generator from the cone vortex. The frequency spectrum of the fluctuating pressure at any location is assumed to scale with the Strouhal number $f\delta^*/U_\infty$ where U_∞ is the free stream flow speed. The pressure spectrum at the mid-section of the cone is shown in Fig. 15.11 . The coherence is expressed in a compact way by the wavenumber spectrum ϕ_3 (k_θ) in the circumferential direction. The following form is chosen for this spectrum at frequency ω

$$\phi_3\ (k_\theta) = \frac{L_3}{2}\ (1 + (k_\theta L_3)^2)^{-\frac{3}{2}} \tag{15.18}$$

with

$$L_3 = 2\delta^*(1 + (\frac{2\omega\delta^*}{U_c})^2)^{-\frac{1}{2}} \ .$$

15.4.1.2.3 Results

The calculation procedure of paragraph 15.4.1.1 was
carried out with the division of the conical shell into seven
sections of equal axial length. The acceleration-response
spectra at the two extreme sections and at the middle section
are shown in Fig. 15.12. Mass-law response is also indicated.
These spectra are normalized with respect to the pressure
spectrum at the midsection as also shown in Fig. 15.11. The
acceleration levels in all frequency bands are seen to increase
from the smaller to the larger end. From Fig. 15.11 the
pressure spectrum increases monotonically with frequency in
the frequency range indicated.

Since the pressure spectrum is assumed to scale on the
local Strouhal number $f\ \delta*/U_\infty$ and since the thickness $\delta*$
increases from the smaller to the larger end of the cone,
the pressure levels increase in any frequency band below the
spectral maximum. This behavior tends to accentuate the difference in the acceleration levels on the different sections of
the cone. The essential mechanism, however, still is the
restricted extent of the vibration patterns of the cone modes.

An explanation of the peaks in acceleration spectra for
the mid-section and the smaller end shown in Fig. 15.12 is of
interest. Consider first the acceleration response of a
cylindrical shell that is excited by a TBL pressure field.
Below the cylinder ring frequency $c_\ell/2\pi a$ where a is the
cylinder radius, a significant portion of vibration is contributed by the membrane-stress-controlled modes. We denote
this type of vibration by MV. This MV contribution is maximum near the ring frequency and drops sharply at higher
frequencies. This is because membrane stresses cease to be
effective above the ring frequency.

For the cone response under consideration, the cone
section near the larger end has a local ring frequency of
about 1000 Hz. The power input from the TBL pressure field
to the MV of this section is maximum at its ring frequency.
Also, the MV is associated with relatively small values of the
modal number n. Furthermore, the modes involving the MV extend
over the entire surface of the cone. Thus, the maximum power
input at 1000 Hz to the MV near the larger end of the cone is
manifested as vibration on the entire cone surface. Let us
recall that the pressure levels in all frequency bands are
maximum near the large end of the cone. Hence, this power
input forms a significant portion of the total power input

to the cone around 1000 Hz. Since the vibration level increases from the smaller to the larger end of the cone, the
presence or absence of the power input to the MV near the
larger end causes the greatest relative change in the
vibration level at the smaller end and the minimum relative
change in the level at the larger end. The sharpness of
the peaks in acceleration spectra at 1000 Hz depends on the
magnitude of the fractional change in the response. Fig.
5.13 presents the response estimates based on two alternate
calculation procedures that involves assumptions that are
relatively less realistic. For the acceleration spectrum
marked A, the structural and the flow models are the same
as those for the results of Fig. 15.12; however, the
assumption of paragraph 15.4.1.1 (that each mode retains
all the power fed into it by the TBL pressure field)
is now replaced with the more naive assumption that the total
power input to the structure in any frequency band is distributed uniformly over the structural surface, yielding a
uniform acceleration level everywhere on the structure. A
third procedure could be based on the assumption that the
modes with resonance frequencies within a particular frequency band share the power (fed into them by the pressure
field in the same frequency band) among each other in such
a way that the time-averaged energy of each mode is the
same. While this assumption is more realistic, it is not
used here because it would entail complicated calculations.
It is interesting to note that the above three assumptions
yield identical estimates of vibration for a rectangular
flat plate or for a cylinder shell, whereas they yield
three distinct estimate for a conical shell.

Acceleration spectrum B in Fig. 15.13 pertains to the
estimated response of a cylindrical shell that has the thickness and the material properties of the conical structure
model and a radius which is the average of the radii of the
two ends, $(a_1 + a_2)/2$. The TBL pressure field is taken to
have the same properties everywhere on the cylinder as those
assumed for the midsection of the structural model.
This estimate B is the simplest to obtain and is seen to
lie within the range of the acceleration spectra of
Fig. 15.12. Therefore, the response of an equivalent
cylinder seems to be useful in yielding an approximate
first-order estimate for the response of a conical shell.

15.4.2 Empirical Prediction of Acceleration Levels

A commonly used empirical prediction scheme is that of
Franken [42]. This approach is not really valid here, however, for two reasons. First, although vehicle diameter is
included in the Franken prediction scheme, the diameter of
our   vehicle is so very much smaller than the diameters
of the vehicles on which Franken took his data that doubts
must be raised about the validity of the extrapolation.
Second, Franken's measurements were taken with acoustic
rather than TBL excitation. The predicted relative
acceleration levels based on Franken's scheme are shown in
Fig. 15.14. In view of the above caveats, the agreement with
the prediction of Fig. 15.13 is surprisingly reasonable at
high frequencies.

15.4.3 Power Input to the Shell

In SEA vibration predictions it is sometimes more useful
to consider the basic input parameter for the problem to be
the power input to part of the system rather than the acceleration level of part of the system. The use of one or
the other of these quantities is discussed in the next paragraphs.

The power input to the shell has been computed in the
approach described in paragraph 15.4.1, although specific
results are not given. It may arise, however, that acceleration
levels are predicted directly, as in the Franken approach. The
total power input to the shell can be found from the acceleration
levels by using the relationship

$$\Pi_{in} = \omega \eta_1 \ E_{1,tot} = \frac{\eta_1 M_1}{\omega} \ <a_1^2> \qquad (15.19)$$

In this equation η_1 is the effective loss factor of the
structure. The other parameter values are $M_1 = 15$ lbs/ft^2
and $\eta_1 = 0.025$ [38]. Values of the input power if required
may thus be computed directly from the predicted values of the
mean square acceleration $<a_1^2>$.

15.5 Vibration Levels of the Instrument Shelf

We are now in a position to use our SEA model of the reentry vehicle to predict the vibration levels on the instrument shelf. The various parameters required for the SEA model have been discussed above. We shall discuss below three ways of using these parameters to predict the vibration levels.

Our goal in the present problem is to predict vibration levels of the instrument shelf within the vehicle when we are given information about the excitation field acting on the skin of the vehicle. Our approach is to predict first either the vibration levels of the skin itself or to predict the power input of the skin from the turbulent boundary layer. Implicit in both these predictions is the neglect of the interior structure of the vehicle on the skin vibration. This effect shows up in two ways: First, as an additional mass and stiffness which may alter the modal resonance frequencies, and second, as an additional path for the loss of energy. The latter effect can be viewed either as an increase in loss factor as far as the skin is concerned, or as a coupling loss factor as far as the relative vibration levels of the internal structure and the skin are concerned.

The power input in a frequency band to a multimodal structure depends on the total mass of the structure and (if we consider only resonant response) on the number of resonant modes in the band. We remember that modal densities are additive, and for shell-like structures at high frequencies tend to be proportional to the surface of the structure. The addition, say, of a stiffener to a shell thus does not greatly affect the modal density. It follows that the total power input to the structure also does not change very much.

On the other hand, the stiffener may be such that it provides energy flow path to other parts of the system. Thus, effective loss factor of the part being considered may increase. As the total energy of the system in a frequency band is inversely proportional to the loss factor (for constant input power) the vibration levels on the structure would decrease. Very often, the coupling between structures is weak. The coupling now has a negligible effect on the vibration of the directly excited structure; both power input and vibration levels remain approximately unchanged.

15.5.1 Prediction Based on Skin Acceleration Level

In this Section, we give SEA predictions on the instrument shelf vibration levels based on:

1. prediction of the acceleration levels of the skin;

2. prediction of the power input to the skin, and

3. direct use of the experimental data.

If we assume an effective loss factor of the vehicle skin and internal structure, then predicted values of the skin vibration levels are appropriate. In terms of these predictions the instrument shelf vibration level is given by

$$<a_2^2> = \frac{M_1}{M_2} \; <a_1^2> \; \frac{\eta_{12}}{\eta_2 + \eta_{21}} \qquad (15.20)$$

The subscripts 1 and 2 refer to the vehicle skin and shelf, respectively. The appropriate values of η_{12} and η_{21} can be used as discussed in paragraph 15.3. Of course, when equations such as (15.14) and (15.15) are used, the parameter η_1 now refers to the modal density of the conical shell (paragraph 15.2). Values of the acceleration level $<a_1^2>$ are obtained from Figs. 15.11 and 15.13. Typical results are discussed in paragraph 15.5.4.

15.5.2 Prediction Based on Input Power

The input power Π_{in} for a turbulent boundary layer can be found from Section I of this report and from Ref. 37. We then have

$$\Pi_{in} = \omega \eta_1 E_{1,tot} + \omega \eta_{12} E_{1,tot} - \omega \eta_{21} E_{2,tot} \qquad (15.21)$$

and

$$E_{2,tot} = E_{1,tot} \; \frac{\eta_{12}}{\eta_2 + \eta_{21}} \; . \qquad (15.22)$$

From these equations we find

$$E_{1,tot} = \frac{M_1}{\omega^2} <a_1^2> = \Pi_{in} \left(\omega\eta_1 + \frac{\omega\eta_2\eta_{12}}{\eta_2 + \eta_{21}}\right)^{-1} . \qquad (15.23)$$

Typical values of the parameters have been given.

A comparison of Eq. (15.23) and (15.19) shows the effect that the internal structure has on the skin vibration. We see immediately that if the coupling loss factors are small, Eqs. (15.23) and (15.19) give approximately the same result. When this is so, then the prediction of paragraphs 15.6.1 and 15.6.2 will be the same. We emphasize, however, that when the coupling is strong, Eqs. (15.23) and (15.22) should be used to predict the vibration levels.

As a final comment on the difference between the two approaches just outlined we note the following: If decay measurements are taken on the skin alone a value of η_1 is obtained. If the internal structure is then put in place and further decay measurements taken, an effective loss factor for the skin of

$$\eta_1 + \frac{\eta_2\eta_{12}}{\eta_2 + \eta_{21}} \qquad (15.24)$$

is obtained. The correct approach thus clearly depends on the experimental configuration used.

15.5.3 Prediction from the Experimental Data

We have discussed methods of obtaining values of the

coupling loss factors from the experimental data at
some length, even though we noted that these values were not
required explicitly as direct vibration predictions are possible.
The Manning experiments were performed on a fixture as similar
as possible to the actual vehicle. When estimating η_{21}, the
coupling loss factor from shelf to skin, we have seen that the
skin can be treated approximately as an infinite structure. As
discussed in earlier parts of this report, the use of the
infinite structure approximation is a common one in obtaining
some SEA parameters. It follows that (to this approximation)
the value of η_{21} is the same for the test fixture as for the
actual vehicle.   On the other hand, values of η_{12} will be
different since the modal density of the test fixture and the
actual vehicle are different.

If we suppose that the skin acceleration levels on the
actual vehicle have been predicted, we can then make an
estimate of the shelf vibration levels. Two equations such as
(15.22) can be written, one describing the test fixture, and
one describing the actual vehicle. Values of η_2 and η_{21} are
the same for each equation. It follows by simple substitution
that

$$\frac{E_{2,tot}}{E_{1,tot}} = \frac{n_1}{n_2} \frac{\tilde{E}_{2,tot}}{\tilde{E}_{1,tot}} \; ,$$

where the variables with a tilde refer to the test fixture,
and those with no tilde to the actual vehicle. The equation
can be written

$$\frac{<a_2^2>}{<a_1^2>} = \frac{\tilde{n}_1 \tilde{M}_1}{n_1 M_1} \frac{<\tilde{a}_2^2>}{<\tilde{a}_1^2>} \tag{15.25}$$

Thus values of the shelf acceleration $<a_2^2>$ relative to the
predicted vehicle skin acceleration level $<a_1^2>$ can be
found directly in terms   of   the transmissibility ratio
$<\tilde{a}_2^2>/<\tilde{a}_1^2>$ found in the experiment. The parameters n_1 and M_1

and $\tilde{n}_1$ and $\tilde{M}_1$ are the modal density and total mass of the skin in the vehicle, and in the test fixture, respectively. Typical values are discussed below.

15.5.4 Results

   Some predictions for the mean square acceleration level on the instrument shelf in the vehicle are presented in this section. We note first that as the coupling loss factors are small, predictions based on paragraphs 15.5.1 and 15.5.2 will be the same. A number of different predictions are possible since in some cases a number of different values have been predicted for each parameter. Some representative predictions are shown in Fig. 15.15.

   Curve A of Fig. 15.15 is obtained from purely theoretical considerations. Values of the coupling loss factors used for the low and high frequency models are summarized in paragraph 15.3.3.5. We have used the equivalent cylinder prediction of the skin acceleration level with $\eta_1 = 0.025$ as shown in Fig. 15.13. The estimated values of η_2 are shown in Fig. 15.9. This curve would be obtained by the prediction schemes of both paragraphs 15.5.1 and 15.5.2 above.

   Curve B of Fig. 15.15 shows the prediction obtained directly from the data by the method described in paragraph 15.5.3.

15.5.5 Other Vibration Parameters

   Fig. 15.15 shows predictions of the acceleration levels of the upper instrument shelf of the vehicle. Other vibration parameters can be obtained directly by the methods of Chapter 7.

15.6 Confidence Limits

   The predictions discussed so far are mean values of the vibration parameters, the mean being obtained from an ensemble of similar systems. As discussed in earlier parts of this report, it is of interest to obtain measures of the deviation

of a realized response from the mean. This can be done in
terms of the standard deviation from the mean square response.

The four major effects producing a variance in response
have been discussed in paragraph 4.2 of Part I. We emphasize
that all four effects are features of the SEA model we have
chosen, and not the actual system under study. Thus, in the
present problem the confidence estimates give no indication
of how accurately our SEA model describes the vibration of the
actual vehicle. What the confidence limits do tell us is
the range of vibration of the response we might expect to
measure if a physical replica of our SEA model were constructed.
However, if data taken on the actual vehicle falls within the
predicted high confidence (say 80%) limits, this is an
indication that our SEA modeling is, at least, adequate.

We shall restrict attention here to "bracketting"
confidence limits for broad band excitation. We have predicted above the mean square acceleration levels in third
octave bands of the upper instrument shelf, $<a_2^2>$. We first
estimate the variance in this response. Eq. (4.2.13) of Part I
gives the ratio of variance to mean square response as

$$\frac{\sigma_a^2}{<a^2>^2} = [n_1 n_2 \frac{\pi}{2} (\omega\eta_1 + \omega\eta_2)\Delta\omega]^{-1} (\frac{<\psi_1^4>^2}{<\psi_1^2>^2})^2 (\frac{<\psi_2^4>}{<\psi_2^3>^2})^2$$

The spatial response factors corresponding to our low and high
frequency models are as follows:

$$\frac{<\psi_1^4>}{<\psi_1^2>^2} \quad = \quad 9/4 \qquad \text{all } f_o$$

$$\frac{<\psi_2^4>}{<\psi_2^2>^2} \quad = \quad \begin{cases} 3/2 & f_o \leq 1000 \text{ Hz} \\ 9/4 & f_o \geq 2000 \text{ Hz} \end{cases}$$

The appropriate values of $N_1 + N_2$ have been discussed above (for example $n_2 \Delta\omega = 1$ for frequencies less than 1000 Hz). Values of the variance are shown in Fig. 15.16.

Confidence limits based on the variance can be obtained from Fig. 4.9 of Part I. The 80% limits for the theoretical prediction given in Fig. 15.15 are shown in Fig. 15.17. The limits tend to bracket the measured values except in the highest frequency band.

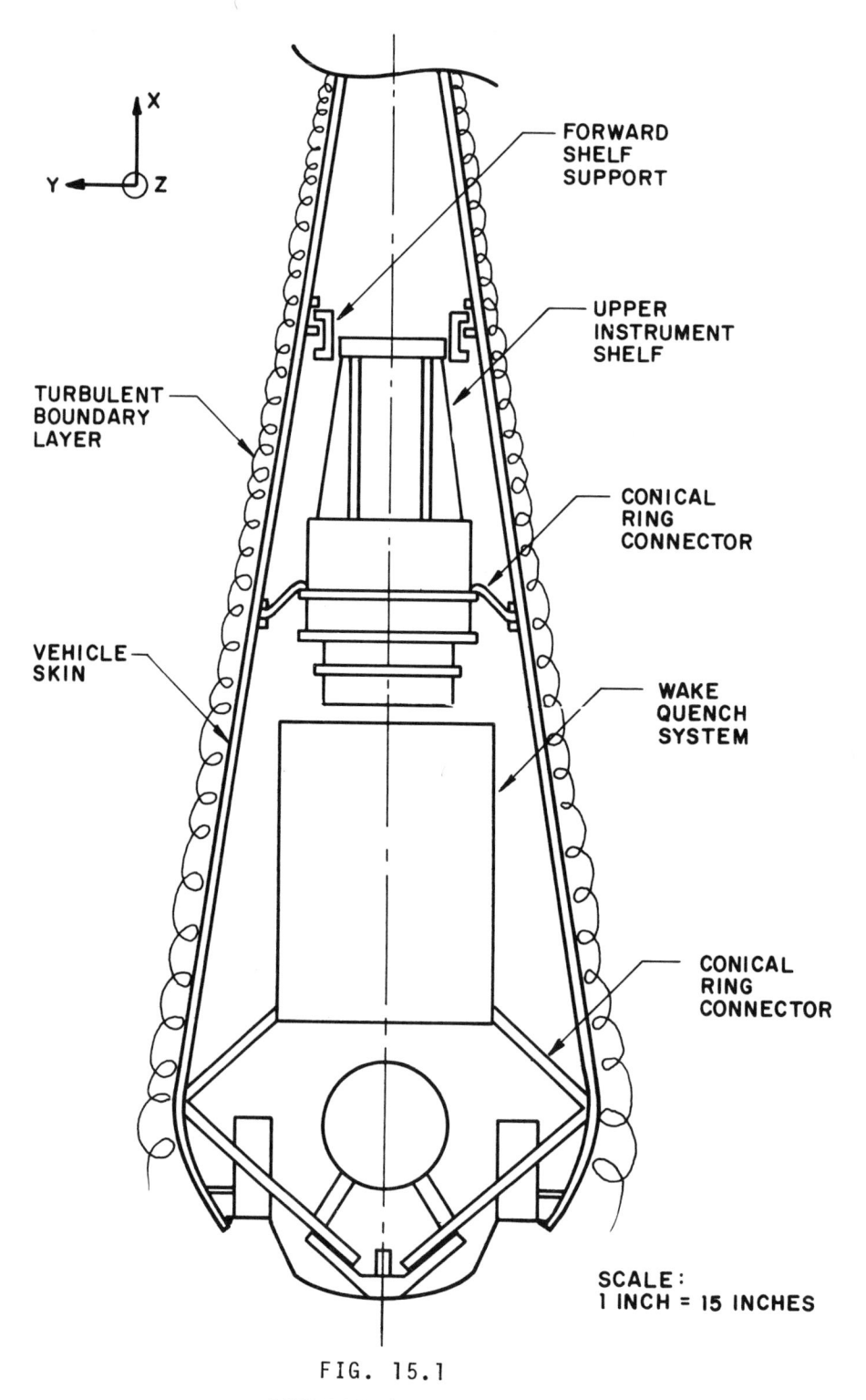

FORWARD
SHELF
SUPPORT

UPPER
INSTRUMENT
SHELF

TURBULENT
BOUNDARY
LAYER

CONICAL
RING
CONNECTOR

VEHICLE
SKIN

WAKE
QUENCH
SYSTEM

CONICAL
RING
CONNECTOR

SCALE:
1 INCH = 15 INCHES

FIG. 15.1

REENTRY VEHICLE

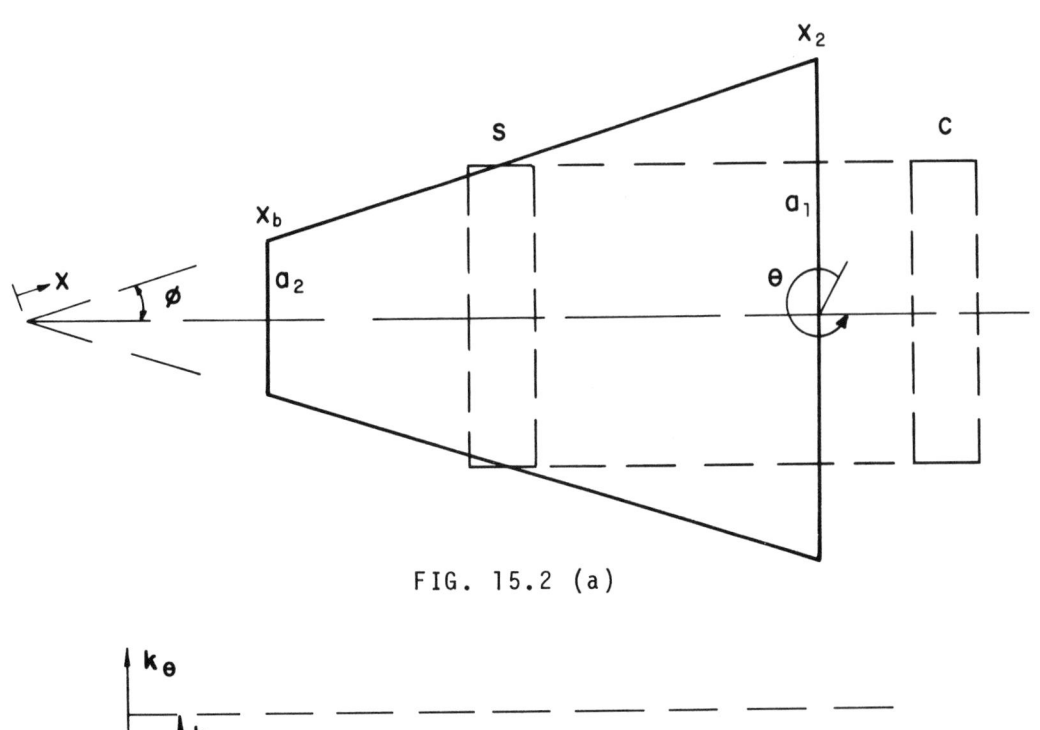

FIG. 15.2 (a)

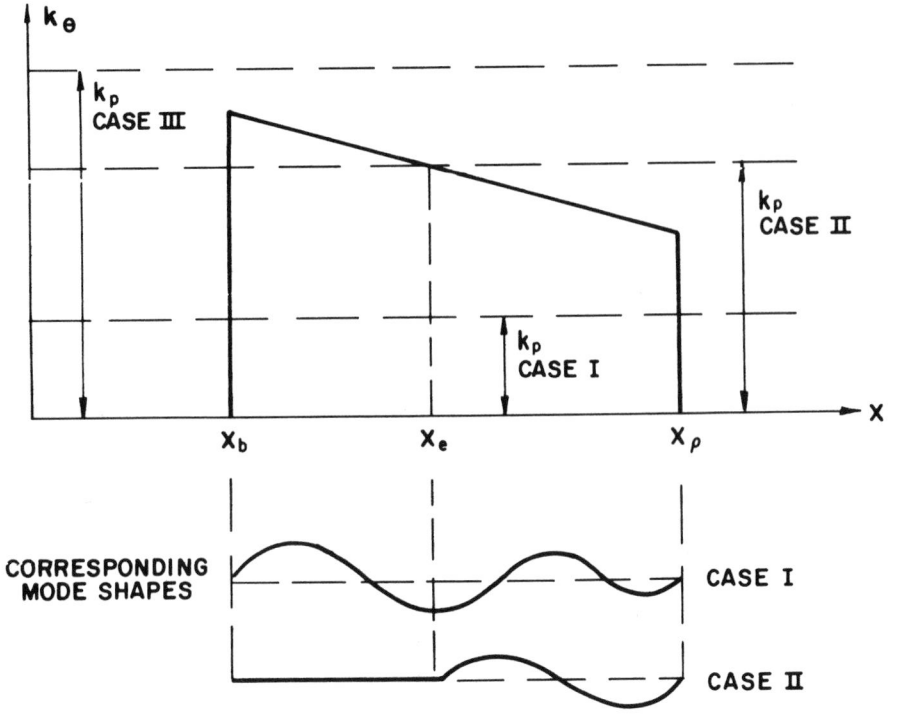

FIG. 15.2 (b)

MODE SHAPES FOR CONE VIBRATION

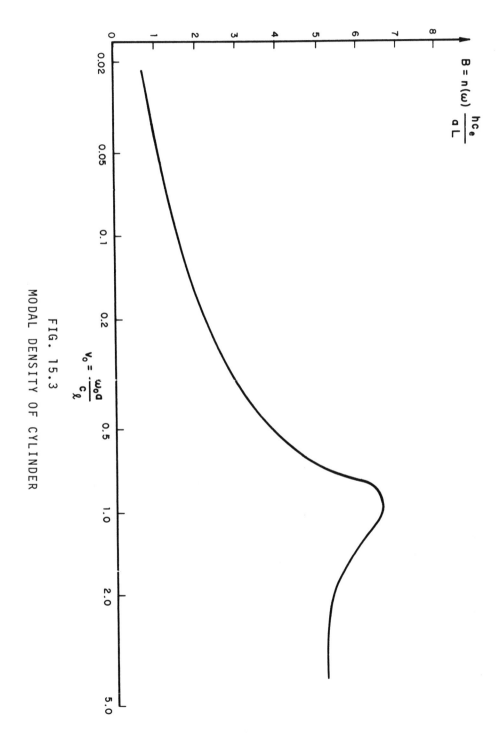

FIG. 15.3

MODAL DENSITY OF CYLINDER

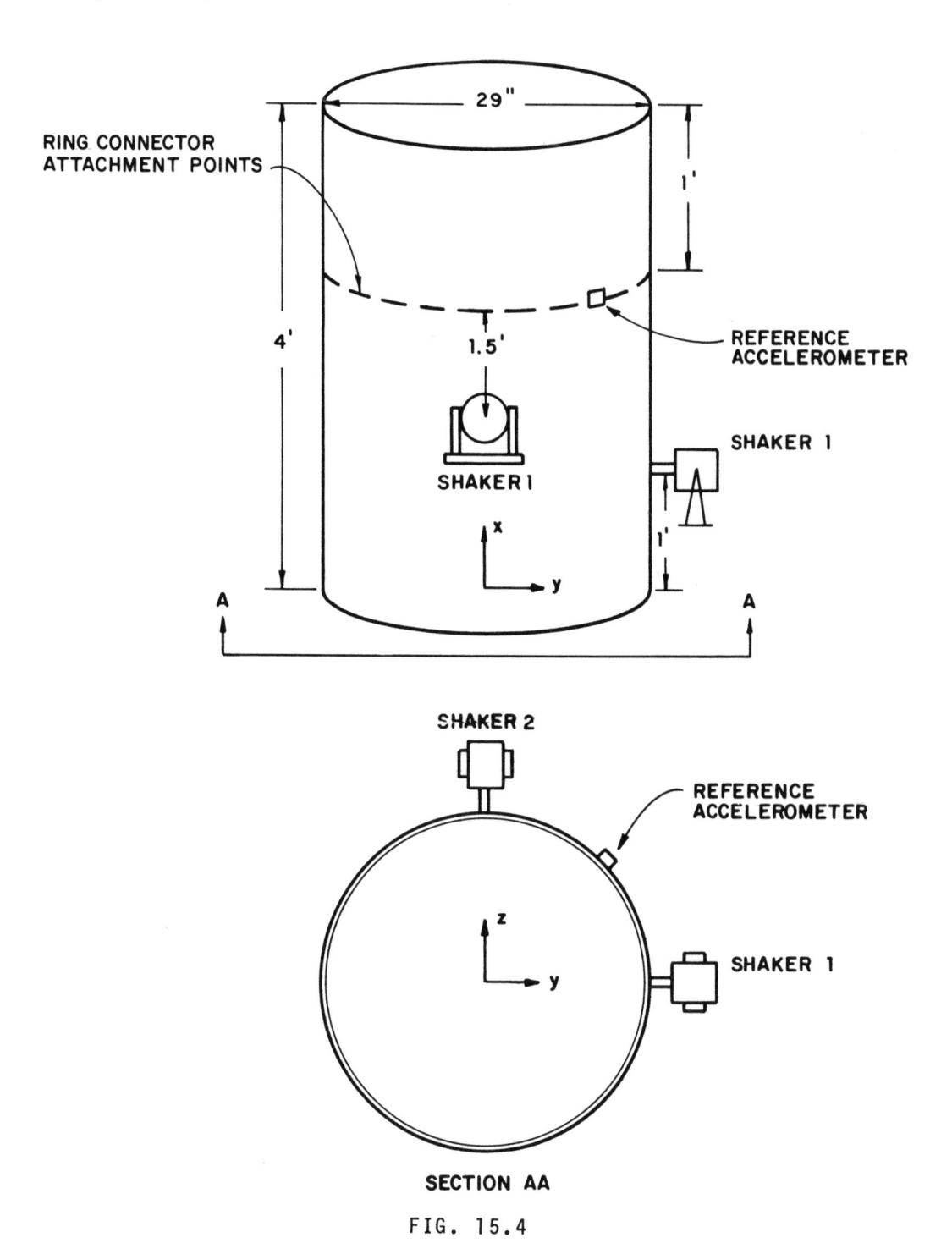

FIG. 15.4

CYLINDRICAL SHELL TEST FIXTURE

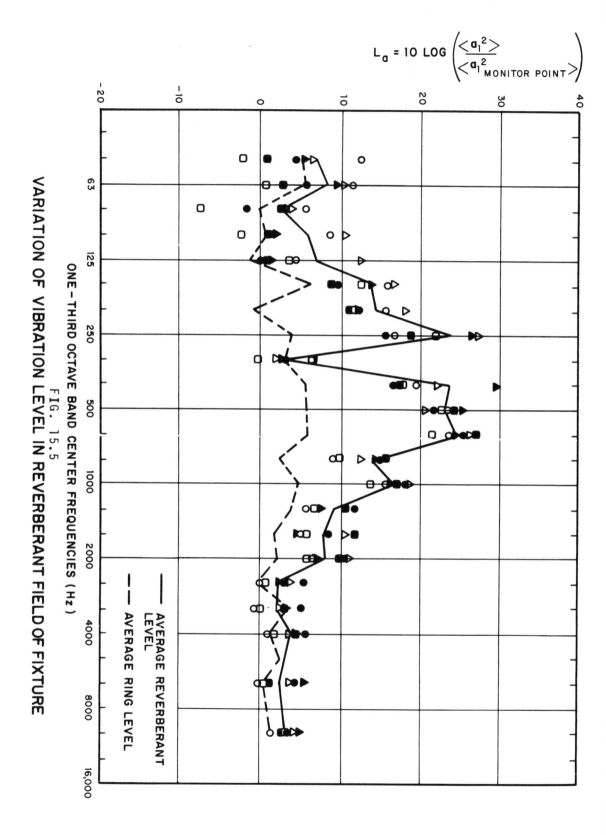

$$L_a = 10 \, LOG \left(\dfrac{\langle a_1^2 \rangle}{\langle a_1^2 \,\text{MONITOR POINT} \rangle} \right)$$

ONE–THIRD OCTAVE BAND CENTER FREQUENCIES (Hz)

FIG. 15.5
VARIATION OF VIBRATION LEVEL IN REVERBERANT FIELD OF FIXTURE

—— AVERAGE REVERBERANT
        LEVEL
— — AVERAGE RING LEVEL

Vibration of a Reentry Vehicle

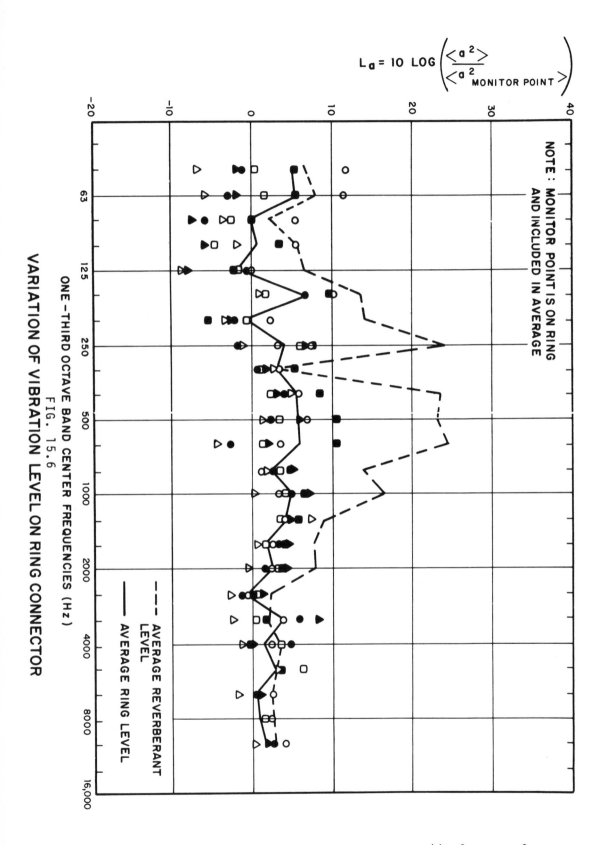

$$L_a = 10 \; LOG \left(\frac{\langle a^2 \rangle}{\langle a^2 \rangle_{MONITOR \; POINT}} \right)$$

NOTE : MONITOR POINT IS ON RING
AND INCLUDED IN AVERAGE

--- AVERAGE REVERBERANT
        LEVEL
——— AVERAGE RING LEVEL

ONE-THIRD OCTAVE BAND CENTER FREQUENCIES (Hz)
FIG. 15.6
VARIATION OF VIBRATION LEVEL ON RING CONNECTOR

L$_a$ RE AVERAGE REVERBERANT FIELD ACCELERATION ON FIXTURE

$$L_a = 10 \, LOG_{10} \left(\frac{\langle a_2^2 \rangle}{\langle a_1^2 \rangle} \right)$$

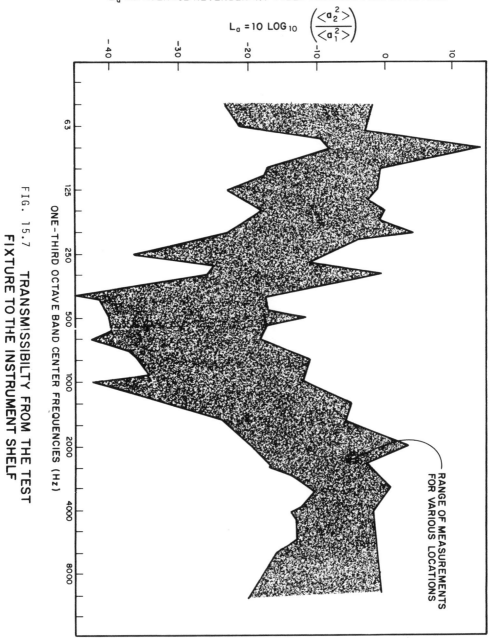

FIG. 15.7 TRANSMISSIBILTY FROM THE TEST FIXTURE TO THE INSTRUMENT SHELF

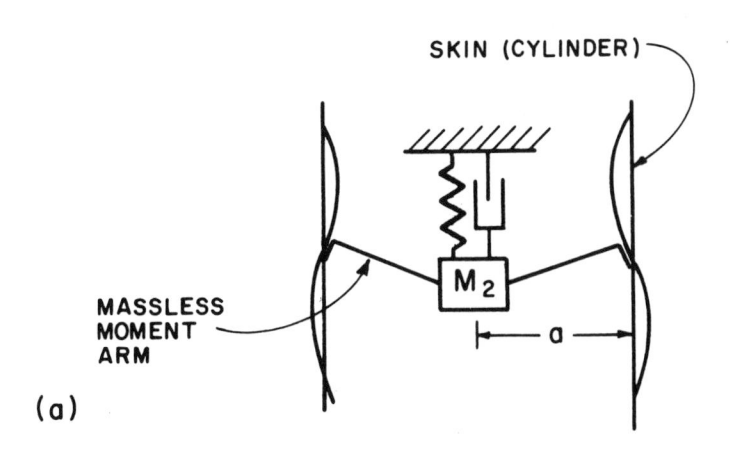

(a)

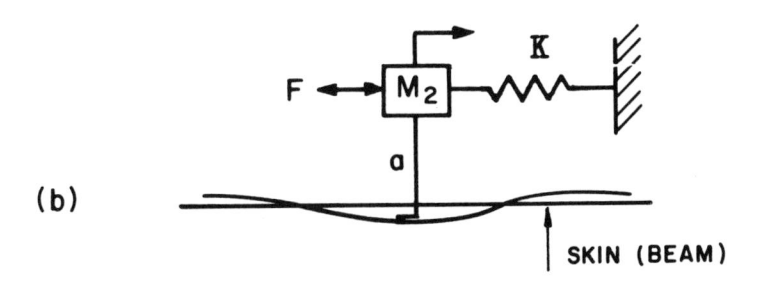

(b)

FIG. 15.8

LUMPED PARAMETER MODEL OF SHELF

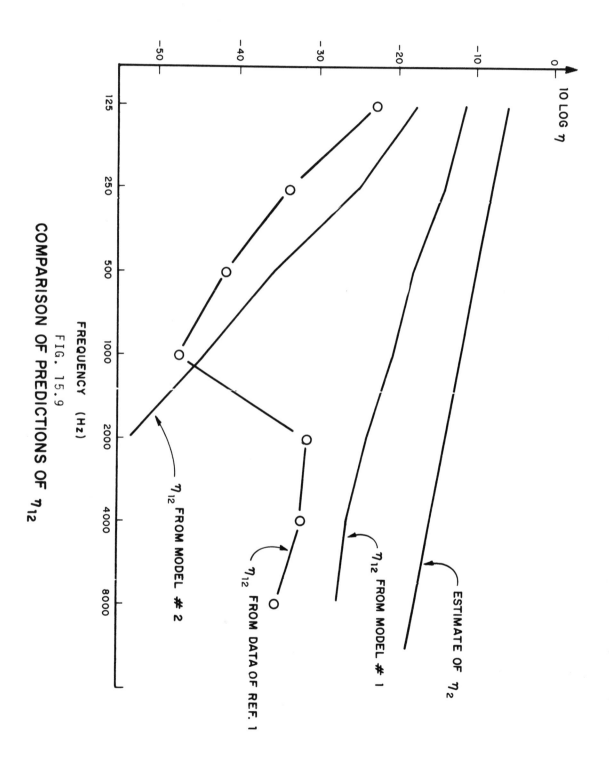

10 LOG η

FREQUENCY (Hz)

FIG. 15.9

COMPARISON OF PREDICTIONS OF η_{12}

η_{12} FROM MODEL # 2

η_{12} FROM DATA OF REF. 1

η_{12} FROM MODEL # 1

ESTIMATE OF η_2

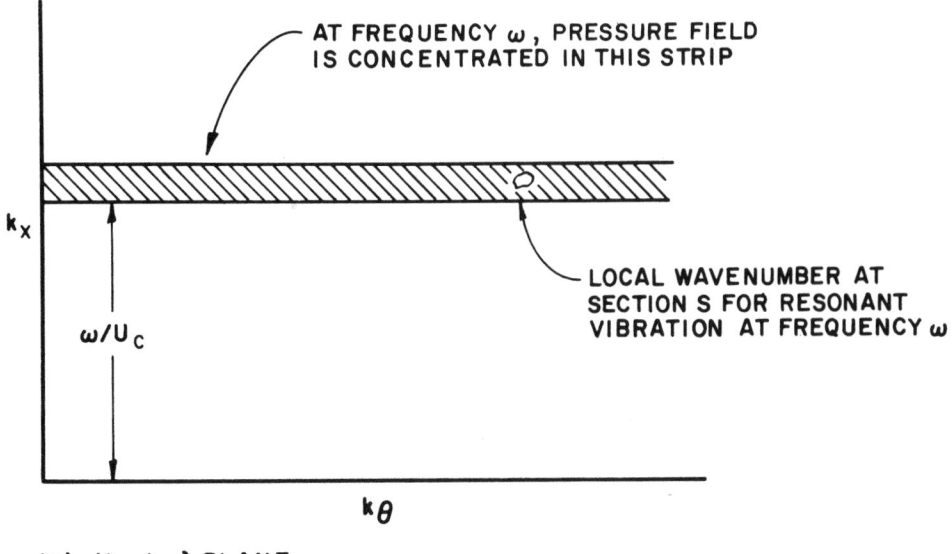

AT FREQUENCY ω, PRESSURE FIELD
IS CONCENTRATED IN THIS STRIP

k_x

ω/U_c

LOCAL WAVENUMBER AT
SECTION S FOR RESONANT
VIBRATION AT FREQUENCY ω

k_θ

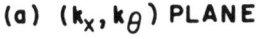

(a) (k_x, k_θ) PLANE

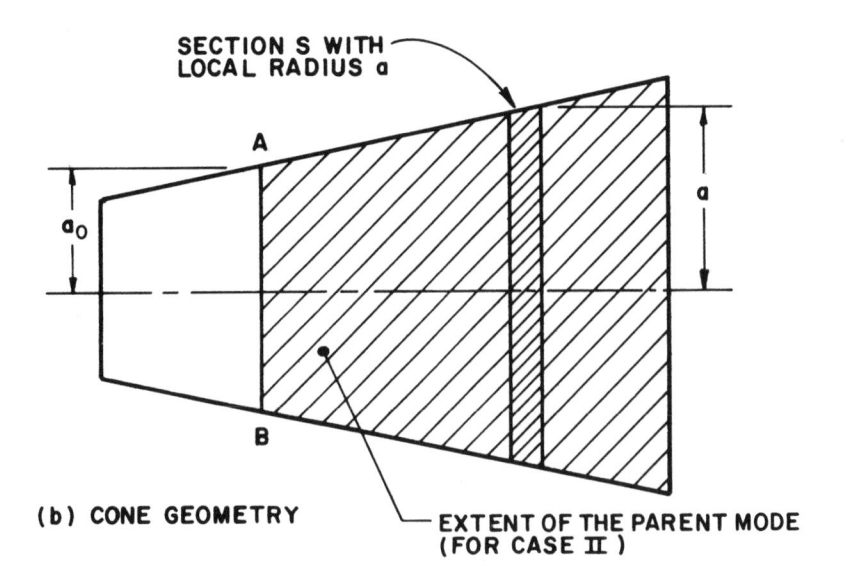

SECTION S WITH
LOCAL RADIUS a

A

a_0

a

B

(b) CONE GEOMETRY

EXTENT OF THE PARENT MODE
(FOR CASE II)

FIG. 15.10

REPRESENTATION IN THE WAVENUMBER
PLANE OF THE TBL PRESSURE FIELD

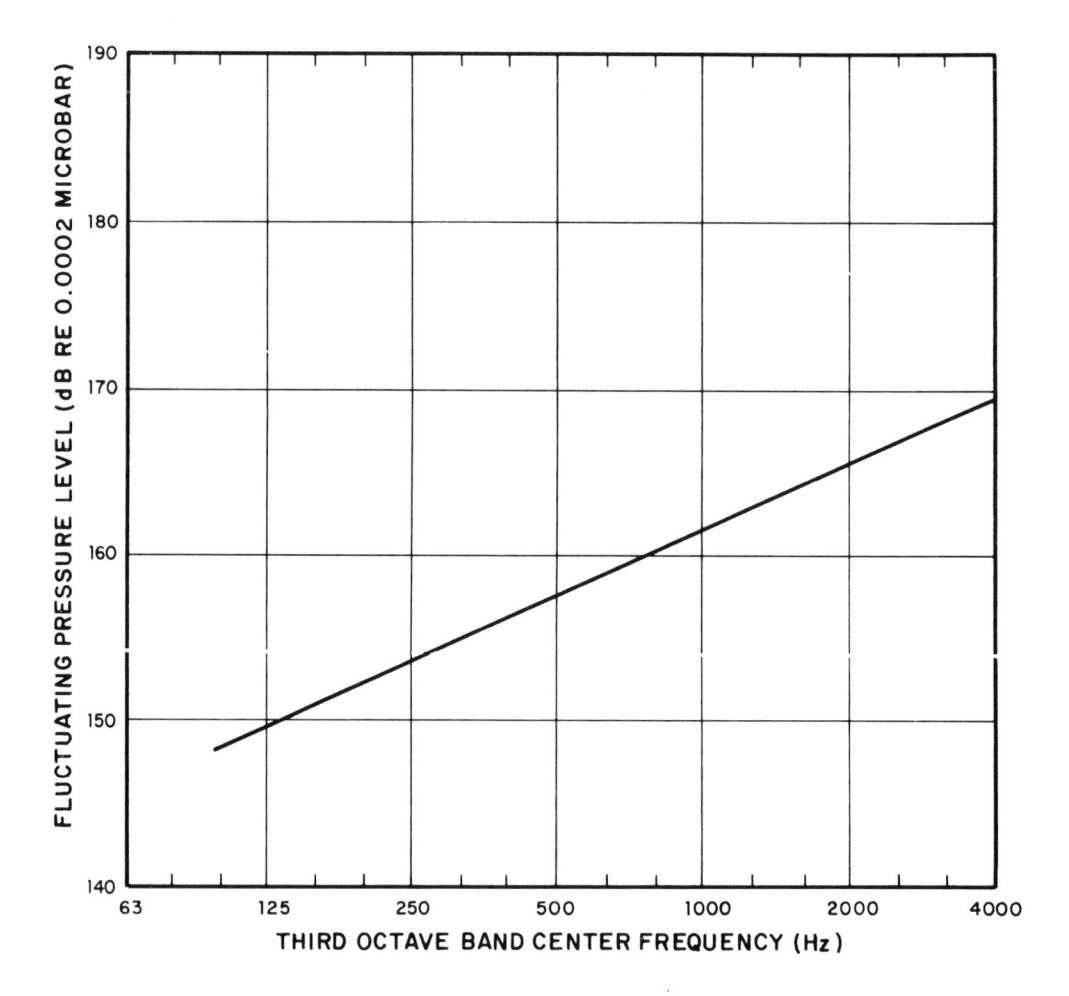

FIG. 15.11

ESTIMATED TBL PRESSURE SPECTRUM AT THE MID-SECTION OF THE
CONICAL SHELL

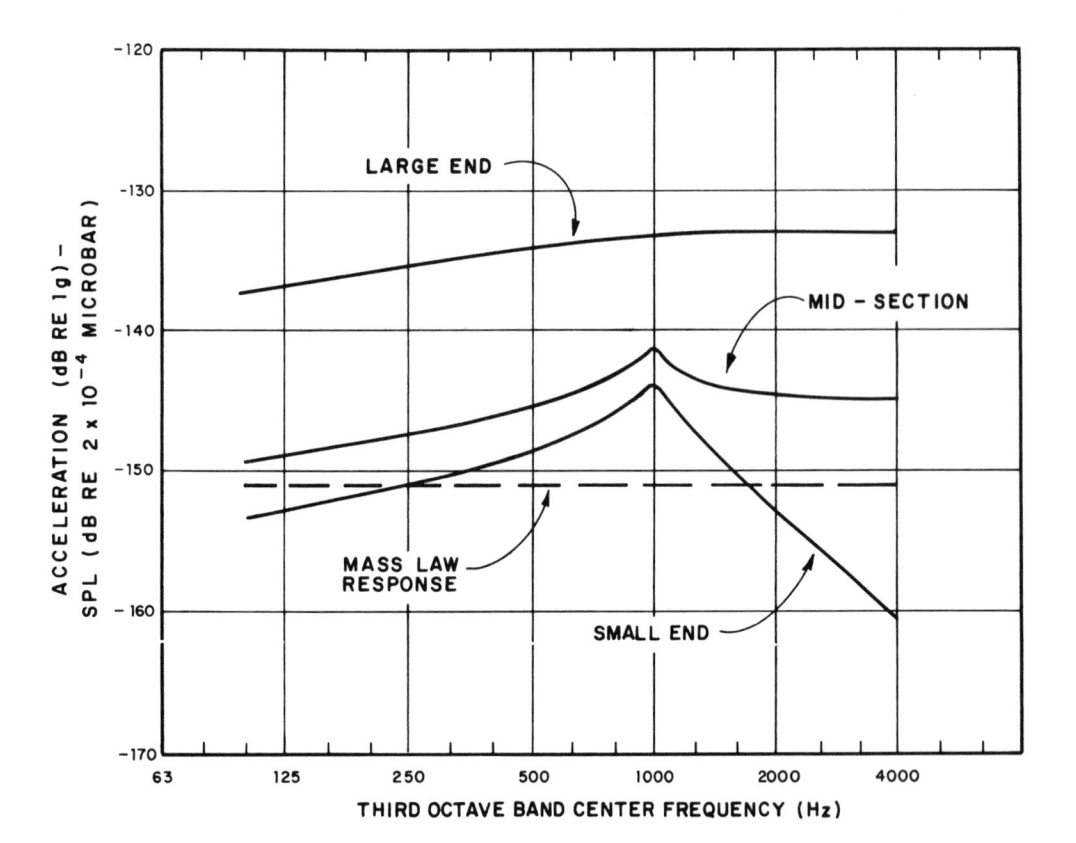

FIG. 15.12

ESTIMATED RESPONSE OF THE CONICAL SHELL TO TBL PRESSURE FIELD

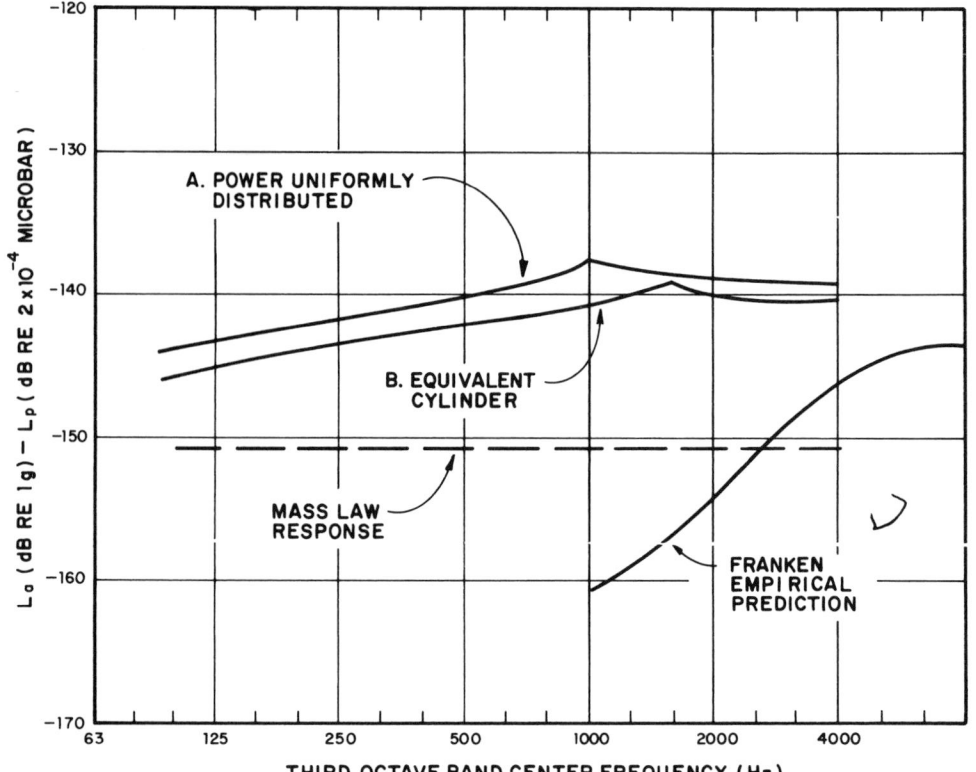

FIG. 15.13

ALTERNATE ESTIMATES OF RESPONSE OF THE CONICAL SHELL,
THE TBL PRESSURE FIELD

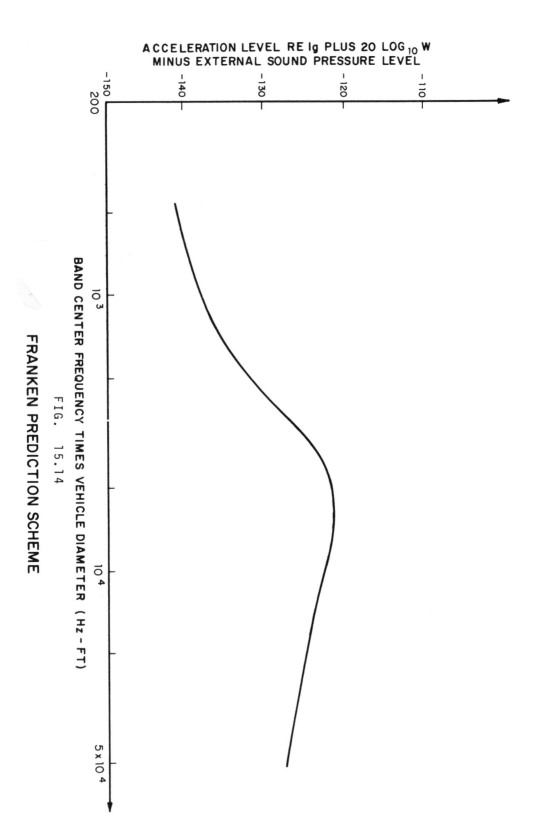

FIG. 15.14

FRANKEN PREDICTION SCHEME

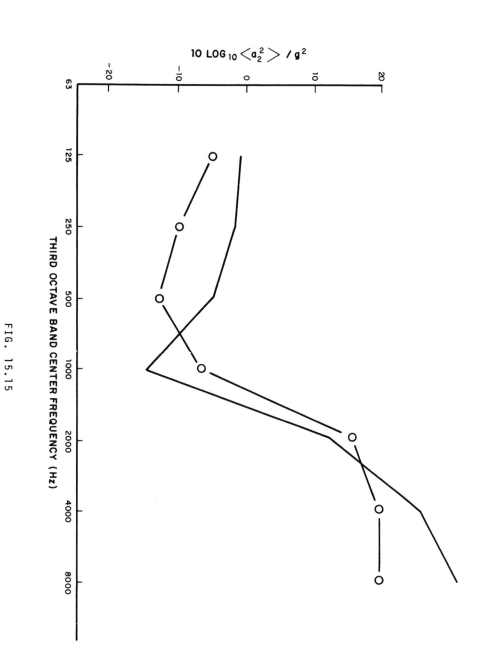

FIG. 15.15

PREDICTION OF VEHICLE SHELF VIBRATION LEVEL

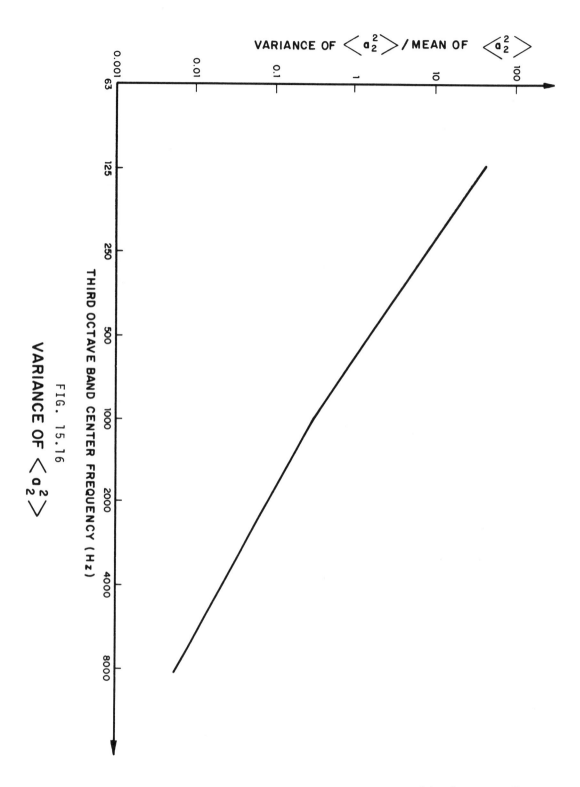

FIG. 15.16
VARIANCE OF $\langle a_2^2 \rangle$

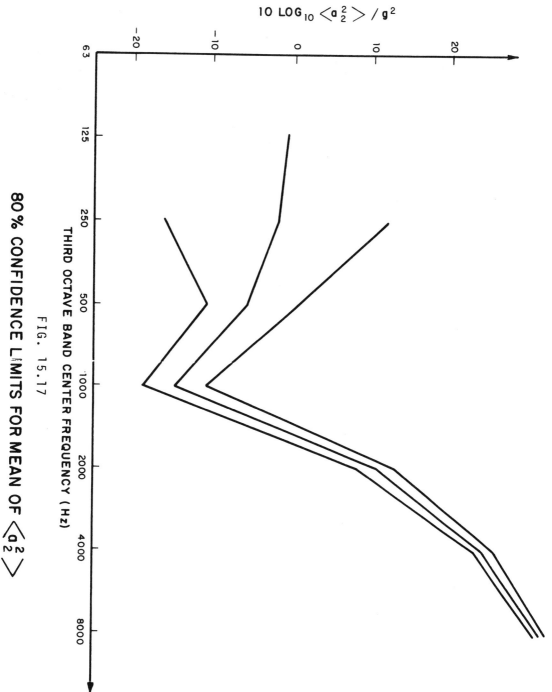

FIG. 15.17

80% CONFIDENCE LIMITS FOR MEAN OF $\langle a_2^2 \rangle$

REFERENCES FOR PART II

1. E. E. Ungar, et al., "A Guide for Predicting the Vibrations of Fighter Aircraft in the Preliminary Design Stage", Tech. Report AFFDL-TR-71-63, April 1973.

2. P. T. Mahaffey and K. W. Smith, "A Method for Predicting Environmental Vibration Levels in Jet-Powered Vehicles" Shock and Vibration Bulletin, No. 28, Pt. 4, pp 1-14 (1960).

3. P. A. Franken, "Sound-Induced Vibrations of Cylindrical Vehicles," J. Acoust. Soc. Am., 34, 453-454 (1962).

4. R. W. White, D. J. Bozich and K. McK. Eldred, "Empirical Correlation of Excitation Environment and Structural Parameters with Flight Vehicle Vibration Response," AFFDL-TR-64-160, Dec. 1964.

5. E. A. Guillemin, Introductory Circuit Theory, Ch. 5, John Wiley and Sons, New York, 1953.

6. R. Lyon, "What Good is Statistical Energy Analysis, Anyway?" Shock and Vibration Digest, Vol. 2, No. 6, pp. 1-9 (June, 1970).

7. J. E. Manning, "A Theoretical and Experimental Model-Study of the Sound-Induced Vibration Transmitted to a Shroud-Enclosed Spacecraft". BBN Report 1891, submitted May 1, 1970 to NASA. Work Performed Under Contract NAS5-10302, and submitted to: NASA Goddard Space Flight Center, Greenbelt, Maryland.

8. P. Kopff, "Note A Mr. Absi" - L.V.B. (Centre Experimental de recherches et d'etudes du batiment et des travaux publics) - September, 1972.

9. E. E. Ungar and K. S. Lee, "Considerations in the Design of Supports for Panels in Fatigue Tests," Tech. Rept. AFFDC-TR-67-86 (1967).

10. R. H. Lyon, "Statistical Analysis of Power Injection and Response in Structures and Rooms" J. Acoust. Soc. Am., Vol. 45, No. 3, pp. 545-565 (1969).

11. R. H. Lyon and E. E. Eichler, "Random Vibrations of Connected Structures," J. Acoust. Soc. Am. 36, pp. 1344-1354 (1964).

12. F. V. Hunt, "Stress and Strain Limits on the Attainable
     Velocity in Mechanical Vibration", J. Acoust. Soc. Am.,
     Vol. 32, No. 9, pp 1123-1128 (1960).

13. E. E. Ungar, "Maximum Stresses in Beams and Plates
     Vibrating at Resonance", J. Eng. Ind. Vol. 84, No. 1
     pp 149-155 (February 1962).

14. R. J. Roark, Formulas for Stress and Strain, McGraw-Hill
     Book Co. (1954).

15. R. H. Lyon, "Boundary Layer Noise Response Simulation
     with a Sound Field", Chapter 10 of Acoustical Fatigue
     in Aerospace Structures, Ed. by W. J. Trapp and
     D. M. Forney (Syracuse University Press, Syracuse,
     New York, 1965).

16. Schroeder, M. R. "The Statistical Parameters of the Fre-
     quency Response Curves of Large Rooms" Acustica 4,
     Beiheft 2, 594, 1954.

17. Lazan, B. J., "Damping of Materials and Members in
     Structural Mechanics" Pergamon Press Ltd., Oxford,
     London, 1968.

18. Snowdon, J. C. "Vibration and Shock in Damped Mechanical
     Systems" John Wiley & Sons, Inc., New York (1968).

19. Mead, D. J., "The Damping Stiffness and Fatigue Properties
     of Joints and Configurations Representative of Aircraft
     Structures", in WADC-Univ. of Minn. Conference on
     Acoustic Fatigue, edited by W. J. Trapp and D. M. Forney,
     Jr. [WADC TR 59-676 (March 1961)].

20. Maidanik, G. "Energy Dissipation Assoc. with Gas-Pumping
     in Structural Joints, J. Acoust. Soc. Am., Nov. 1966,
     Vol. 40, p. 1064.

21. Ungar, E. E., "A Guide to Designing Highly Damped
     Structures .... Using Layers of Viscoelastic Material"
     Machine Design, February 14, 1963, p. 162-168.

22. Szechenyi, E. "Modal Densities and Radiation Efficiencies
     of Unstiffened Cylinders Using Statistical Methods",
     J. Sound Vib. 19, No. 7, pp. 65-81 (1971).

23. Wilkinson, J. P. D., "Modal Densities of Certain Shallow
     Structural Elements", J. Acoust. Soc. Am. 43, No. 2 pp. 245-
     251 (February 1968).

24. Erdelyi, A., et al., Higher Transcendental Functions, Vol. 2 (McGraw-Hill Book Co., Inc., New York, 1953) p. 313 (March 1971).

25. Pierce, A. D. "Physical Interpretation of the WKB or Eckonal Approximation for Waves and Vibrations in In-homogeneous Beams and Plates", J. Acoust. Soc. Am. 48, No. 1 (Pt. 2) pp. 275-284 (July 1970).

26. Gemogenova, O. A., "Formalism of Geometrical Optics for Flexural Waves", J. Acoust. Soc. Am. 49, No. 3 (Pt. 2) pp 776-780.

27. Berendt, Raymond D.; Winzer, George E.; Burroughs, Courtney B., "A Guide to Airborne, Impact and Structure-Borne Noise-Control in Multi-Family Dwellings" U. S. Department of Housing and Urban Development, Washington, D. C., September, 1967.

28. Crocker, M. J. and Battarcharya, M. C. and Price, A. J., "Sound and Vibration Transmission through Panels and Tie-Beams Using Statistical Energy Analysis" Jour. of Eng. Ind., Trans. ASME Vol. 93, Series B No. 3, August 1971.

29. Beranek, L. L., Noise and Vibration Control, (McGraw-Hill Book Co., New York 1971) p. 283.

30. This result from a report by J. E. Manning is similar in form to a result by L. Cremer, Akustische Zeitschrift, Vol. 7, p. 81 (1942).

31. Maidanik, G., "Response of Ribbed Panels to Reverberant Acoustic Fields", J. Acoust. Soc. Am. 34 (6) p. 809, (1962).

32. Fahy, F. J. "Vibration of Containing Structures by Sound in the Contained Fluid" J. Sound Vib. (1969) 10 (3) pp 490-512.

33. Fahy, F. J. "Response of a Cylinder to Random Sound to Contained Fluid", J. Sound Vib., (1970) 13 (2) pp 171-194.

34. Lyon, R. H. and Eichler, E., "Random Vibration of Con-nected Structures", J. Acoust. Soc. Am., Vol. 36, No. 7, pp 1344-1354, July 1964.

35. Heckl, M. "Wave Propagation in Beam-Plate Systems" J. Acoust. Soc. Am., Vol. 33, No. 5, p. 640-651.

36. Ungar, Eric E.; Koronaios, Nicholas; and Manning, J. E.,
     "Application of Statistical Energy Analysis to Vibrations
     of Multi-Panel Structures" BBN Report No. 1491,
     Submitted to: Air Force Flight Dynamics Laboratory,
     Directorate of Laboratories, Air Force Systems Command,
     Wright-Patterson Air Force Base, Ohio 45433 (May 1967).

37. Cremer, L.; Heckl, M.; and Ungar, E., Structure-Borne
     Sound, (Springer, New York, 1973) p. 281.

38. Manning, J.E., Private Communication based on unpublished
     consulting work.

39. Ungar, E. E., et al., "A Guide for Predicting the
     Vibrations of Fighter Aircraft in the Preliminary Design
     Stages", AFFDL-TR-71-63, (April, 1973).

40. See reference number 7.   Also, see page 37, reference 44.

41. Heckl, M. A., "Compendium of Impedance Formulas",
     BBN Report #774 (1961). Work performed under
     NONR 2322 (00); see also Chapter 4 of reference 37.

42. Franken, P. A., "Sound-Induced Vibrations of Cylindrical
     Vehicles," J. Acoust. Soc. Am. 34, p. 453 (1962).

43. Chandiramani, K. L., et al., "Structural Response to
     Inflight Acoustic and Aerodynamic Environments", BBN Report
     #1417, (1966). NAS 8-20026, NASA George C. Marshall
     Space Flight Center, Huntsville, Alabama.

44. R. H. Lyon, Random Noise and Vibration in Space Vehicles.
     Shock and Vibration Information Center, U. S. Dept. of
     Defense.

GLOSSARY OF SYMBOLS

VARIABLES AND PARAMETERS

a   strength of impulse of force (2.1)

A   proportionality factor for power flow (3.1),
            wall area (3.3)

ΔA_k area associated with lattice point in
            k-space (2.2)

b   modal susceptance (2.2)

B   mechanical susceptance (2.2), bending rigidity
            (2.3), power flow-proportionality factor (3.1)

c   wave speed (see subscript section)

CC   confidence coefficient (4.3)

d   mean free path (2.4), determinant of power
            flow equations (3.1)

D   system determinant for pure tone excitation
            (3.1)

$D(\Omega)$ directivity function for reverberant field (3.3)

E   total vibrational energy - kinetic plus potential

$\mathcal{E}$ energy density (3.3)

f   cyclic frequency

G   mechanical conductance (2.2), gyroscopic coupling
            parameter (3.1)

h   plate or beam thickness (2.2)

H   frequency response function (3.1)

I   wave intensity (2.4)

$J_o(x)$ bessel function (2.4)

k   wavenumber = 2π/wavelength

K	mechanical stiffness
KE	kinetic energy (2.1)
ℓ	dimension of beam or rectangular plate (2.2)
$\ell(t)$	time varying load (2.1)
L	amplitude of sinusoidal load (2.1), projected length of junction (3.3)
m	mean value (4.2), surface density of wall (3.4)
M	mass
n	modal density
N	mode count in band $\Delta\omega$
p	pressure, usually of sound field
P	system perimeter (2.4), fourier transform of p (2.3)
PE	potential energy (2.1)
Q	quality factor (2.1)
r	damping of continuous system (2.2), distance from origin (4.2), ratio of limit of confidence interval to mean (4.3)
r_D	distance from source to reverberant field (2.4)
R	mechanical resistance, vector distance (2.4)
S	spectral density
t	time
T_R	reverberation time
U	volume velocity
v	mechanical velocity

V	system volume (3.4), velocity amplitude (2.1)
w	beam width (4.1)
x	spatial coordinate
X	mechanical reactance
y	resonator displacement (2.1), distributed system displacement (2.2)
Y	mechanical admittance (2.2), modal displacement (2.2)
z	displacement of coupled mode (3.1), distance from neutral axis of beam (4.1)
Z	mechanical impedance
α	complex natural frequency (2.1)
γ	normalized gyroscopic coupling parameter (3.1), ratio of specific heats (4.1), incomplete gamma function (4.3)
Γ	gamma function (4.3), proportionality constant (3.2)
δ_{mn}	Kronecker delta; = 1 for m=n, 0 otherwise (2.2)
$\delta(x)$	delta function (2.2)
Δ	frequency bandwidth (2.1)
$\Delta\omega$	noise or averaging bandwidth (2.1)
ϵ	mechanical strain (4.1)
ϕ	probability density (4.3), wave orientation (4.1), vibration phase angle (2.1)
η	loss factor (2.1), coupling loss factor (3.2)
κ	radius of gyration (2.3), normalized stiffness coupling (3.1)

λ   coupling parameter (3.1), square root of mass
             ratio (3.1)

μ   inverse of normalized variance (4.3), normalized
             mass coupling (3.1)

ω   radian frequency

Ω   wave orientation (3.3), solid angle (3.3),
             natural frequency of coupled mode (3.1)

Π   power

ψ   mode shape (2.2)

ρ   surface density (2.2), system density (2.2)

σ   standard deviation (4.2)

τ   transmissibility (3.3)

ξ   normalized frequency variable (2.2)

OPERATORS

Arg(...)   phase angle of

∂   partial differential

∇^2 Laplacian or nabla

max(···)   maximum value of

$\Lambda(\partial/\partial x)$ spatial differential operator representing
             elastic restoring forces

$\langle ... \rangle_\nu$ average with respect to parameter ν

i   rotator in complex plane by $\pi/2$

SUPERSCRIPTS

M	moment
(b)	blocked
s	source

SUBSCRIPTS

ϕ	phase
b	beam, bending
p	plate
σ, α	modal order indices
1,2	system order indices
R	room, reverberant
g	group
o	ambient
coup	coupling
s	surface, source
D	direct

INDEX

AUTHOR INDEX (Numbers in parenthesis are number of citations)

H. Andres 167

R. Bamford 153
M.C. Battarcharya 367
L.L. Beranek 154 (2), 367
R.D. Berendt 160, 367
J.L. Bogdanoff 10,157,160,166
V.V. Bolotin 156,192,204
R.H. Bolt 153
Bolt Beranek and Newman
    Inc. (BBN)7
D.J. Bozich 365
L.M. Brekhovskikh 163
L. Brillouin 162
C.B. Burrough 160,367

K.L. Chandiramani 9, 156,368
H. Cramer 166
S.H. Crandall 153
L. Cremer 156,367,368
M.J. Crocker 10,157,297,367

H.G. Davies 166
J.P. DenHartog 153
I. Dyer 159

D.A. Earls 159
E. Eichler 8,9,155(2)306,365,
    367
K.McK. Eldred 365
A. Erdelyi 168,367

F. Fahy 165(2),302,367(2)
H. Feshbach 162
P.A. Franken 161,339,365,368

O.A. Germogenova 288,367
W. Gersch 158(2)
E.A. Guillemin 162,365

F.D. Hart 156
M.A. Heckl 9,156(2),157,164,307,
    367,368(2)
C.I. Holmer 159(2)

F.V. Hunt 160,366

E.H. Kennard 154
P. Kopff 365
N. Koronaios 368
L. Kurzweil 9,156

B.J. Lazan 366
K.S. Lee 182,192,193,194,365
Y. K. Lin 153
R. Lotz 10,158
D. Lubman 167
R.H. Lyon 7,8,9,153,154,155(5),
    158(2),159(2),160,161(5),162,
    164,165,168,169,306,365(3),
    366,367,368

P.T. Mahaffey 168,365
G. Maidanik 7,9,154,155,156,
    161,366,367
J.E. Manning 9,156(2)157,
    158(2),317,318,319,320,
    324,343,365,367,368
W.D. Mark 153
D.J. Mead 366
M.L. Mehta 6,154
D.K. Miller 9,156
A.M. Mood 160
C.T. Morrow 160
P.M. Morse 153,162,163

D.E. Newland 159,164
National Aeronautics and
    Space Administration (NASA)
    10

E. Parzen 168
A.D. Pierce 288, 367
A. Powell 10, 157
A.J. Price 367

E.J. Rathe 164
P.J. Remington 10, 158